Human Factors in Privacy Research

Nina Gerber • Alina Stöver • Karola Marky
Editors

Human Factors in Privacy Research

Springer

Editors
Nina Gerber
Institute for Psychology
Technical University of Darmstadt
Darmstadt, Germany

Alina Stöver
Institute for Psychology
Technical University of Darmstadt
Darmstadt, Germany

Karola Marky
Institute for IT Security
Gottlieb Wilhelm Leibniz University
Hannover, Germany

This work was supported by GRK2050 Privacy and Trust for Mobile Users

ISBN 978-3-031-28642-1 ISBN 978-3-031-28643-8 (eBook)
https://doi.org/10.1007/978-3-031-28643-8

This Springer imprint is published by the registered company Springer Nature Switzerland AG
The registered company address is: Gewerbestrasse 11, 6330 Cham, Switzerland

Foreword

Every day, there is some new kind of privacy incident reported in the media. It might be a data breach, some kind of smartphone app or new device that's collecting too much information, or some kind of scandal about how personal data is being abused. The news articles about privacy are legion, ranging from the large scale like Cambridge Analytica and Equifax data breach, to the small scale like people stalking their ex-partners using smart home technologies or a priest outed as gay due to Grindr selling location data of its users.

The thing is, it doesn't have to be this way.

The good news is that things are starting to move in the right directions, slowly but surely. There are new laws that govern how companies and large organizations must handle data. There are new kinds of technologies, tools, standards, and guidelines for helping developers create privacy-sensitive apps. Lastly, and the focus of this book, there are new kinds of human-centered methods and empirical results to help researchers and practitioners design better user interfaces and systems.

This book is a treasure trove for researchers and practitioners interested in usable privacy. If you are interested in designing and building interactive systems that everyday people can use and would want to use, or want to know best practices in evaluating these kinds of systems in an ethical manner, this book is for you.

From a theoretical perspective, this book offers a foundation about theories, both philosophical and behavioral, that can help explain people's attitudes and behaviors towards privacy. For example, this book touches on Nissenbaum's conceptualization of contextual integrity, as well as how the Technology Acceptance Model might influence people's willingness to adopt new privacy enhancing technologies.

From a more pragmatic perspective, this book also offers a number of tools to help with practical concerns, ranging from survey scales to assess people's privacy concerns to human-centered design processes, from designing effective privacy notices to applying nudges to influence people's behaviors. These chapters contain especially useful overviews of a wide range of topics related to privacy, regardless of whether your background is in computer science, psychology, or design.

This book also offers something unique for researchers and practitioners, namely a deep discussion of the challenges that corporations face with compliance with laws and the conflicts they face when rolling out privacy measures.

There are some books you skim over, and then put away, never to be looked at again. There are other books you keep on your shelf just to look smart (yes, admit it, you do it too). And then there are books like this one, which contain so much useful information, that you will keep coming back to it time and time again.

October 2022 Jason Hong

Acknowledgements

This work has been co-funded by the Deutsche Forschungsgemeinschaft (DFG, German Research Foundation, grant number 251805230/GRK 2050) and by the German Federal Ministry of Education and Research and the Hessen State Ministry for Higher Education, Research, Science and the Arts within their joint support of the National Research Center for Applied Cybersecurity ATHENE.

We would also like to thank everyone who contributed to the creation of this book—first and foremost, of course, the great authors of the chapters including Jason Hong for contributing the foreword, as well as the thorough reviewers who provided valuable feedback for the chapters, the coordination team of the RTG "Privacy and Trust for Mobile Users" for their efforts to make funding for the book possible, the Springer Nature team for their support, and last but not least our families and friends who ensured that we (and our children) did not have to starve during the final hot phase.

About This Book

This book tackles the topic of human factors in privacy research from four different angles: theoretically, methodically, specifically with reference to various application areas, and solution-oriented by considering approaches for user privacy support.

Theory We start the book with the theoretical foundations of usable privacy research. For this purpose, the chapter "Data Collection Is Not Mostly Harmless: An Introduction to Privacy Theories and Basics" gives an introduction to the most important concepts of the privacy topic. Subsequently, the chapter "From the Privacy Calculus to Crossing the Rubicon: An Introduction to Theoretical Models of User Privacy Behavior" shows how theoretical behavioral models from psychology, health or economics can be used to explain human privacy behavior.

Methodology After that, we approach the topic of usable privacy research from a methodological point of view. First, the chapter "Empirical Research Methods in Usable Privacy and Security" gives an overview of the different research methods that are commonly used in usable privacy research. Furthermore, an introduction to ethical considerations is given. Then, the chapter "Toward Valid and Reliable Privacy Concern Scales: The Example of IUIPC-8" takes a closer look at the quantitative approach, using the IUIPC-8 as an example to describe how questionnaires can be examined for their validity and reliability as quantitative measurement instruments. This is followed by a consideration of the more qualitative approach in the chapter "Achieving Usable Security and Privacy Through Human-Centered Design" by describing how research and design in the privacy context can be conducted using methods of human-centered design. Here, approaches such as mental models, user requirements, user group profiles, and personas are presented as examples. Then, in the chapter "What HCI Can Do for (Data Protection) Law—Beyond Design", a bridge is built between law and HCI and the authors discuss how both can be combined in order to do truly user-centered, law-compliant privacy research. In the chapter "Expert Opinions as a Method of Validating Ideas: Applied to Making GDPR Usable", this combination is then presented through a case study by describing the implementation and results of a study that uses expert interviews from the legal and HCI sciences, among others, to investigate how requirements of the General Data Protection Regulation (GDPR) can be implemented in a user-friendly way.

Application Areas Subsequently, we consider different application areas of privacy research. Here, we start in the chapter "Privacy Nudges and Informed Consent? Challenges for Privacy Nudge Design" with the question to what extent nudges can be used to help users to make better informed decisions when handling their private data. The discussion is complemented by reflections on how a general use of nudges should be designed from an ethical point of view. The chapter "The Hows and Whys of Dark Patterns: Categorizations and Privacy" then discusses the use case of dark patterns, in which the psychological principles used in nudging for positive purposes are used to trick users into disclosing more of their data than initially intended. The chapter "'They see me scrollin'—Lessons Learned From Investigating Shoulder Surfing Behavior and Attack Mitigation Strategies" discusses the specific application of shoulder surfing studies and how Virtual Reality (VR) can be used as a study methodology and what ethical aspects should be considered. The chapter "Privacy Research on the Pulse of Time: COVID-19 Contact-Tracing Apps" gives an overview of the current research on contact tracing apps, which rapidly gained relevance in the research field of usable privacy during the COVID-19 pandemic starting in 2020. Here, the issue of using different measurement tools based on different privacy concepts is exemplified and thus the connection to the methodological foundations discussed in the chapter "Toward Valid and Reliable Privacy Concern Scales: The Example of IUIPC-8" is made. Finally, the chapter "Privacy Perception and Behavior in Safety-Critical Environments" presents various case studies investigating privacy perceptions and behaviors within safety critical environments and elaborates on the relationship between security and privacy behavior.

Solutions In the last part of the book, we look at various approaches that are intended to support users in finding a meaningful, self-determined way of dealing with their private data. For this, we first turn to the concern of obtaining user consent for data collection and processing in a legally compliant and user-friendly way. Here, the chapter "Generic Consents in Digital Ecosystems: Legal, Psychological, and Technical Perspectives" discusses the extent to which users could generically give their consent and the legal principles and challenges that need to be considered in this regard. The chapter "Human-Centered Design for Data-Sparse Tailored Privacy Information Provision" then describes a possible solution in which transparency for users is increased through context-sensitive, tailored privacy information provision. Thus, the chapter "Human-Centered Design for Data-Sparse Tailored Privacy Information Provision" discusses transparency-enhancing technologies (TETs), which are a subcategory of privacy-enhancing technologies (PETs), which are described in the following chapter "Acceptance Factors of Privacy-Enhancing Technologies on the Basis of Tor and JonDonym" using the examples of Tor and JonDonym. For this purpose, the chapter "Acceptance Factors of Privacy-Enhancing Technologies on the Basis of Tor and JonDonym" presents empirical results on which factors influence the acceptance of users for such

PETs. The chapter "Increasing Users' Privacy Awareness in the Internet of Things: Design Space and Sample Scenarios" spans the design space for privacy solutions aimed at increasing user awareness in Internet of Things (IoT) environments about the presence of sensors, such as cameras, and presents PriView, a concrete solution for this.

Finally, we turn to the enterprise context, with the chapter "Challenges, Conflicts, and Solution Strategies for the Introduction of Corporate Data Protection Measures" discussing how data protection measures can be introduced in companies and what social aspects need to be considered in this process. The chapter "Data Cart: A Privacy Pattern for Personal Data Management in Organizations" then presents data cart, an example of a concrete solution for the corporate context, which enables data protection-compliant processing of personal data in companies and was developed according to the principles of human-centered design, as described in the chapter "Achieving Usable Security and Privacy Through Human-Centered Design".

Contents

Part III Application Areas

Part IV Solutions

Part I
Theory

Data Collection Is Not Mostly Harmless: An Introduction to Privacy Theories and Basics

Karola Marky

1 Introduction

The contributions presented in this book belong to the broader field of *human factors in privacy*, *usable privacy research*, or generally deal with the concept *privacy*. Usable privacy, in particular, is situated at the intersection of cybersecurity with a focus on privacy and human–computer interaction [9] specifically considering the users' capabilities and knowledge when interacting with a technology.

The remainder of **this chapter** particularly focuses on the digital life of individuals and interactions with a digital system, such as a smartphone, personal computer, or Internet-of-Things (IoT) devices. Before we dive into why we need privacy, especially in our digital lives, we first take a look at different privacy definitions and theories that have been described in the literature.

2 Privacy Theories

This section details core privacy theories in the scientific literature. We start historically with the "The Right to Privacy" [23]. Next, the theories of Westin [24], Altman [1, 2], and Solove [22] are summarized. From these theories and further scientific literature, we learn specific properties of privacy and highlight why privacy is a highly individual concept.

K. Marky (✉)
Ruhr-University Bochum, Bochum, Germany
e-mail: karola.marky@rub.de

N. Gerber et al. (eds.), *Human Factors in Privacy Research*,
https://doi.org/10.1007/978-3-031-28643-8_1

The Right to Be Let Alone An early mention of privacy in the literature is the article "The Right to Privacy" by Warren and Brandeis in 1890 [23]. In this early work, the authors informally define privacy as "*the right to be let alone*" [23, p. 195].

Warren and Brandeis [23] cite the judge Thomas M. Cooley when making this statement and refer to a section on bodily integrity in his book [6, p. 29] where the original quote reads "*The right to one's person may be said to be a right of complete immunity: to be let alone*" [6, p. 29]. However, Cooley mainly refers to the integrity of the human body, specifically to instances of battery, while Warren and Brandeis take "*the right to be let alone*" to the social domain. Further, Cooley does not attempt to provide a notion of privacy. Also Warren and Brandeis do not attempt to provide a definition of the right to privacy [18], and they argue that privacy should be "*part of the more general right to the immunity of the person, – the right to one's personality*" [23, p. 207].

Warren and Brandeis specifically mention early technical devices that allow pictures of individuals to be taken as well as devices that allow eavesdropping conversations from afar mostly referring to the press that might invade people's private lives. Yet, this leaves room for interpretation what the "*the right to be let alone*" entails [21]. Nevertheless, this article had quite an impact by motivating privacy laws in the USA because it showed that the tort law did not protect privacy adequately at that time and because privacy violation is an injury to feelings and not to the body [21, 23].

Westin's Privacy Theory "*The right to be let alone*" [23, p. 195] was later on extended to individuals that determine what information about themselves should be known to others [24]. The political scientist and lawyer Alan F. Westin influenced how we understand privacy today.

His privacy theory defines privacy as "*the claim of individuals, groups, or institutions to determine for themselves when, how, and to what extent information about them is communicated to others*" [24, p. 7]. To show the different reasons "why" individuals might want privacy, Westin describes four privacy functions, which are detailed below:

Westin's Four Privacy Functions

1. *Personal autonomy* is the desire of individuals to not be manipulated, dominated, or exposed by others.
2. *Emotional release* describes a time-out from social demands, such as role demands.
3. *Self-evaluation* considers processing experiences.
4. *Limited and protected communication* sets interpersonal boundaries, while protected communication exchanges information with trusted peers.

see [24]

Westin also details different ways on the "hows" to achieve privacy that he denotes as states of privacy [24]. Below, we apply these four states to the analog and digital life and give some examples:

Westin's Four Privacy States This box gives an overview of Westin's four privacy states completed with examples from the analog life (denoted as) and digital life (denoted as):

1. *Solitude* means that information is not shared with others, similar to the "right to be let alone" [23, p. 195].
   There is a possibility to physically separate from others.
   A technology provides access control to keep information private.
2. *Intimacy* refers to information being shared only with specific humans.
   Close relationship between peers based on information exchange.
   A technology provides options to share information only with specific humans, e.g., specific posts can only be shared with "friends" in an online social network.
3. *Anonymity* means that information cannot be connected to an individual.
   The desire of public privacy.
   A technology offers the possibility to store or submit anonymized data, e.g., in an online election, the identities of the voters are not disclosed.
4. *Reserve* describes that information disclosures to others are limited.
   The creation of a psychological barrier against unwanted intrusion.
   A technology offers options to limit information disclosures, e.g., IoT devices do not capture specific information.

see [24]

Altman's Privacy Regulation Theory Similar like Westin, the social psychologist Irwin Altman also impacted our understanding of the concept privacy. He concisely defines privacy as "*the selective control of access to the self*" [1, p. 24], yet also captures more nuanced aspects of privacy in his work.

Altman states that privacy involves a dynamic process of boundary control between individuals [1]. Within this process, the desired level of privacy wanted by an individual might not match the achieved level in reality. To better describe this, he models privacy as a non-monotonic function with three different privacy levels: (1) optimal level where the desired level matches reality, (2), too much privacy, i.e., the desired level is lower than reality, and (3) too little privacy, i.e., the desired level is higher than reality. This function also shows several important aspects that Altman detailed in his later work: privacy, in principal, is a social process, which is why an in-depth understanding of psychological aspects is needed [2]. Too much privacy might result in social isolation, while too little might alter the behavior of individuals. We will talk about that in more details in the next section. An interesting extension of Altman's theory that specifically considers online communication is the Communication Privacy Management (CPM) by Petronio [17].

Solove's Privacy Taxonomy While Westin and Altman discuss privacy as a rather positive concept that enables individuals to exert control, Solove specifically considers the negative side of privacy invasions [22]. He first dives into different existing privacy theories mainly demonstrating that those are "too narrow, too broad, or too vague" [22, p. 8]. Then, he identifies four types of privacy problems that he uses to build a four-layered taxonomy. Each layer contains a different number of specific activities that can be done to harm the privacy of individuals:

Solove's Taxonomy

1. *Information collection*: surveillance and interrogation
2. *Information processing*: aggregation (combining different data pieces), identification (linking information to individuals), insecurity (not protect stored information adequately), secondary use (using collected information for a different purpose), and exclusion (not informing individuals properly about data handling)
3. *Information dissemination*: breach of confidentiality, disclosure, exposure (revealing nudity, grief, or bodily functions of individuals), increased accessibility, blackmail, appropriation (identity misuse), and distortion (propagating false information)
4. *Invasions*: intrusion (i.e., disturbing one's tranquility or solitude), and decisional interference (i.e., impact on private decisions by governments)

see [22]

It should be noted that each action by itself might not impose any harm on individuals as long as consent is given [22].

2.1 How (Not) to Define Privacy

Even though several attempts have been made to define privacy later on, no overall definition has been agreed on so far. Solove discussed different existing privacy theories concluding that they mainly are "too narrow, too broad, or too vague" [22, p. 8], and later in his book, he compares the term privacy to the ambiguity of the term animal to highlight how problematic ambiguity can be [22]. The reason for that lies in the complexity of privacy as an umbrella term for different concepts within different disciplines and scopes [22]. Further, privacy has a quite challenging property: it is a highly *individual and elastic concept* meaning each individual decides what kind of information they wish to keep private [15]. Something that is private information for one individual might be happily shared by another.

Further, there are differences in privacy perceptions based on specific contexts, such as culture [15]. Hence, there are different spheres that can impact privacy norms on different levels, such as political, socio-cultural, and personal levels [25].

The definition considers the possibility for individuals to exert control on when and how personal information about them is collected and processed by others [7, 8, 23, 24]. Consequently, it is a personal good that also protects individuals. One must also mention that sometimes, privacy is considered as a value that can be traded against specific benefits [5], such as financial benefits or services that are free of charge. The chapter "From the Privacy Calculus to Crossing the Rubicon: An Introduction to Theoretical Models of User Privacy Behavior" specifically describes theories and behavioral models that aim to explain privacy behavior.

Finally, it is also challenging to separate privacy from related concepts, such as secrecy or anonymity. Especially in the legal context, privacy can be defined as secrecy, and there are several disagreements on the specific boundaries between privacy and its related concepts [12]. A core aspect of privacy, however, is that it is a highly *individual concept*. Individual differences also make it particularly challenging to implement one specific overall solution that fits the needs of each and every individual. Consequently, specific technologies ideally offer a possibility for individuals to configure it according to their privacy needs. Privacy, furthermore, can fulfill different functions.

3 Why Do We Need Privacy?

Now that we introduced the concept of privacy, different theories, and its functions, we discuss why privacy is needed in the first place. Solove's taxonomy detailed above already provides a list of negative consequences of privacy invasions [22]. In the remainder of this chapter, we provide three specific reasons why privacy is important:

1. **Missing Privacy Can Bias Decisions**: Early research in the field of psychology showed that sacrificing privacy is not a viable solution. It has repeatedly been demonstrated that people alter their behavior when observed by others [3, 10, 19]. For instance, Asch studied the extent to which the opinions and behavior of a majority could affect individual decisions and judgments of individuals [3]. Therefore, he performed a series of experiments that became known under the terms *elevator test* and *line test*. Both experiments share that one participant is confronted with a group of actors. In the elevator test, the group performs unexpected actions, such as facing the elevator's wall instead of the door. In the line test, the participants received a card with a line and have to pick a line that matches the line length on the received card from a set. The actors chose a line from the set that was obviously not matching the one on the card. Asch's

results indicate that individuals conformed to the majority's opinion even when the correct answer was obvious. Thus, social influence can make people question their own decision under the supervision of a contradicting majority. This is also one reason for a central principle of modern democracies: vote privacy. In summary, the need for privacy comes from the presence of society and other individuals around us [14]. Without that, we would not need privacy [14].

2. **Missing Privacy Allows Others to Control Us**: The amount of information that another entity holds about individuals can also be used to influence that specific individual without the presence of other humans. This also relates to Westin's privacy function *personal autonomy* described above [24]. Zuboff coins the term *surveillance capitalism* [29] to describe the influence on humans by massively using data captured about them. More specifically, she describes it as a "new economic order that claims human experience as free raw material for hidden commercial practices of extraction, prediction, and sales" [29, p. 1]. The idea behind this is that any kind of data created by human experiences, such as sharing pictures or purchasing products, is fed into algorithms that aim to subtly influence actions of humans, e.g., going to a specific restaurant. Such influence can occur via targeted advertisements, but also via coupons or even games. While individuals might benefits from such data analysis, many mechanisms are designed in a way that do not keep individuals in control, and there is a fine line between benefit and exploitation. A possible solution to that would be not to process data about individuals.
3. **Missing Privacy Can Impact Mental Health**: Privacy is an integral human need. Each individual has different kinds of personal boundaries. In this context, privacy serves as a boundary control that enables individuals to regulate contact and information exchange with other individuals on several levels. Too much information (or contact) is perceived as an invasion of the self [16]. Complete withdrawal of others, however, can result in feelings of loneliness [16]. Therefore, privacy regulation is essential for mental health [11].

The reasons outlined above are just a fraction of the reasons to motivate a need for privacy. Privacy in the digital world is particularly challenging. In the analog world, humans can use physical restrictions to protect personal information from others. Until the early two-thousands, the majority of information had been in analog format. To interact with analog information, humans either needed to be in the vicinity of the information or had to make a physical copy. To enforce restrictions based on privacy preferences, humans could physically limit access to analog information about them. In doing so, humans can decide which information they share with others. Translating such physical limitations into the digital world, however, is not trivial.

The ongoing digital transformation is fundamentally changing how humans interact with information and the kind of information they share with others. At the beginning of the digital transformation, computers were obvious standalone devices, and users always *intentionally* interacted with them. Thus, privacy did not require much added extra effort. Just two decades later, in 2023, the majority of information is digital data. Networks, such as the Internet, serve as an infrastructure to interact with data that are stored remotely. Computational capabilities and sensors for collecting data are integrated into everyday objects connected to the Internet—the so-called Internet of Things (IoT) [4]. This has numerous benefits for users, such as availability or convenience of everyday life [13, 26]. However, the ubiquitous abilities of digital services and the IoT devices they are connected with raised several privacy challenges because digital services generate, collect, store, and analyze data about people's private lives (cf. [20, 27, 28]). As Warren and Brandeis already feared in 1890, technology can now penetrate our very private places and eavesdrop on our private conversations [23].

In summary, privacy is a highly individual concept. Missing privacy can impact mental health, social decisions, and our lives in general. Privacy in the digital world is challenging for several reasons demanding a need for more in-depth research in this field and novel solutions that better help protecting the essential need of our society.

References

1. Altman, I. (1975). *The environment and social behavior: Privacy, personal space, territory, and crowding*. ERIC
2. Altman, I. (1990). Toward a transactional perspective. In *Environment and behavior studies* (pp. 225–255). Springer.
3. Asch, S. E. (1956). Studies of independence and conformity: I. A minority of one against a unanimous majority. *Psychological monographs: General and Applied, 70*(9), 1.
4. Atzori, L., Iera, A., & Morabito, G. (2017). Understanding the Internet of Things: Definition, potentials, and societal role of a fast evolving paradigm. *Ad Hoc Networks, 56*, 122–140.
5. Bennett, C. J. (1995). *The political economy of privacy: A review of the literature*. Center for Social and Legal Research.
6. Cooley, T. M. (1879). Callaghan and Company, Chicago.
7. Culnan, M. J., & Armstrong, P. K. (1999). Information privacy concerns, procedural fairness, and impersonal trust: An empirical investigation. *Organization Science, 10*(1), 104–115.
8. Fried, C. (1968). Privacy. *Yale Law Journal, 77*, 21.
9. Garfinkel, S., & Lipford, H. R. (2014). Usable security: History, themes, and challenges. *Synthesis Lectures on Information Security, Privacy, and Trust, 5*(2), 1–124.
10. Jenness, A. (1932). The role of discussion in changing opinion regarding a matter of fact. *The Journal of Abnormal and Social Psychology, 27*(3), 279.
11. Johnson, C. A. (1974). Privacy as personal control. *Man-Environment Interactions: Evaluations and Applications: Part, 2*, 83–100.
12. Margulis, S. T. (2003). Privacy as a social issue and behavioral concept. *Journal of Social Issues, 59*(2), 243–261.
13. Marikyan, D., Papagiannidis, S., & Alamanos, E. (2019). A systematic review of the smart home literature: A user perspective. *Technological Forecasting and Social Change, 138*, 139–154.
14. Moore, B. (1984). *Privacy: Studies in social and cultural history*.
15. Nissenbaum, H. (2020). Protecting privacy in an information age: The problem of privacy in public. In *The ethics of information technologies* (pp. 141–178). Routledge.
16. Pedersen, D. M. (1997). Psychological functions of privacy. *Journal of Environmental Psychology, 17*(2), 147–156.

17. Petronio, S. (2002). *Boundaries of privacy: Dialectics of disclosure*. Suny Press.
18. Schoeman, F. (1984). Privacy: Philosophical dimensions of the literature. *Philosophical Dimensions of Privacy: An Anthology, 1*, 33.
19. Sherif, M. (1935). *A study of some social factors in perception*. Archives of Psychology (Columbia University).
20. Sivaraman, V., Gharakheili, H. H., Vishwanath, A., Boreli, R., & Mehani, O. (2015). Network-level security and privacy control for smart-home IoT devices. In *2015 IEEE 11th International Conference on Wireless and Mobile Computing, Networking and Communications (WiMob)* (pp. 163–167).
21. Solove, D. J. (2002). Conceptualizing privacy. *California Law Review* 1087–1155.
22. Solove, D. J. (2008). *Understanding privacy*. Harvard University Press
23. Warren, S., & Brandeis, L. (1890). The right to privacy. In *Harvard law review* (pp. 193–220).
24. Westin, A. (1967). *Privacy and freedom*. Atheneum.
25. Westin, A. F. (2003). Social and political dimensions of privacy. *Journal of Social Issues, 59*(2), 431–453.
26. Wilson, C., Hargreaves, T., & Hauxwell-Baldwin, R. (2017). Benefits and risks of smart home technologies. *Energy Policy, 103*, 72–83.
27. Ziegeldorf, J. H., Morchon, O. G., & Wehrle, K. (2014). Privacy in the Internet of Things: Threats and challenges. *Security and Communication Networks, 7*(12), 2728–2742.
28. Zimmermann, V., Dickhaut, E., Gerber, P., & Vogt, J. (2019). Vision: Shining light on smart homes—supporting informed decision-making of end users. In *Proceedings of IEEE European Symposium on Security and Privacy Workshops* (pp. 149–153).
29. Zuboff, S. (2019). *The age of surveillance capitalism: The fight for a human future at the new frontier of power*. Profile Books.

From the Privacy Calculus to Crossing the Rubicon: An Introduction to Theoretical Models of User Privacy Behavior

Nina Gerber and Alina Stöver

1 Introduction

A plethora of empirical studies aims to explain human privacy behavior, of which many focus on the so-called *privacy paradox*, i.e., the discrepancy between stated privacy concerns and privacy behavior (for an overview, the reader is referred to, e.g., [15]). Several theoretical explanations have been proposed for this phenomenon so far, however, there is no agreed-upon theoretical framework that correctly predicts and explains human privacy behavior. While the privacy paradox received considerable attention in the usable privacy research field, this attitude-behavior gap is not new in psychological research and has been investigated in other application areas such as health behavior for decades [3, 8, 25, 28, 30]. Hence, it could be worthwhile to consider theoretical models of human behavior stemming from other research contexts for privacy research as well, as these might provide novel explanations for the privacy-specific attitude-behavior gap (i.e., the privacy paradox), and add valuable factors for predicting privacy behavior, which can serve as a basis for designing privacy-supportive interventions.

The aim of this chapter is thus to summarize theoretical frameworks for explaining and predicting human behavior that could add to our understanding of user privacy behavior. Some of these concepts were already investigated in depth in the privacy context, while others originate from other contexts, such as health or working psychology, and have not been applied in the privacy context yet. The list of models is not exhaustive, rather, we selected such models that have either been

N. Gerber (✉) · A. Stöver
Technical University of Darmstadt, FAI, Darmstadt, Germany
e-mail: nina.gerber@tu-darmstadt.de; alina.stoever@tu-darmstadt.de

N. Gerber et al. (eds.), *Human Factors in Privacy Research*,
https://doi.org/10.1007/978-3-031-28643-8_2

applied extensively in privacy research or provide promising potential to add to the existing privacy models and stimulate novel insights.

2 Homo Economicus

The concept of the homo economicus [13, 33], originating from economic theory, forms the basis for the among privacy researchers well-known privacy calculus model [24]. This behavioral model is based on the idea that people act purely rationally and pursue the goal of maximizing their benefit in all their actions. To this end, the advantages and disadvantages of a decision are weighed up against each other, and in each case the behavior is chosen which has more positive than negative consequences (see Fig. 1). In the case of privacy, for example, social benefits like feeling connected to one's social contacts can outweigh the downsides of using a privacy-threatening messenger or social network. On the other hand, potentially severe consequences of a privacy breach, such as the possibility of sensitive health information getting publicly known, can discourage information sharing in this area.

Although the privacy calculus or the underlying model of the homo economicus, respectively, offer intuitive and easy-to-understand explanations for human behavior, they fall short in explaining how and which consequences are evaluated by the users. While positive consequences, such as social inclusion or free and easy access to services, might be relatively easy to identify in a specific context, negative consequences are usually fuzzier and harder to pinpoint. For example, these can include the psychological burden of feeling surveilled, a vague perception of various risks that might become relevant in the future, or additional costs in terms of time and money for using more privacy-preserving technologies. Furthermore, the model does not make any assumptions about how the benefits and disadvantages are weighted by the individual users. Hence, the model of the homo economicus seems intuitive for explaining behavior retrospectively—e.g., the user decided to participate in a social network because the advantages of doing so were perceived to be greater than potential negative consequences—but it fails in predicting future behavior. More refined models are needed to also map the various factors that determine how positive and negative consequences are perceived by different users in different situations and how these translate into actual behavior.

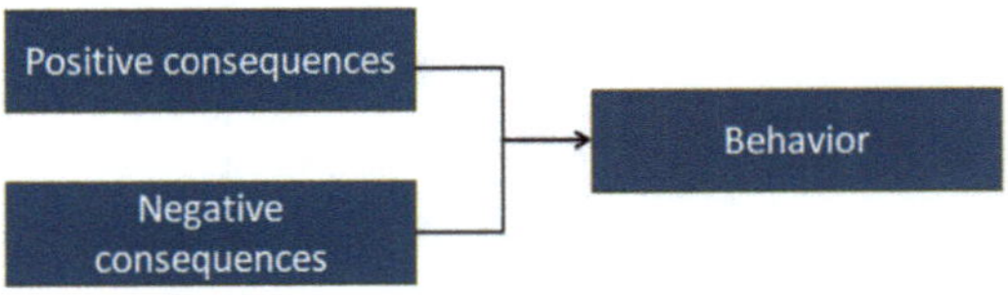

Fig. 1 The homo economicus model

3 Antecedents → Privacy Concerns → Outcomes (APCO) Model

The APCO model [29, 36] was developed based on reviews of the privacy literature [7]. It focuses on privacy concerns (also referred to as beliefs, attitudes, perceptions), which are directly and independently influenced by antecedents (see Fig. 2). According to the model, *privacy concerns* are a function of *previous privacy experiences* (e.g., users who have had bad experiences in the past tend to have greater privacy concerns), *privacy awareness* (e.g., if users are not at all aware that data is being collected from them in a certain situation, they will have fewer privacy concerns), *demographic factors* such as age or gender (the empirical evidence on the relationship between demographic factors and privacy is very mixed, however, so we will refrain from specifying a concrete direction of effect here), *personality factors* (here, too, the evidence is rather mixed), and culture or corporate climate (in some cultures, for example, more value is placed on privacy protection than in others). The privacy concerns of the users in turn affect the outcomes in the form of *regulations*, *behavior* (including data disclosure), and *trust* (e.g., towards the data collecting entity). Trust depends on the content provided in the privacy notice, which, for example, provides information about what data the entity claims to collect and how the collected data is protected. Furthermore, the privacy calculus is considered for the concrete decision for or against a certain behavior (see Sect. 2), i.e., the weighing of costs and benefits of this behavior.

The APCO model has been widely criticized since its publication [7], among other things because important psychological processes such as cognitive biases and bounded rationality are not taken into account. Hence, revised versions of the APCO model have been proposed [7, 10]. Still, the APCO model considers various factors that are internal and external to the user. In this respect, the APCO model is superior to the privacy calculus, of which it is a direct extension, but it is nevertheless only suitable to a very limited extent for explaining or predicting privacy behavior, since important factors are not considered and the current state of studies on the various influencing factors is rather inconclusive.

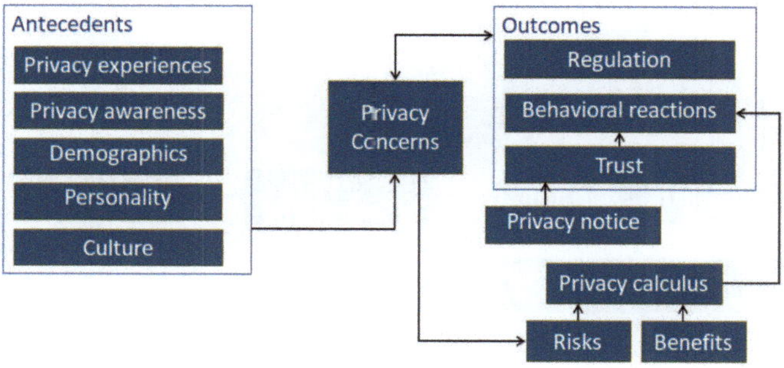

Fig. 2 The APCO model

4 Theory of Planned Behavior

The theory of planned behavior is among the most popular models that explain human behavior in psychological research [4, 9, 26, 37]. As the name already suggests, this theory aims to explain only deliberate, i.e., planned behavior, and is less suited to explain automatic or reflexive behavior. It postulates that the users' behavioral intention, e.g., to provide their data, is mainly affected by their *attitude* towards this behavior (i.e., do these users think it is a good idea to provide their data in general), the *perceived social norm* of this behavior (i.e., does their close social circle think it is a good idea to provide this data or to provide data in general), and their *perceived behavior control* (see Fig. 3). The latter distinguishes this theory from its predecessor, the theory of reasoned action [5], in which this factor was not considered.

However, behavioral control might be an important factor—for instance, it seems reasonable that someone who thinks it might be a good idea to protect their private communication by using end-to-end encryption (E2EE) and who further thinks the people close to them also think this is a great idea is still not likely to use E2EE if they feel they are not able to implement E2EE at all. The perceived behavioral control can depend on internal factors, such as knowledge or self-efficacy, but also on external resources, such as time, money, or autonomy. For example, someone who is employed in an organization may not have the authorization to decide whether their colleagues should also implement E2EE. Yet, they cannot send encrypted mails without the receivers also having implemented E2EE. The head of the company, on the other hand, might dictate their employees to use E2EE, in which case these have little perceived behavioral control to decide *against* using E2EE. Thus, perceived behavioral control is assumed not only to affect behavioral intention but also to have a direct effect on behavior, as sometimes users are forced to go against their intention due to external factors.

The theory of planned behavior further considers the *attitude-behavior gap* referred to as *privacy paradox* [15, 22], i.e., the fact that people are often willing to do something (e.g., better protect their privacy, delete their Facebook account) but fail to actually do so, whether out of apathy or, for example, because they are still postponing the respective action for other reasons. The fact that this phenomenon,

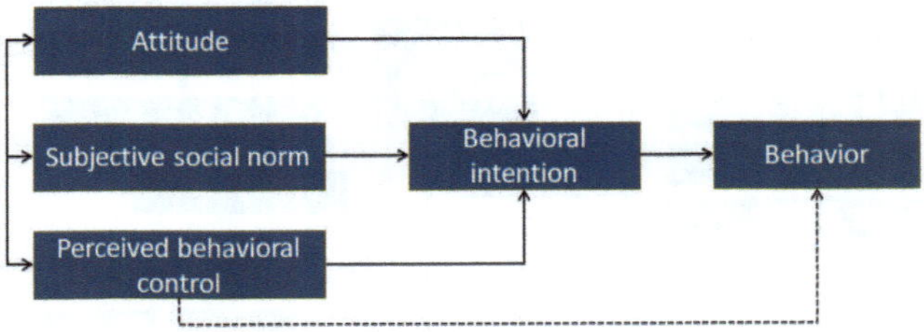

Fig. 3 The theory of planned behavior

which is richly explored in other areas such as health behavior [3, 8, 25, 28, 30], was rarely known among privacy researchers may have led to an overestimation of this phenomenon in the field of privacy research. Hence, while it is important to include the attitude-behavior gap in models aimed at explaining privacy behavior, there might be several other influencing factors in the case of privacy behavior apart from the attitude-behavior gap. We will thus explore further behavioral models, which, to the best of our knowledge, have not yet been widely applied to privacy research, in the following sections.

5 Cognitive Consistency Theories

Cognitive consistency theories [1, 11, 12] describe the fact that people strive to avoid inconsistencies in their attitudes, beliefs, intentions, and actions, i.e., they strive for consistency among these factors. According to these theories, contradictions between behavior and attitudes lead to *cognitive dissonance*, which is perceived as unpleasant. To resolve this dissonance, people therefore adjust either their behavior or their attitude (see Fig. 4).

In terms of privacy behavior, this could, e.g., look as follows: Users feel a general level of privacy concerns, which is reflected in specific privacy concerns (related to the data disclosed during the interaction) when interacting with a concrete application. This leads to a negative perception of the application due to privacy concerns, i.e., the privacy-threatening potential of the application. On the other hand, the application may also provide various positive features, e.g., offer very useful functions or a good user experience. This leads to a positive perception of the application. The two contradictory perceptions of the application (negative because privacy-threatening and positive because of the functionalities provided) leads to cognitive dissonance. The users do not know how to act in the situation, since the two contradictory perceptions suggest to them at the same time that they should use the application and that they should refrain from using it. To resolve this dissonance,

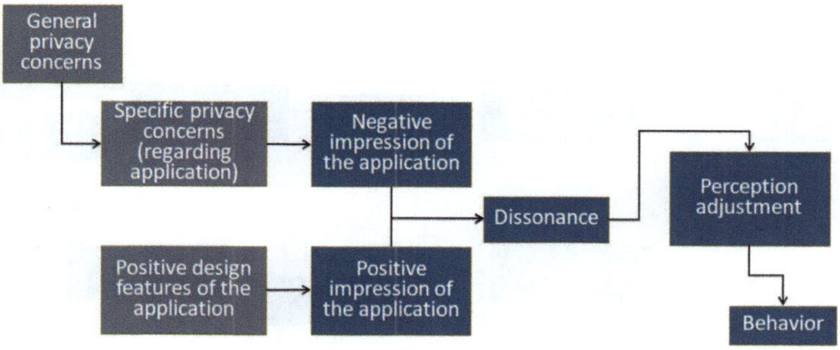

Fig. 4 Cognitive consistency theories

the perception of the application is adjusted by either relativizing the specific privacy concerns (along the lines of "Although I am in principle against the disclosure of this kind of information, it is not so bad with the present application because. . .") or by correcting the positive perception of the application functions downward ("The application is not really that great after all, because. . ."). The impression of the application adjusted in this way can now be transferred directly into consistent behavior.

In the long term, however, a decision to use an application or to disclose data, i.e., behavior that is not privacy-preserving, can also lead to a dissonance between behavior and general privacy concerns. In these cases, either the behavior or the general privacy concerns can be adjusted, which can potentially lead to a gradual weakening of existing privacy concerns.

Like the homo economicus model, cognitive consistency theories explain privacy behavior in a rather post hoc manner. The theories offer an explanation beyond the rational model of homo economicus for seemingly inconsistent expressions of general privacy concerns and concrete privacy behavior, i.e., the privacy paradox, by taking well-studied psychological processes into account. Nevertheless, this model does also not allow for the prediction of privacy behavior, since adjustments for the purpose of establishing consistency may refer to different cognitions as well as behaviors. Moreover, in contrast to the theory of planned behavior, no concrete external factors such as social influence are considered.

6 Transactional Model of Stress and Coping

The transactional model of stress and coping [23] aims to explain in which circumstances a person experiences stress. Behind this lies the classic working psychology assumption that not every person reacts in the same way to a stressor and that this stressor may or may not lead to stress depending on personal conditions (see Fig. 5).

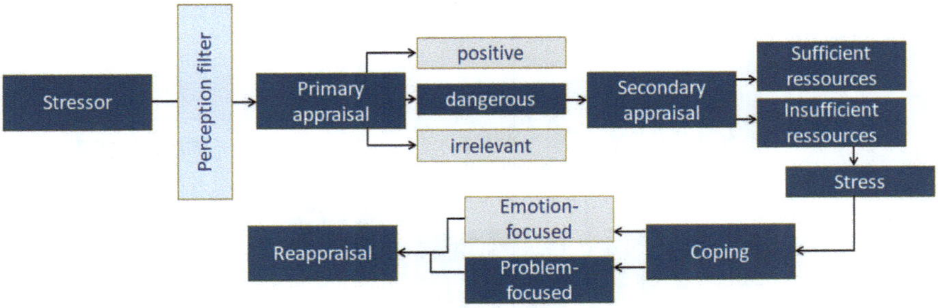

Fig. 5 The transactional model of stress and coping

A *stressor* occurring in the environment, for example, the collection of behavioral data via the use of cookies for the purpose of playing out personalized advertising, must first pass through the users' *perception filter*, i.e., be registered by them in the first place. Here, for example, users who visit websites from an EU country will have a higher probability of perceiving this stressor, as they are usually made aware of the use of these cookies via a cookie consent notice. However, individuals who are fundamentally more privacy-aware are also more likely to register such a stressor than individuals who are more indifferent to online tracking. If the stressor is registered by the users, a *primary appraisal* takes place: the stressor is classified as positive, dangerous, or irrelevant. This classification depends, of course, both on the nature of the stressor (e.g., using cookies for the purpose of serving personalized advertising results in the collection of far more sensitive data than using cookies for the purpose of generating statistical analyses of website usage) and on the attitudes of the individual (in this case, for example, the importance given to the protection of personal data). If the stressor is evaluated as positive, e.g., because the users would like to receive personalized advertising, or as irrelevant, because the users are neither positive nor negative about the process, the process ends at this point. Only if the stressor is evaluated as dangerous, a *secondary appraisal* follows, in which the users check to what extent they have resources to react to the stressor. If the users have sufficient resources (e.g., technical knowledge, time, an interface that allows them to refuse the use of cookies for the purpose of displaying personalized advertising), they neutralize the stressor by using these resources. However, if the secondary appraisal turns out to be negative, i.e., the users conclude that they do not have enough resources (for example, because the cookie consent notice is a content blocker that requires consent to use all cookies in order to visit the desired page, too little time or knowledge is available to make the required settings, and/or additional dark patterns—see also the chapter "The Hows and Whys of Dark Patterns: Categorizations and Privacy"—have been used), *stress* evolves. The users now attempt to deal with this stress by following either an emotion-focused/appraisal-focused or a problem-focused *coping strategy*. In the former, an attempt is made to reduce the negative emotions, e.g., by distraction, or to change the reference to the situation, e.g., by the users convincing themselves that the acquisition of their data in this case is bearable. Only problem-focused coping leads to privacy-protecting action. Here, an attempt is made to build up additional resources, e.g., by asking another person for help, acquiring additional knowledge through research, or installing a technical assistance tool. Finally, a *reassessment* of the stressor takes place, reflecting on how successful the coping performed was. Based on this, it is possible that the same stressor will no longer be perceived as generating stress in the future if the users realize that they now have sufficient resources to counter it.

Although the transactional model of stress and coping does not directly seek to explain behavior, but only the genesis of and coping with (negative) stress, we believe it adds value to privacy research. It emphasizes the otherwise possibly easily overlooked fact that a stimulus—such as the collection of private data—must first pass through a person's perceptual filter, which does not happen as a matter of course in times of ubiquitous data collection. Subsequently, this data collection must be classified as dangerous—this factor is also present in many other behavioral models. However, the model offers new input about dealing with insufficient resources. On the one hand, it illustrates that users experience stress in the nowadays common situation of being overwhelmed when dealing with the collection of their private data that is taking place—a circumstance that is potentially detrimental to health and has so far received little attention in the public debate on data protection. On the other hand, it captures what is in many cases ineffective coping via reassessment of the situation in a formal model. However, it does not explain how users can be persuaded to use a more goal-directed, problem-focused coping strategy. At this point, it should be emphasized that emotion-focused coping is by no means a bad option per se, because by reducing emotional stress, it allows users to shift into a more positive mindset that can facilitate better, problem-focused coping with the stressor. It is only harmful if emotion-focused coping is not combined with a problem-focused coping mechanism, because then the problem itself cannot be solved.

7 Rubicon Model

The Rubicon model [2] is a classic motivation model from psychology that distinguishes between different phases of action. Similar to the theory of planned behavior, the choice of action goals ("Behavioral Intention" in the theory of planned behavior) and the realization of these action goals ("Behavior") are considered separately (see Fig. 6).

The first phase (evaluation) describes the *weighing* of different action goals. Here the model assumes that people have more potential action goals than they can realize and therefore must weigh up which goals (a) are particularly desirable and (b) have a good chance of being achieved with a realistic use of resources. Hence, this is a primarily motivational phase. The conclusion of this evaluation phase is the formulation of a concrete goal for action—the rather general desire to better protect one's data could, for example, give rise to the possible goal for action of switching digital communication in a private context to privacy-friendly channels wherever possible. This transition from desire to concrete action goal is referred

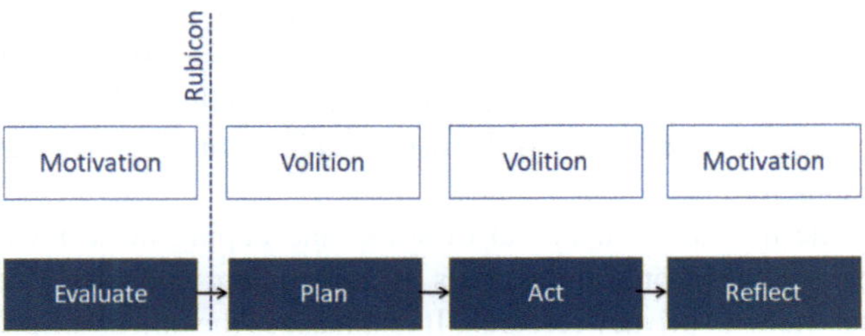

Fig. 6 The Rubicon model

to as "crossing the Rubicon," in analogy to Caesar's crossing of the Rubicon in 49 B.C., with which he instigated a civil war and after which there was literally no turning back. In our everyday life, of course, crossing the Rubicon is far less dramatic; here, finality refers to the fact that by setting one's goal for action, users create a commitment to themselves to reach that goal.

This is followed by the *planning phase*, in which the weighing of action goals is completed, and consideration is given to how the action goal formulated in the previous phase can best be achieved. Hence, according to theory, this is no longer a motivational phase, but a volitional phase. No action is taken at this stage; the user merely makes resolutions to act and considers at which points in the implementation of the goal difficulties could arise and how these can best be addressed. It is assumed that users do not act immediately because they first have to wait for favorable opportunities. When potentially favorable opportunities occur (*favorable* means compared to other past and anticipated future opportunities), action initiation occurs, drawing on the pre-determined strategies. For example, users may consider which channels or messengers they no longer want to use in the future and which they should be replaced with. In addition, they consider the people with whom they would like to communicate via these alternative channels and how the change of communication channel can best be implemented—for example, by selecting messengers that are available free of charge and easy to use. One potential difficulty could be that certain key communication partners may not want to switch channels voluntarily. In this case, the users would be well advised to consider in the planning phase how these communication partners can best be convinced—perhaps by providing them with a newspaper article that deals with what consequences the exploitation of data from private communications can have for private individuals.

Once an action has been initiated, the users are in the *actional phase*, in which they attempt to realize the action goal by implementing the actions and strategies defined in the previous phase. Depending on the complexity of the goal and the occurrence of difficulties, it is necessary here to accept considerable efforts and to resume interrupted actions several times in order to successfully achieve the goal. For example, users could fail here in the action of no longer using certain messengers if they find that a communication partner with whom they would like to remain in contact digitally in the future is not willing to change channels. In this case, persistent attempts at persuasion may be necessary if the goal is ultimately to be achieved. The effort that users are willing to make results from the commitment to the goal of action, which in turn depends on the attractiveness and feasibility of the goal.

In the last phase, the users *reflect* on the extent to which they have achieved the set action goal, also taking into account the extent to which the intended positive consequences have occurred as a result. In this phase, it may become apparent, for example, that despite successful achievement of the goal, not all the intended positive effects or additional negative effects not considered in advance have occurred. Here, motivational factors are again in the foreground. If the action goal is evaluated as achieved and the subsequent consequences as satisfactory, the action goal is mentally deactivated. If the action goal is judged as not or only insufficiently fulfilled, either the level of ambition is lowered and then the goal is deactivated or the action goal is maintained and new actions are planned, which are to make the achievement of the action goal possible after all. In the privacy context, this reveals a difficulty: While it is comparatively easy for the users to assess whether the action goal—in our example, the switch to privacy-friendly communication channels—has been achieved, it is almost impossible to assess the associated positive consequences. To do this, users would first have to have an overview of what data is collected about them through the use of privacy-unfriendly and privacy-friendly communication channels, how this data is processed, and what tangible consequences this has for their life. We know from research that users find it very difficult to assess the last aspect [6, 14, 16–21, 40], while at least users from the scope of the General Data Protection Regulation (GDPR) would in principle be entitled to information on the first two points—but here, too, anecdotal research shows that practice in many large companies is unfortunately currently far from providing users with comprehensive information on these aspects, even upon request [31, 38, 39]. Often, therefore, users at this stage are left to speculate about the positive effect of their actions, while the negative effect, for example, in the form of reduced usability or not reaching out to certain people via digital channels such as messenger, is clearly evident and thus can potentially lead to a change in the goal of action towards less privacy-preserving behavior. Similarly, some users find that they have only been able to partially achieve their goal but, in the absence of alternative promising action strategies, find themselves unable to continue pursuing their goal (with the prospect of successfully achieving it) and therefore lower their aspiration level. Especially in the context of messengers, this case often occurs when users are faced with the seemingly insurmountable obstacle that important communication partners cannot be reached via alternative channels (the so-called walled garden phenomenon [32]).

With its distinction between motivational and volitional phases, the Rubicon model also provides an explanation for the *privacy paradox*. In addition, it provides a suitable framework for designing interventions that are intended to support users in achieving their goals, such as a more conscious approach to their digital privacy. The model does not make any concrete assumptions about the occurrence of different desires or action goals, desired consequences, successful action strategies, and potential difficulties. It is therefore not suitable for predicting or explaining privacy behaviors. In our opinion, this model is helpful in principle, but it addresses a different context of application than, for example, the theory of planned behavior.

8 Capability, Opportunity, Motivation → Behavior (COM-B) System

The COM-B system is a behavior system in which capability, opportunity, and motivation interact and lead to a certain behavior that in turn influences these components (see Fig. 7) [27]. For example, a user wants to protect their privacy by using a more privacy-friendly channel to communicate with their friends. According

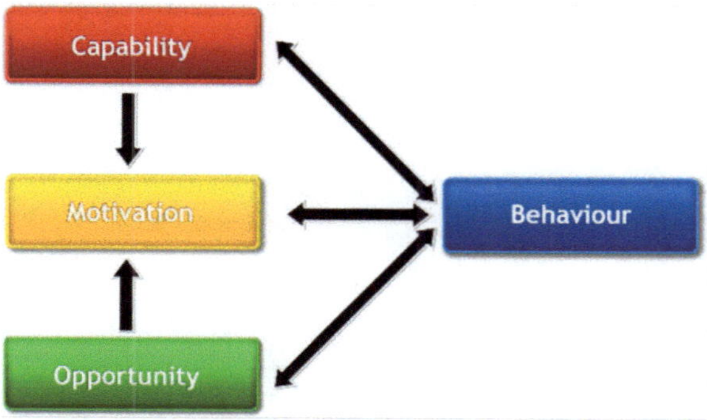

Fig. 7 The COM-B system, figure by Michie et al. [27] licensed under CC BY 2.0

to the COM-B model, the person needs the *psychological* and *physical ability* to perform the activity. This includes the required knowledge and skills. If we revisit the example from the previous section, this can mean that the person knows which privacy-friendly channels are available and how they are installed or used. Furthermore, the person must be *motivated*. Motivation is defined as all brain processes that stimulate and control behavior, not just goals and conscious decisions. It includes habitual processes, emotional responses, and analytical decisions. Applied to our example, this can mean that if the person wants to communicate with a friend in a privacy-friendly way, on the one hand, they have to believe that alternative messengers protect privacy (reflective motivation) and then automatically select the privacy-friendly channel in the process of communication (automatic motivation). Last but not least, there have to be *appropriate opportunities* for a certain behavior to be shown. This includes all factors that are external to the individual and enable or trigger the behavior. For our example, this may mean that the person is directly offered a privacy-friendly channel as the first choice for communication.

While the COM-B model is a behavioral model, it also provides a basis for designing interventions aimed at changing behavior. According to Michie et al. [27], a particular intervention can change one or more components of the behavioral system. The causal links within the system can reduce or increase the effect of certain interventions by leading to changes elsewhere. The task is to consider what the behavior should be and what components of the behavioral system need to be changed to achieve this. Thus, the model can also serve as a basis for developing interventions to promote privacy-friendly behavior.

9 Health Action Process Approach

Originating from health research, the health action process approach (HAPA) [34, 35] describes the factors that bring about a change toward healthier behavior (either by taking up health-promoting activities such as exercise or by quitting unhealthy activities such as smoking). Like the Rubicon model, a motivational phase and

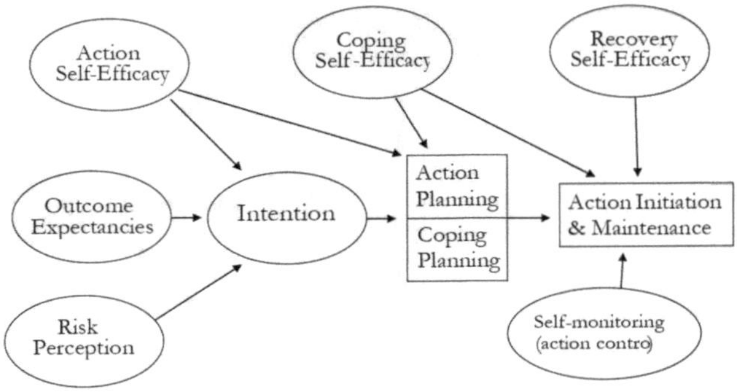

Fig. 8 The HAPA model, figure taken from Schwarzer [35] licensed under CC BY-NC-ND 4.0

a volitional phase are distinguished here, with intention forming the transition between the two phases (see Fig. 8).

Motivation may start with a *perceived risk* (in the privacy context, this could come, e.g., from a conversation with privacy-aware individuals or from hearing media reports about adverse consequences of data disclosure). In the further course, however, perceived risk plays a rather subordinate role and thus serves primarily as a trigger for building motivation. The motivational strength is mainly influenced by the other two variables, i.e., similar to the Rubicon model, *outcome expectancies* (which in this model would again be a weighing of potential advantages and disadvantages of a behavior), and *self-efficacy* (i.e., the extent to which the users are convinced that they can actually perform the behavior). Once the intention for a behavior has been formed, a planning phase follows first in this model as well. However, HAPA differentiates between figuring out strategies to perform the actual planned behavior (to revisit the previous example again, this could be, e.g., installing and using privacy-preserving messengers and uninstalling privacy-threatening messengers), called *action planning*, and figuring out replacement strategies or strategies to deal with potential obstacles (e.g., "If my mother is not willing to switch to another messenger, I will communicate with her by phone call and email in the future instead"), called *coping planning*. Again, self-efficacy plays a crucial role at this point in terms of the extent to which behavior can be maintained and potential difficulties dealt with, assuming that coping self-efficacy is distinct from action self-efficacy, i.e., users who have high action self-efficacy do not necessarily have high coping self-efficacy. Conceptually closely related to this is recovery self-efficacy, which describes the extent to which users can recover from possible setbacks and resume the desired behavior. Also important for maintaining the desired behavior is *action control*, which is usually achieved via self-monitoring—i.e., the users monitor their own behavior and check whether it is consistent with the targeted behavior.

HAPA combines the distinction between motivational and volitional phases with explanatory factors, providing both an approach to explaining behavior (and the *intention-behavior gap*) and a theoretical framework for designing interventions.

For the latter, the model further distinguishes between individuals who are in the motivational phase (non-intenders), those who have already formed an intention but are still engaged in planning (intenders), and those who are already acting (actors). For non-intenders, successful interventions should focus on risk and resource communication, while intenders are best served by interventions designed to help them plan specific strategies (for primary behavior and dealing with obstacles), and actors benefit most from interventions designed to protect them from potential relapse, for example, by avoiding risky situations.

10 Conclusion

As of yet, there is no theory or behavioral model, which includes all factors that contribute to user privacy and can be used to perfectly predict privacy behavior. Still, aligning one's research with theoretical behavior models adds validity and can inspire novel avenues for future work. Particularly models that originate from other contexts than privacy, e.g., originally stemming from health research, can provide valuable input to trigger new perspectives. The transactional model of stress and coping, for example, shines light on the fact that users being constantly overwhelmed by the management of their digital privacy can experience stress, which might lead to severe mental and/or physical health problems; a fact that should receive more attention in the public debates around the worth of data protection. Further, classic and concise psychological theories such as the theory of planned behavior can provide an intuitive to understand framework for conducting empirical studies. Models focusing on behavior change, such as the HAPA model, can contribute to our understanding of when, why, and how users can alter their privacy behavior towards a more deliberate handling of their private data, and thus form a decent basis for developing privacy interventions.

References

1. Abelson, R. P., Aronson, E. E., McGuire, W. J., Newcomb, T. M., Rosenberg, M. J., & Tannenbaum, P. H. (1968). *Theories of cognitive consistency: A sourcebook.* Rand-McNally.
2. Achtziger, A. & Gollwitzer, P. M. (2008). Motivation and volition in the course of action. In *Motivation and action* (2nd ed., pp. 272–295). Cambridge University Press.
3. Acock, A. C., & DeFleur, M. L. (1972). A configurational approach to contingent consistency in the attitude–behavior relationship. *American Sociological Review, 37*(6), 714–726.
4. Ajzen, I. (1991). The theory of planned behavior. *Organizational Behavior and Human Decision Processes, 50*(2), 179–211. Theories of Cognitive Self-Regulation.
5. Ajzen, I. (2012). Martin Fishbein's legacy: The reasoned action approach. *The ANNALS of the American Academy of Political and Social Science, 640*(1), 11–27.
6. Aktypi, A., Nurse, J. R. C., & Goldsmith, M. (2017). Unwinding Ariadne's identity thread: Privacy risks with fitness trackers and online social networks. In *Proceedings of the Multimedia Privacy and Security Workshop (MPS) at the 24th ACM Conference on Computer & Communication Security (CCS).*

7. Buck, C., Dinev, T., & Anaraky, R. G. (2022). Revisiting APCO. In *Modern socio-technical perspectives on privacy* (pp. 43–60). Springer International Publishing.
8. Claudy, M. C., Peterson, M., & O'Driscoll, A. (2013). Understanding the attitude-behavior gap for renewable energy systems using behavioral reasoning theory. *Journal of Macromarketing, 33*(4), 273–287.
9. Conner, M., & Armitage, C. J. (1998). Extending the theory of planned behavior: A review and avenues for further research. *Journal of Applied Social Psychology, 28*(15), 1429–1464.
10. Dinev, T., McConnell, A. R., & Smith, H. J. (2015). Research commentary—Informing privacy research through information systems, psychology, and behavioral economics: Thinking outside the "APCO" box. *Information Systems Research, 26*(4), 639–655.
11. Festinger, L. (1962). Cognitive dissonance. *Scientific American, 207*(4), 93–106.
12. Festinger, L. (1962). *A theory of cognitive dissonance* (Vol. 2). Stanford University Press.
13. Flender, C., & Müller, G. (2012). Type indeterminacy in privacy decisions: The privacy paradox revisited. In J. R. Busemeyer, F. Dubois, A. Lambert-Mogiliansky, & M. Melucci (Eds.), *Quantum interaction* (pp. 148–159). Springer.
14. Garg, V., Benton, K., & Camp, L. J. (2014). The privacy paradox: A Facebook case study. In *Proceedings of the 42nd Research Conference on Communication, Information and Internet Policy*.
15. Gerber, N., Gerber, P., & Volkamer, M. (2018). Explaining the privacy paradox: A systematic review of literature investigating privacy attitude and behavior. *Computers & Security, 77*, 226–261.
16. Gerber, N., Reinheimer, B., & Volkamer, M. (2018). Home sweet home? Investigating users' awareness of smart home privacy threats. In *Proceedings of an Interactive Workshop on the Human aspects of Smarthome Security and Privacy (WSSP)*. USENIX.
17. Gerber, N., Reinheimer, B., & Volkamer, M. (2019). Investigating people's privacy risk perception. *Proceedings on Privacy Enhancing Technologies (PoPETs), 3*, 267–288.
18. Gerber, N., Zimmermann, V., & Volkamer, M. (2019). Why johnny fails to protect his privacy. In *Proceedings of the 4th European Workshop on Usable Security (EuroUSEC)*. IEEE.
19. Harbach, M., Fahl, S., & Smith, M. (2014). Who's afraid of which bad wolf? A survey of IT security risk awareness. In *2014 IEEE 27th Computer Security Foundations Symposium* (pp. 97–110).
20. Jakobi, T., von Grafenstein, M., Smieskol, P., & Stevens, G. (2022). A taxonomy of user-perceived privacy risks to foster accountability of data-based services. *Journal of Responsible Technology, 10*, 100029.
21. Karwatzki, S., Trenz, M., Tuunainen, V. K., & Veit, D. (2017). Adverse consequences of access to individuals' information: An analysis of perceptions and the scope of organisational influence. *European Journal of Information Systems, 26*(6), 688–715.
22. Kokolakis, S. (2017). Privacy attitudes and privacy behaviour: A review of current research on the privacy paradox phenomenon. *Computers & Security, 64*, 122–134.
23. Lazarus, R. S., & Folkman, S.1984. *Stress, appraisal, and coping*. Springer.
24. Lee, N., & Kwon, O. (2015). A privacy-aware feature selection method for solving the personalization-privacy paradox in mobile wellness healthcare services. *Expert Systems with Applications, 42*(5), 2764–2771.
25. Leone, L., Perugini, M., & Ercolani, A. P. (1999). A comparison of three models of attitude–behavior relationships in the studying behavior domain. *European Journal of Social Psychology, 29*(2-3), 161–189.
26. Machuletz, D., Laube, S., & Böhme, R. (2018). Webcam covering as planned behavior. In *Proceedings of the 2018 CHI Conference on Human Factors in Computing Systems (CHI)* (pp. 180:1–180:13). ACM.
27. Michie, S., Van Stralen, M. M., & West, R. (2011). The behaviour change wheel: A new method for characterising and designing behaviour change interventions. *Implementation Science, 6*(1), 1–12.
28. Mittal, B. (1988). Achieving higher seat belt usage: The role of habit in bridging the attitude–behavior gap. *Journal of Applied Social Psychology, 18*(12), 993–1016.

29. Ozdemir, Z. D., Smith, H. J., & Benamati, J. H. (2017). Antecedents and outcomes of information privacy concerns in a peer context: An exploratory study. *European Journal of Information Systems, 26*(6), 642–660.
30. Park, H. J., & Lin, L. M. (2020). Exploring attitude–behavior gap in sustainable consumption: Comparison of recycled and upcycled fashion products. *Journal of Business Research, 117*, 623–628.
31. Pins, D., Jakobi, T., Stevens, G., Alizadeh, F., & Krüger, J. (2022). Finding, getting and understanding: The user journey for the GDPR's right to access. *Behaviour & Information Technology, 41*(10), 2174–2200.
32. Riley, C. (2019). *Meet the newest walled garden*. https://blog.mozilla.org/netpolicy/2019/03/11/meet-the-newest-walled-garden. Accessed October 14, 2022.
33. Rittenberg, L., & Trigarthen, T. (2012). *Principles of microeconomics*. Flat World Knowledge.
34. Schwarzer, R. (1999). Self-regulatory processes in the adoption and maintenance of health behaviors. *Journal of Health Psychology, 4*(2), 115–127.
35. Schwarzer, R. (2016). Health action process approach (HAPA) as a theoretical framework to understand behavior change. *Actualidades en Psicología, 30*(121), 119–130.
36. Smith, H. J., Dinev, T., & Xu, H. (2011). Information privacy research: An interdisciplinary review. *MIS Quarterly, 35*(4), 989–1015.
37. Sniehotta, F. (2009). An experimental test of the theory of planned behavior. *Applied Psychology: Health and Well-Being, 1*(2), 257–270.
38. Tolsdorf, J., Fischer, M., & Iacono, L. L. (2021). A case study on the implementation of the right of access in privacy dashboards. In N. Gruschka, L. F. C. Antunes, K. Rannenberg, & P. Drogkaris (Eds.), *Privacy technologies and policy* (pp. 23–46). Springer International Publishing.
39. Urban, T., Tatang, D., Degeling, M., Holz, T., & Pohlmann, N. (2019). A study on subject data access in online advertising after the GDPR. In *Data privacy management, cryptocurrencies and blockchain technology* (pp. 61–79). Springer.
40. Zeng, E., Mare, S., & Roesner, F. (2017). End user security and privacy concerns with smart homes. In *Proceedings of the 13th Symposium on Usable Privacy and Security (SOUPS)* (pp. 65–80). USENIX.

Part II
Methodology

Empirical Research Methods in Usable Privacy and Security

Verena Distler, Matthias Fassl, Hana Habib, Katharina Krombholz, Gabriele Lenzini, Carine Lallemand, Vincent Koenig, and Lorrie Faith Cranor

1 Introduction

Researchers in the usable privacy and security (UPS) field study privacy- and security-relevant perceptions and behaviors and aim to design systems that simultaneously address requirements for usability/user experience, security, and privacy. Human-computer interaction (HCI) and social science research methods are well-suited to study many of the types of questions that are relevant in UPS, which often involve concepts such as subjective experience, attitudes, understanding, behavior and behavior change. However, there are many challenges specific to UPS that are not usually described in more generic methods textbooks. We highlight techniques

V. Distler (✉)
University of Luxembourg, HCI Research Group, Esch-sur-Alzette, Luxembourg

University of Bundeswehr, Munich, Germany
e-mail: verena.distler@unibw.de

M. Fassl · K. Krombholz
CISPA Helmholtz Center for Information Security, Saarbrücken, Germany
e-mail: matthias.fassl@cispa.de; krombholz@cispa.de

H. Habib · L. F. Cranor
Carnegie Mellon University, Pittsburgh, PA, USA
e-mail: htq@cs.cmu.edu; lorrie@cs.cmu.edu

G. Lenzini · V. Koenig
University of Luxembourg, HCI Research Group, Esch-sur-Alzette, Luxembourg
e-mail: gabriele.lenzini@uni.lu; vincent.koenig@uni.lu

C. Lallemand
University of Luxembourg, HCI Research Group, Esch-sur-Alzette, Luxembourg

Department of Industrial Design, Eindhoven University of Technology, Eindhoven, Netherlands
e-mail: carine.lallemand@uni.lu

N. Gerber et al. (eds.), *Human Factors in Privacy Research*,
https://doi.org/10.1007/978-3-031-28643-8_3

for risk representation, options for participant recruitment, ethics-related topics in study design, and biases that may play a role in UPS studies with human participants.

Structure of this Chapter We first highlight specific challenges in applying research methods with human participants to the field of UPS (section "Common Challenges in UPS Research"). We then describe the most frequently used research methods, providing a general overview for each method, potential challenges when applying the method to UPS, examples of prior UPS work, and references for further reading on the methods (Sect. 2). Next, we describe methodological "techniques" that can be used in conjunction with these methods (Sect. 3). We then describe options for recruiting participants (Sect. 4). We also discuss important ethical considerations (Sect. 5), and psychological biases (Sect. 6).

Common Challenges in UPS Research UPS studies often set out to study motivations, perceptions, or reactions related to security or privacy threats. The methodological challenges in UPS differ from other areas of human-computer interaction, as including privacy and security threats in an ecologically valid way significantly increases complexity of study designs. While lab studies, for example, can help control the environment to exclude the influence of external variables, the threat participants perceive may be vastly different than what they experience in their everyday lives. When study participants are given fictitious personal information and credentials for the purposes of the study so as to keep their personal data confidential, participants are likely to feel and behave differently than they would if their own data was at risk. In addition, when researchers simulate attacks in a lab, these attacks typically happen at a higher rate than outside of a study context, which further jeopardizes ecological validity [34]. Another challenge is that when people interact with technology, they usually aim to complete tasks that are not related to security or privacy. Thus, mentioning security or privacy may prime participants to think about these topics and cause them to behave differently than they normally would [24]. Privacy and security-related behaviors can also be perceived as socially desirable by participants, and they are likely to adapt answers and behaviors accordingly [25] (more on the social desirability bias in Sect. 6). While self-reported data is crucial to explore participants' experiences and perceptions, it is also not sufficient to quantify and fully understand past behaviors, which are difficult to remember and report accurately. UPS researchers have followed a variety of approaches to tackle these challenges, and to create a realistic experience of risk in their studies, including the use of role-playing to simulate realistic situations [90], the use of deception to simulate the presence of an adversary [16], or launching simulated attacks on research participants [24]. These approaches can bring highly valuable insights, but they need to be balanced with ethical and feasibility considerations.

2 Research Methods in UPS Studies

This section provides readers with an overview of a selection of research methods. These methods are described in more detail in many methods books. Broadly, we can think of empirical methods on a continuum of qualitative to more quantitative methods. Qualitative methods generally have the objective of exploring topics in depth and generating hypotheses inductively, while quantitative methods often build upon such qualitative research to generate research results that generalize to larger samples, or test hypotheses. This distinction is relatively fluid. For example, while interviews are typically used to gain deep understanding of a topic and generate hypotheses, they could also be used in the evaluation of a technology. Similarly, questionnaires (a quantitative method) can be used for descriptive, relational, and experimental objectives.

Regardless of method, studies generally build on previous work. Reviewing existing literature is an essential process. We begin by describing a specific type of literature review: systematic literature reviews. We then describe empirical methods, starting from qualitative methods that are used more frequently to explore and understand topics in depth, and then moving on to quantitative methods that are used more frequently to evaluate hypotheses. In each subsection, we provide a short description of the method, give examples of use of this method in UPS, and discuss how risk is represented. We then point out references that provide detailed descriptions of how to apply these methods. We include methods that are used frequently in UPS, as well as methods that are used less frequently, but seem promising.

2.1 Systematic Literature Reviews

Systematic literature reviews refer to an approach that reviews the literature by adopting explicit procedures [12]. It provides a log of the researcher's decisions, procedures, and conclusions [98] and can help avoid biases on the part of the researcher conducting the literature review [12]. Broadly, systematic literature reviews follow the process of (1) defining the purpose and scope of the review, (2) seeking out relevant studies according to keywords and inclusion criteria (e.g., articles containing the keywords within certain scientific databases), (3) narrowing down the selection of studies from step 2, and (4) analyzing each study based on the study objectives [12].

The authors of this chapter have previously conducted a systematic literature review of recent UPS conference papers [20]. We will describe the method we used and some trends in the UPS research literature we identified through this process. The objective of this systematic literature review was to understand the methods used in the UPS community, how researchers represent risk, and how deception is used in study protocols. We included papers from five top-tier UPS venues if they mentioned security or privacy in addition to one user-related term in the title or abstract. After filtering out papers, we conducted a detailed review of the papers, excluding additional papers that did not meet our criteria. We analyzed the remaining papers according to an analysis structure to respond to our research questions.

The analysis structure researchers use to make sense of a large corpus of work can be a contribution in itself. For example, to investigate how risk was represented to research participants, we used a categorization of risk representation that is useful when considering user research methods in UPS. *Naturally occurring risk* refers to studies that rely on risk perceptions that already exist in a participant, for instance, when observing or self-reporting on naturally occurring behavior. *Simulated risk* is used in studies in which participants are asked to situate themselves in a scenario that would trigger a perception of risk (e.g., role-playing a security-critical interaction), as well as in studies that use deception to make participants believe that a simulated risk is real. *Mentioned risk* is used in studies that describe risk to research participants (e.g., "some people's partners install tracking apps on their phone"). Mentioned risk can be compared to presenting a hypothetical situation to participants to "anchor" [32] their responses in this context, but without asking them to place themselves in the situation. Some studies also *do not represent risk* at all, for instance, when simply investigating the usability of a new tool. In this chapter, we use this categorization when describing study methods.

In our literature review we found that the included papers predominantly used experiments, surveys, and interviews. The majority of papers involved naturally occurring or simulated risk. The literature review also provided an analysis structure that could be used by venues to encourage better reporting of user studies, and a checklist for researchers and reviewers to use for detailed reporting of methods. We discussed some methods that were used rarely in the sample (e.g., co-creation and participatory design, group methods such as focus groups), some participant groups that were rarely included in the analyzed sample (non-Western populations, people with disabilities, members of the LGBTQ+ community, older adults, and minors). We invite the reader to read the original paper for details [20].

2.2 Interviews

Interviewing is the most widely used method in qualitative research and also one of the top-three methods used in empirical studies in UPS. In our previous systematic literature review, 13% of papers used interviews on their own [20]. An interview typically takes the form of a conversation between an interviewer, who asks questions, and an interviewee, who answers these questions. An individual interview usually lasts between 45 and 90 min. Interviews are a powerful, yet labor-intensive data collection technique [53]. While it seems easy to have a "conversation" with a participant, research interviews require skills and the awareness of best practices to follow, especially related to the formulation of unbiased questions (see Sect. 6) and qualitative data analysis.

The most common interviewing format in HCI and UPS is the semi-structured interview, which offers a compromise between the level of flexibility to reflect the interviewee's perspective and the comparability between interviewees. In a semi-structured interview, the interviewer has a list of topics to address, formulated as a set of questions compiled in an interview guide. There is some flexibility in the sequence and in the way the interviewee can respond, but all questions will be asked in a similar way to all interviewees. In-between the questions from the interview guide, the interviewer will use follow-up questions aimed at clarifying or exploring some points further.

Interviews most frequently use naturally occurring risks to investigate people's real-life privacy and security experiences [20]. For example, Rashidi et al. [76] interviewed undergraduates to understand their real-life privacy workarounds in the context of pervasive photography. Similarly, Ahmed et al. [1] enquired about the real-life privacy concerns of people with visual impairments. Interviews can also be used to explore naturally occurring risks in organizational settings [15, 41]. Interviewers sometimes make use of material or situations to support the interview [53], especially when involving specific users. McReynolds et al. [59] invited kids to play with connected toys in a lab setting, with their parents present. Both parents and children were interviewed about their mental model of the toys and privacy perceptions. Less frequently, interview studies use scenarios to simulate risk. For example, Vaniea et al. [99] used a set of hypothetical scenarios to elicit stories about software update experiences.

Many texts offer guidelines about preparing, conducting, and analyzing interviews [5, 12, 50, 53].

2.3 Focus Groups

Focus groups are a form of interviewing where several participants are invited to share and discuss their ideas and opinions around a predefined subject in a "group interview" [12]. The sessions are run by a moderator, who guides the participants in a rather flexible and non-intrusive way. Group interaction is essential: it helps participants to explore, clarify, and build their points of view, revealing insights to researchers in the process. Focus groups might involve a combination of individual and group tasks as well as discussion. The main challenges lie in the adequate setup of the sessions and the transcription and analysis of the data [53].

Focus groups have been used only rarely in UPS (less than 1% of papers analyzed in our previous literature review used focus groups on their own [20]), generally in studies employing naturally occurring risk. In most cases, focus groups aim at gathering qualitative in-depth insights from lay participants. For instance, Freed et al. [31] conducted 11 focus groups with 39 survivors of intimate partner violence to understand how abusers exploit technology for harmful purposes. While discussing sensitive topics in a group setting can sometimes trigger discomfort, group interviews were useful in this context. Focus groups questions usually revolve around high-level topics, in this case how technology was used in these abusive relationships, the strategies used by participants to defend themselves, and ideas to cope with technology-enabled abuse. The findings unveiled abuser strategies and provided a nuanced, yet detailed view of these attacks, helpful for designing mitigation strategies. Sambasivan et al. [86] conducted focus groups to gain an understanding of how South Asian women perceive, manage, and control their personal privacy on shared phones. The authors identified five performative practices that participants employed to maintain individuality and privacy in contexts where devices were frequently borrowed and monitored by social relations.

Focus groups can also be conducted with security and privacy experts. Following interviews and a drawing task with end-users, Murillo et al. [62] ran two focus groups with seven data deletion experts. The aim of the focus groups was to set a baseline to compare the interview findings against. Prior to the session, the experts were asked to draw how they think online data deletion works. The drawings were discussed during the focus group and acted as a support to decide the most important aspects that a lay person should know to have a good understanding about deleting online data. Finally, many disciplines have shown a growing interest in conducting focus groups online, and we expect this practice to develop in the UPS field as well. Studying privacy trade-offs, Distler et al. [22] used private social media groups to conduct asynchronous online focus groups in combination with in-person focus groups. Confronted with several scenarios, participants first noted advantages and shortcomings individually, and then discussed and confronted their opinions in the online focus group setting.

Focus groups in UPS are an underused yet insightful method, holding the potential of gathering qualitative in-depth insights into privacy and security attitudes that might help the community obtain rich qualitative results. Methodological research in social sciences has produced worthwhile recommendations to guide the setup of focus groups. Questions about number and type of participants to involve, how to moderate focus groups, and how to analyze focus group data are addressed in textbooks [5, 12, 36, 53].

2.4 Co-Creation Methods

Co-creation refers to a wide range of practices that usually bring together both designers and people who were not trained in design [54, 87]. The involvement of users goes from a temporary involvement during a workshop to an in-depth participation across all stages of the design process. Co-creation approaches treat users as co-creators and partners who are experts in their lived experience [87, 88]. In contrast, user-centered design broadly sees users as more passive subjects whose role is to inform the designers about requirements or to give feedback on their ideas. Co-creation involves users during knowledge development, idea generation, and concept development, with designers facilitating the process and providing tools for expression. Participatory design is a related approach that attempts to understand a problem from the participants' perspective and sees methods as means for users to gain influence in the design process [40]. Participatory design allows participants to contribute to building solutions [53]. Originally, participatory design was a form of co-creation with political goals, but today, the term participatory design is often used as a synonym for co-design [9].

As there are no fixed rules for conducting co-creation studies, it is crucial to document the co-creation process in detail and justify choices. In UPS research, user understanding and acceptance of security mechanisms are reoccurring issues. Co-creation methods may contribute to improving them. Design research often focuses on providing design outcomes, but researchers can also use research-through-design approaches [109] throughout the design process itself, e.g., by engaging with wicked problems to reframe problem statements. When recruiting participants who are not experts in UPS for co-creation studies, researchers should consider additionally obtaining security and UX experts' input to create user-centered, appropriate, and functional designs.

Currently, co-creation methods are still used rarely in UPS. Only two papers in the corpus of our UPS literature review [20] used them: Egelman et al. [23] crowd-sourced 274 sketches of potential privacy icons, and Adams et al. [19] asked VR developers to contribute to a VR code of ethics on a shared online document. Outside the scope of our literature review, Weber et al. [105] used a participatory approach to create SSL warning messages. Fassl et al. [26] used collaborative design workshops to ideate potential authentication ceremonies for secure messengers. In Distler et al.'s [21] study, security and HCI experts co-created textual and visual representations of security mechanisms.

2.5 Surveys

Surveys (also called questionnaires) are one of the most common research methods in social sciences, as well as HCI and UPS [20]. 12% of papers in our previous systematic literature review used surveys on their own [20]. Surveys consist of a well-defined and carefully written set of questions to which individuals are invited to respond [53]. They are mainly used to measure people's attitudes, to gauge the existing knowledge around a topic, to profile users, or to gather feedback about user experiences [53]. Surveys are mostly administered in a written form (on paper or online) but can also be conducted face-to-face or by telephone [77]. Surveys are an effective and flexible technique that can address several research objectives and reach a larger audience than most other empirical methods. Surveys can be used as a standalone data collection method or in combination with other techniques. Noteworthy, the validity of the data produced by surveys highly depends on the rigor of the survey design: often misperceived as easy to create, poorly designed surveys can provide meaningless or inaccurate information [5]. Methodological textbooks provide extensive guidance about survey design [77, 97].

In our systematic literature review [20], papers in UPS that used a survey-based approach most frequently used naturally occurring risk or multiple risk representations but were less likely to mention or simulate risk. In the case of naturally occurring risk, participants are asked about real-world behaviors in actual situations. Redmiles et al. [78], for example, investigated how users' profile, security beliefs, and knowledge influenced security behaviors. Surveys have been used in UPS to explore user's understanding, behaviors, and attitudes towards a myriad of topics, from password expiration policies and authentication [14, 39], data security practices [66], data breaches [47], private browsing [37], security indicators [28], targeted ads [72], privacy perception [85, 96], or encryption [82] to only mention a few. Researchers often compare several user profiles (using demographics questions) or populations (through sampling), for instance, expert and non-expert security practices or mental models [18, 46]. Surveys can, for example, be combined with data logs (see Sect. 2.6) or interviews (Sect. 2.2), both methods being complementary to the type of data produced by survey research. When risk is simulated in a survey study, a prototype or scenario is usually used [20]. For instance, Karunakaran et al. [47] used a scenario to invite participants to imagine that they were victims of a data breach.

2.6 Analyzing Measurement Data (Data Logs)

Measurement data can be analyzed to provide important insights into the use of a particular system, as well as the response of a system to user behaviors. In UPS research, this may involve the measurement of a specific privacy/security-related user behavior or evaluation of a privacy/security enhancing technology. Research studies using measurement data are typically conducted as field studies to analyze events of interest as they occur in the real world, where data is collected through an application developed specifically for data collection—such as a browser extension—or logs integrated into a developed system. Measurement data captured over a period of time can then be analyzed to report on frequency of occurrence of certain behaviors, longitudinal trends in the data, or performance metrics.

As with all research methods, analysis of measurement data offers advantages and disadvantages. Such studies offer the opportunity to capture users' behavior in their ordinary use of a system. This mitigates data quality issues associated with other study methods, such as participant behavior being influenced by a lab setting or poor recall of past behaviors in self-report methods. Furthermore, measurement studies analyze behaviors in environments where users encounter privacy and security risk as they naturally occur, rather than in settings where such risk must be mentioned or simulated. Logs and other data measurement tools can be instrumented to collect data from a large sample over a period of time. While this can allow for important quantitative and longitudinal insights, measurement data may need to be accompanied with other research methods (e.g., a survey) to explain the context behind the observed data. Additionally, developing and maintaining a data collection infrastructure for measurement studies may require more resources and technical expertise relative to other methods. While there may be opportunities to repurpose existing logs and measurements to answer new research questions, any data sharing and analysis should be done following guidelines for ethical research.

Logs and measurement data have been previously used to explore several usable privacy and security topics. In our literature review, 24 papers included the use of log analysis, and 4 papers used datalogs on their own. These papers usually used naturally occurring risk. Data logs from deployed systems have been analyzed to explore users' adoption and usage of multi-factor authentication [14, 80], user responses to quarantining actions by Internet Service Providers [13], and reidentifiability risk of browsing data [7, 70]. Field studies using measurement data have also been used to evaluate prototype systems, such as those helping users manage Android permissions [69, 91]. Others have developed applications and browser extensions to measure privacy and security behaviors occurring in users' daily online activities, including characteristics about passwords used in real-world accounts [55, 71, 104], usage of private browsing mode [38], and effectiveness of phishing attacks [68]. In addition to user behavior, previous work has used measurement data to evaluate system behavior and characteristics related to different privacy and security topics, such as adoption of HTTPS [27], phishing detection [45, 67], and unsolicited phone calls [73]. Altogether, this past work—which is only a small sample of recent measurement studies in UPS—demonstrates that measurement and log data can be used to explore a wide range of research questions.

2.7 Extracting Online Datasets

In some social sciences, it is common for researchers to conduct research using datasets that are collected by government agencies or other official entities [53]. This is not (yet) a common practice in HCI or UPS, and most researchers collect their own data. Some researchers use creative approaches to extract online datasets. For example, Fiesler and Hallinan [29] collected and analyzed public comments to news articles discussing online data sharing and privacy controversies in the media. Researchers can also use other online data sources to extract and analyze datasets, for instance, from forums [74] or Twitter. In our previous systematic literature review, we found that 4% of papers analyzed datasets. Ethical concerns should be carefully considered when using "public" data for research purposes, as it is often infeasible to obtain informed consent from the people whose data are being studied and these types of studies are often not subject to ethics board review [30]. Indeed, there seems to be a lack of consensus on the role of ethical review boards in this type of research [101]. It seems promising to evaluate research approaches that use online data through the lens of contextual integrity [65, 108].

2.8 Experience Sampling Method

The experience sampling method (ESM) was developed in psychology during the 1960s and 1970s [44, 51]. It focuses on the individual participant's ongoing behavior and everyday experiences in their natural environment rather than laboratory settings. While the method was developed in psychology, it is applicable to a wide variety of questions. ESM studies usually take several weeks to collect sufficiently large samples. Participants receive a signal and then have a limited time to complete a short questionnaire about their current situation and experience. The method to receive these signals evolved over time: Earlier approaches used timers and beepers, while more recent approaches employ mobile phone apps and desktop software. Researchers regularly use time-based, situation-based, or random sampling strategies—depending on the study purpose. The primary benefit of experience sampling is the increased ecological validity compared to lab studies. Other methods also investigate naturally occurring situations, e.g., participant observation and diary studies, a qualitative approach where participants record entries about their daily lives. In contrast, ESM requires less effort from researchers and participants. However, the resulting data usually contains fewer details, and researchers often have limited control over the sampled situations and must rely on participants' situation descriptions. In practice, some qualitative diary studies and mixed-methods ESM studies may look very similar.

Experience sampling is an infrequently used method in UPS. Our UPS literature review [20] uncovered eight studies that used experience sampling (3% of the corpus). The following examples illustrate how experience sampling can provide valuable insights into naturally occurring situations: Bonné et al. [11] explored user decisions in runtime permission dialogs on Android. They used experience sampling to survey participants after each permission dialog. Reeder et al. [79] investigated users' experience with browser security warnings, using a browser extension to survey users immediately after the warnings appeared. Gallagher et al. [33] studied users' problems and frustrations with Tor Browser. They asked participants to explain their reasons every time users ended a browsing session or switched browsers. Most of these studies in UPS use a context-specific trigger for sampling experiences, but there are examples of random sampling as well: Yang et al. [107] evaluated a smartphone authentication mechanism under natural conditions, asking participants to complete a short questionnaire after each authentication task. All of the approaches repeatedly sample user experiences over long study periods, asking users for immediate and necessarily short feedback in situ.

2.9 Experiments

Experiments are an important method to identify the causes of a situation or a set of events [53], in which participants are randomly assigned to experimental conditions. The researchers then compare the effects of these conditions on one or multiple dependent variables. Experiments are important tools to test hypotheses and research questions, and they are a highly flexible method that can be used to address many types of research questions. Studies in HCI often use experiments to compare types of devices or versions of an interface [53], using both measured variables (e.g., time to complete) and subjective variables such as satisfaction, user experience, perceived security, and acceptance of a technology. While the term experiment is sometimes used more loosely in HCI to describe any type of user test, the random assignment of participants to experimental conditions is typically a defining characteristic of an experiment [12].

Experiments are commonly used in UPS research. Our previous systematic literature review found that experimental approaches were used in more than a third of the analyzed papers [20]. An example of an experiment is a study that asked computer science students to role-play and imagine that they were responsible for creating the code for user registration of a social networking platform. Students were randomly assigned one of four experimental conditions that varied on the instructions they were given (e.g., level of security priming). The researchers compared the effects of the conditions on dependent variables such as functionality or use of secure coding practices [63]. Experimental research can also investigate perceptions of security and privacy-relevant topics. A recent study [21] used a vignette experiment to investigate various approaches to displaying encryption to non-experts. Participants were randomly assigned to conditions that varied in the visual and textual representation of encryption. The authors then compared the effects of these representations on participants' user experience, perceived security and understanding of encryption.

Our 2021 systematic literature review found that papers focusing on experiments used simulated risk representation to a large extent (35% of analyzed papers used experiments on their own [20]). As exemplified by the studies mentioned previously, risk is often simulated using scenarios that participants should situate themselves in.

Insightful experiments are challenging to design, and research on experimental design is constantly progressing. Textbooks provide relevant instructions for conducting experiments [12, 35, 53].

3 Techniques that Can Be Used in Combination with Methods

Researchers often creatively combine the methods in Sect. 2 with a variety of techniques or "tools" that can help study security- and privacy-relevant interactions and represent risk to research participants in UPS.

Assigned Tasks UPS researchers sometimes ask participants to complete tasks that can be related or unrelated to security or privacy. Examples of such security/privacy-related tasks could be attempting to send an encrypted email [81] or asking participants to try using a new authentication method [17]. Sometimes researchers also ask participants to complete tasks that are not directly related to security or privacy so that they can observe routine tasks or tasks that incidentally have a security component, without drawing participants' attention to the security or privacy aspect.

Prototypes Many studies make use of prototypes developed by the study authors. These are often new low- or high-fidelity interface designs, icons, notices, or instructions [20]. Many of these studies use simulated risk.

Scenarios Many studies include scenarios in which researchers ask participants to imagine themselves being in a certain situation [20]. Such scenarios can be helpful to simulate risk by describing situations that would be risky to research participants. For example, some passwords research papers involve a scenario in which researchers ask participants to imagine that their email account had been hacked, and to change their password in response [49, 60]. Researchers also sometimes ask participants to role-play. For example, in one study participants were recruited together with a friend. These pairs of participants were placed in separate rooms and asked to communicate with each other using a secure email system. They were asked to role-play a scenario about completing taxes. They were instructed to share sensitive information (e.g., social security number) and treat it as if it was their own sensitive information [100].

Educational Interventions Some studies use interventions designed to educate participants about security/privacy topics. For example, Wash and Cooper educated their participants on how to detect phishing attempts [103], and Lastdrager et al. [52] evaluated how effective anti-phishing training can be with children. Warshaw et al. [102] tried to improve adults' understanding about how companies collect and make inferences about their data.

Financial Incentives While the field of behavioral economics frequently uses financial incentives in experiment designs to model real-world incentives, these approaches are still relatively rare in UPS studies [20]. In UPS, researchers often provide some compensation to research participants, but they do rarely provide additional Financial incentives can be used, for example, to provide encouragement for participants to complete privacy and security tasks well by making a part of their compensation contingent upon successful completion of a task (bonus in addition to base compensation). The consequences a user might face in real life for badly performing security and privacy tasks are different and go beyond financial disadvantage (e.g., account compromise, having to reset a password after forgetting it). However, a financial incentive can provide some additional motivation or perception of risk to research participants in a research setting. For example, Tan et al. [94] conducted an online experiment in which participants were asked to play the role of an accountant who requested confidential information from employees using a secure messaging system. The participants were asked to complete fingerprint verification for each request, and the researchers investigated whether the participants were able to recognize mismatched fingerprints. To simulate on-the-job pressures to get the task done quickly, the fastest 15% of participants were promised a $1 bonus in addition to their base compensation ($3).

Deception When using technology, users often have primary objectives that are unrelated to security or privacy. Letting participants know that a study's objective concerns privacy and security would prime them to think about these topics, when they might not have done so in a non-research context ("security priming" [93]). For additional realism, some studies use deception. There is a grey area between avoiding priming participants and the use of deception [20]. Deception is typically defined as deliberately misleading participants or not informing them about the purpose of the investigation, usually to avoid participants giving answers that they perceive are expected by the researchers [3]. In UPS studies, deception is often used to make participants believe a risk to their security or privacy is real. Studies involving deception need to carefully consider ethics. We describe these considerations in Sect. 5.2.

4 Participant Recruitment

There are a variety of approaches to participant recruitment, and different approaches can be used in combination. In addition to study objectives, logistical concerns play a role when deciding on a recruitment strategy. It is important to report on participant recruitment and compensation transparently in research papers, as this helps readers situate study results and aids replicability. We have previously provided a checklist that can help report all relevant information [20]. We describe common sampling approaches below.

Convenience Samples Convenience samples refer to populations that the researchers can easily access, and no systematic selection criteria are applied. Examples of convenience sampling include recruiting students at the researchers' own institutions, posting flyers in an institution's neighborhood, or recruiting colleagues or friends. Currently, a large proportion of studies in UPS are based on convenience samples [20]. Note that the question of whether a convenience sample will suffice for a study depends on the research question the researchers set out to address. For example, if the objective is to understand whether a novel gesture authentication mechanism is, in principle, usable while walking, a sample of students might be sufficient. However, if the objective is to determine the proportion of adults in the United States who hold certain views on privacy, convenience sampling is not appropriate. Snowball sampling is often closely related to convenience sampling. In this approach, research participants introduce the researcher to other potential participants, who fulfil certain criteria [8].

Representative Samples If the study objective is to generalize results to a larger population, researchers need to recruit a sample of participants representative of that population. Recruiting a randomly selected, representative sample is ideal.

Representative sampling is currently rare in UPS [20] but is now easier to (partially) achieve through new crowdsourcing solutions. Some crowdsourcing solutions, for instance, offer representative sampling options based on indicators such as gender, ethnicity, and age [75]. It is important to remember that these samples may not generalize to less tech-savvy parts of the population.

Crowd Workers Platforms such as Prolific, Mechanical Turk, or Crowdflower provide access to crowd workers who volunteer to take part in paid research studies. They allow researchers to filter potential participants regarding specific characteristics or recruit large, unfiltered samples easily. They also allow researchers to recruit for a variety of methods, from remote interviews to survey experiments. When recruiting on these types of platforms, it is important to keep in mind that participants are likely more tech-savvy than the average person in the population [75].

Specific Groups of People Some studies also include specific groups of people, such as employees in an organization, experts in a relevant field (e.g., [21]), or focus on specific groups such as women, children, or older adults. Some groups, for example, minors, are considered vulnerable populations, and researchers need special ethics board approvals for their participation (also refer to Sect. 5.3).

5 Basics of Ethical Research Design with Human Participants

5.1 Ethical Core Principles

There are numerous frameworks for ethical research, such as APA's ethical principles of psychologists [2] or the Belmont Report [95], which share common principles. The ethics statements of major security conferences reference the 2012 Menlo Report [48], which we use here to illustrate four ethical core principles: (1) Respect for Persons, (2) Beneficence, (3) Justice, and (4) Respect for Law and Public Interest. While the full report provides guidelines to operationalize these core principles, this section summarizes issues that frequently arise in UPS research.

Respect for Persons Human research subjects are autonomous persons; researchers should treat them as such. They have the right to decide if and in what research they want to participate. To ensure this, researchers inform them of the type of research and ask for explicit consent to participate. Written consent forms are a common way to document this procedure. Similarly, human subjects always have the right to withdraw from a research study—researchers must remind them of this and provide an easy way to do so, e.g., contact information or link to an opt-out website. However, some research designs make a detailed informed consent procedure difficult. First, it may be an unreasonable effort for researchers to collect informed consent from all participants (e.g., analyzing existing large datasets or when participants are hard to identify or contact). Second, research designs that deceive participants cannot collect meaningful informed consent. Researchers should debrief participants to prevent mistrust in researchers [2]. Schechter et al. [89] recommend rigorous reporting on debriefing, participants' reactions and data protection measures in these cases. In addition, we recommend allowing participants to withdraw from the study after debriefing them.

Beneficence Research should have the best interest of the participants and society in mind. Identifying benefits and potential harms is the initial step to deliberating a research project's beneficence. While participants' welfare is the primary focus, researchers should also consider other stakeholders' welfare. In the UPS field, this often concerns participants' family members, malicious actors, security product vendors, or the research community itself. These stakeholders may be involved in or affected by harm in different ways: research may reveal too much personal information or harm their social standing with their relatives or workplace. Research into users' security behavior may also reveal issues with security products and inform future attacks. Deceptive research designs may also impact researchers' reputation, making future research projects on the same topic difficult. Researchers should avoid any potential harm resulting from their research. If some harm is unavoidable, researchers must do their best to mitigate it, e.g., by anonymizing personal information, rephrasing quotes, or debriefing participants after deceptive research.

Justice Research is a burden for participants—they invest time to inform researchers about their privacy and security experiences. There is a danger of exploitative researcher-participant relationships when vulnerable participants bear the burden of research to benefit a different group of people. Hence, research participants should benefit from the research in one way or the other. Fair monetary compensation for the participants' time is a common way to ensure that. In UPS research, this is especially important when using crowdsourcing platforms (e.g., Prolific) for recruitment. Crowdworkers participate in research to make income and might not benefit in any other way. Paying workers at least the minimum wage at their location is a regular (currently not met [42]) demand of workers and researchers alike [92]. Noteworthy, participants may also reap the benefits of security research in other ways, such as personalized security education or improved security and privacy of the tools they use. We recommend reflecting and reporting on these issues when conducting studies.

Respect for Law and Public Interest Researchers need to identify and adhere to laws that apply to their research. Legal issues might be hard to navigate when security research involves reverse engineering or vulnerability disclosure. However, in UPS research, data protection laws and terms of service for crowdsourcing platforms, survey platforms, and transcriptions services are the most common issues. Collecting participants' informed consent about data collection, storage, sharing, and deletion is a best practice. In addition, data protection laws (e.g., GDPR in the European Union) give data subjects certain rights, e.g., information about data practices and data deletion upon request. At most research institutions, support for legal issues is available. To provide transparency and accountability, researchers should responsibly share research methods, ethical evaluations, collected data, and results with researchers or policy-makers.

5.2 Ethical Considerations for Deceptive Research

In rare cases, deceiving participants about the research purpose may be necessary to address underlying research questions, e.g., knowing a study purpose might influence user behavior under investigation. Deception in research designs is a contentious issue and should remain a highly regulated exception. Guidelines recommend debriefing participants as soon as possible after the deception to avoid any psychological harm (e.g., feelings of shame, guilt, and worrying about negative impacts from assumed but not real risks). Debriefing avoids participants losing trust in researchers [2]. Research protocols involving deception usually require approval from an institution's ethics board.

Studies may deceive participants about the study objective or the presence of risk. For instance, Samat and Acquisti [84] told participants that they would share their information with a specific group of people, when in reality they did not. Marforio et al. [57] asked participants to test a banking prototype for one week and simulated a phishing attack on the prototype. Our 2021 literature review [20] shows that 6% of the sampled UPS papers used deception. Researchers mostly used deception in experimental studies on social engineering and privacy inform-and-consent mechanisms. Two-thirds of these papers mention a debriefing process, and most have IRB-approval.

5.3 Ethical Review Boards

Many universities and other organizations have an Ethical Review Board (ERB) or an Institutional Review Board (IRB) to support researchers with ethical deliberations and legal compliance. While the application process (incl. duration and required level of detail) differs between countries and universities, most applications require at least a description of the research process and the identified ethical concerns. Applying for an ERB/IRB decision should be one of the first steps of starting a research project. Continued communication is necessary when changing the research design.

The call for papers of top-tier security conferences (IEEE S&P, USENIX Security, ACM CCS, and NDSS) includes ethics statements. They require that all submissions that experiment on human subjects, analyze data derived from human subjects, or put humans at risk should discuss their research ethics and disclose whether they have an ERB/IRB-approval or exemption. These ethics statements describe minimum requirements for human-subjects' studies; ethical community standards and review processes are continuously under scrutiny and updated accordingly.

In our 2021 review of UPS literature [20], we found that 56% of the papers obtained approval or received exempt approval, 35% did not mention the topic, and 4% were from institutions without approval procedure. The remaining papers either described a corporate internal review process or claimed to be exempt from needing approval.

When a researcher's institution does not have its own approval process, we advise stating so in the paper and transparently discussing potential ethical issues and how they were justified and mitigated. Especially in sensitive cases (e.g., studying the privacy and security concerns of political activists or survivors of intimate partner violence (IPV), or designing privacy/security solutions for an elderly population) an informal advisory board with research peers and subject-matter experts may be useful for providing feedback to ethical questions as they arise.

6 Biases in Research with Human Participants

The validity and reliability of what we measure in user studies can be compromised by systematic errors or biases. We highlight some of the most common biases that can impact UPS research.

The Method The choice of an appropriate method or the construction of an ad-hoc method bears numerous opportunities for biasing measures. For instance, issues can arise from the question order, priming effects, inappropriate response types or scale formats.

Biases Related to Study Procedures Biases are grounded in insufficient procedures, too. For instance, a lack of care in the design of a study procedure can lead to problems of *validity* (e.g., when instructions are unclear, excessively complicated, or tasks too numerous) and *reliability* (when a missing or insufficiently defined protocol leads one or several researchers to administer a method variably across participants). A possible remedy is to train and prepare researchers for data collections in a standardized way. Researchers should also pre-test all parts of their procedure.

Biases Related to the Experimental Environment Ecological validity in data collection refers to the extent to which the experimental environment is similar to the real world. It is much easier to obtain high levels of ecological validity in a field study than a lab study, but researchers can take steps to introduce enough realism to have some confidence that results may generalize to real-world conditions.

Biases Related to Sampling It is a frequent limitation that researchers unknowingly overestimate how representative of the population their sample is, thus introducing bias. This can lead to underestimating problems in UPS as these participants are on average higher educated than the general population. Additional aspects of sample bias can result from specific geographic or time frames. It is important to be aware of the selection and the self-selection bias. The selection bias describes how any sample that was selected by rules that are not fully random sampling do not accurately describe the true population [43], and participants who voluntarily self-select to take part in research studies may represent a subset of the population that differs in terms of privacy and security perceptions and behaviors from the general population. When these biases cannot be avoided, it is important to be transparent about the sampling procedure and to discuss findings in relation to the sample.

Biases Related to the Participant Biases can influence how research participants behave and respond to questions. Typical response biases include, for example, the tendency to acquiescence, i.e., participants tend to give a positive answer more frequently than a negative one [61, 106]. Participants can also alter their behavior, e.g., towards higher performance, when they are aware of being observed. The Hawthorne effect refers to the effect of participants' awareness of study participation on their behavior [58]. Participants also tend to provide answers or show behavior they feel is positively connoted by the researchers or social norms; this is the social desirability bias. For example, in a non-anonymous setting, participants may prefer not to say that they use simplistic passwords rather than complex ones [10, 61, 106]. The halo effect describes the propensity of participants to transfer their impression (positive or negative) to a wider set of properties that they have not experienced before. Participants who feel sympathy for a brand may extend their trust to a specific service [4, 6, 106]. The recall or retrospection bias refers to memory limitations leading to vague or wrong responses. Distortion of recollections may also occur under the influence of emotional tainting [83, 97]. A specific problem in UPS research is that of security priming, which can lead to increased risk awareness in participants and can trigger unnatural responses or behavior.

Biases Related to the Researcher When conducting research with human participants, researchers can underestimate the influence of their own subjectivity when applying procedures, formulating questions, taking notes, or coding answers. Researchers should be aware of the confirmation bias, which refers to the propensity of humans to seek evidence in support of their hypotheses or beliefs, compromising neutrality in identifying all relevant data points [56]. Especially in quantitative studies, pre-registration of the research questions and methodology can be helpful for researchers to document their initial assumptions, and transparently describe how they might have adapted their analysis procedure during the study, and why. The cultural bias describes a propensity to underestimate the diversity in language, behavior, values, or conventions among the participants in a study. It can lead to inappropriate wording or color codes, or a lack of support for right-to-left languages [64]. When interacting with a participant, a frequent problem is the use of leading questions, which increase the risk of socially desirable responses [61, 77].

7 Conclusion

Studying behaviors and subjective experiences in the context of UPS is complex. Researchers need to balance realistic risk representation with practical, ethical, and legal concerns. This chapter describes social science and HCI research methods that are relevant for UPS. We describe a variety of methods, discuss how they can be used in UPS, and direct the reader to relevant resources. We discuss additional methodological tools to enhance the methods, participant recruitment, ethical challenges, and biases that could play a role in UPS research. While there is no such thing as a one-size-fits-all method, this chapter aims to provide readers with the tools to weigh the advantages and shortcomings of the described methods. We hope that UPS, as an inherently interdisciplinary field, can use the rich insights in methods research to conduct valid, ethical, and replicable science.

References

1. Ahmed, T., Hoyle, R., Connelly, K., Crandall, D., & Kapadia, A. (2015). Privacy concerns and behaviors of people with visual impairments. In *Proceedings of the 33rd Annual ACM Conference on Human Factors in Computing Systems* (pp. 3523–3532). ACM.
2. American Psychological Association. (2017). Ethical principles of psychologists and code of conduct (2002, amended effective June 1, 2010, and January 1, 2017). http://www.apa.org/ethics/code/index.html
3. American Psychological Association. (2022). Deception research—APA dictionary of psychology. https://dictionary.apa.org/deception-research
4. Balzer, W. K., & Sulsky, L. M. (1992). Halo and performance appraisal research: A critical examination. *Journal of Applied Psychology, 77*(6), 975.
5. Baxter, K., Courage, C., & Caine, K. (2015). *Understanding your users* (2nd ed.). Morgan Kaufmann.
6. Bellé, N., Cantarelli, P., & Belardinelli, P. (2017). Cognitive biases in performance appraisal: Experimental evidence on anchoring and halo effects with public sector managers and employees. *Review of Public Personnel Administration, 37*(3), 275–294.
7. Bird, S., Segall, I., & Lopatka, M. (2020). Replication: Why we still can't browse in peace: On the uniqueness and reidentifiability of web browsing histories. In *Sixteenth Symposium on Usable Privacy and Security (SOUPS 2020)* (pp. 489–503).
8. Blandford, A., Furniss, D., & Makri, S. (2016). Qualitative HCI research: Going behind the scenes. *Synthesis Lectures on Human-Centered Informatics, 9*(1), 1–115.
9. Bødker, S., Dindler, C., Iversen, O. S., & Smith, R. C. (2022). *Participatory design*. Number 1946-7680 in synthesis lectures on human-centered informatics. Springer .
10. Bogner, K., & Landrock, U. (2016). *Response biases in standardised surveys (version 2.0)*. GESIS—Leibniz-Institut für Sozialwissenschaften.
11. Bonné, B., Peddinti, S. T., Bilogrevic, I., & Taft, N. (2017). Exploring decision making with Android's runtime permission dialogs using in-context surveys. In *Thirteenth Symposium on Usable Privacy and Security (SOUPS 2017)* (pp. 195–210). USENIX Association.
12. Bryman, A. (2015). *Social research methods*. Oxford University Press.
13. Cetin, O., Ganán, C., Altena, L., Tajalizadehkhoob, S. & van Eeten, M. (2018). Let me out! Evaluating the effectiveness of quarantining compromised users in walled gardens. In *Fourteenth Symposium on Usable Privacy and Security (SOUPS 2018)* (pp. 251–263).

14. Colnago, J., Devlin, S., Oates, M., Swoopes, C., Bauer, L., Cranor, L., & Christin, N. (2018). "it's not actually that horrible" exploring adoption of two-factor authentication at a university. In *Proceedings of the 2018 CHI Conference on Human Factors in Computing Systems* (pp. 1–11).
15. Conway, D., Taib, R., Harris, M., Yu, K., Berkovsky, S., & Chen, F. (2017). A qualitative investigation of bank employee experiences of information security and phishing. In *Thirteenth Symposium on Usable Privacy and Security (SOUPS 2017)* (pp. 115–129). USENIX Association.
16. Cranor, L. F., & Buchler, N. (2014). Better together: Usability and security go hand in hand. *IEEE Security Privacy, 12*(6), 89–93.
17. Das, S., Laput, G., Harrison, C., & Hong, J. I. (2017). Thumprint: Socially-inclusive local group authentication through shared secret knocks. In *Proceedings of the 2017 CHI Conference on Human Factors in Computing Systems* (pp. 3764–3774).
18. De Luca, A., Das, S., Ortlieb, M., Ion, I., & Laurie, B. (2016). Expert and non-expert attitudes towards (secure) instant messaging. In *Twelfth Symposium on Usable Privacy and Security (SOUPS 2016)* (pp. 147–157).
19. Devon A., Bah, A., & Barwulor, C. (2018). Ethics emerging: The Story of privacy and security perceptions in virtual reality. In *SOUPS '18: Proceedings of the Fourteenth USENIX Conference on Usable Privacy and Security* (p. 17).
20. Distler, V., Fassl, M., Habib, H., Krombholz, K., Lenzini, G., Lallemand, C., Cranor, L. F., & Koenig, V. (2021). A Systematic literature review of empirical methods and risk representation in usable privacy and security research. *ACM Transactions on Computer-Human Interaction, 28*(6), 1–50.
21. Distler, V., Gutfleisch, T., Lallemand, C., Lenzini, G., & Koenig, V. (2022). Complex, but in a good way? How to represent encryption to non-experts through text and visuals—Evidence from expert co-creation and a vignette experiment. *Computers in Human Behavior Reports, 5*, 100161.
22. Distler, V., Lallemand, C., & Koenig, V. (2020). How acceptable is this? How user experience factors can broaden our understanding of the acceptance of privacy trade-offs. *Computers in Human Behavior, 106*, 106227.
23. Egelman, S., Kannavara, R., & Chow, R. (2015). Is this thing on?: Crowdsourcing privacy indicators for ubiquitous sensing platforms. In *Proceedings of the 33rd Annual ACM Conference on Human Factors in Computing Systems* (pp. 1669–1678). ACM.
24. Egelman, S., King, J., Miller, R. C., Ragouzis, N., & Shehan, E. (2007). Security user studies: Methodologies and best practices. In *CHI '07 Extended Abstracts on Human Factors in Computing Systems - CHI '07* (p. 2833). ACM Press.
25. Egelman, S., & Peer, E. (2015). Scaling the security wall: Developing a security behavior intentions scale (SeBIS). In *Proceedings of the 33rd Annual ACM Conference on Human Factors in Computing Systems* (pp. 2873–2882). ACM.
26. Fassl, M., Gröber, L. T., & Krombholz, K. (2021). Exploring user-centered security design for usable authentication ceremonies. In *Proceedings of the 2021 CHI Conference on Human Factors in Computing Systems*, CHI (pp. 1–15). ACM.
27. Felt, A. P., Barnes, R., King, A., Palmer, C., Bentzel, C., & Tabriz, P. (2017). Measuring https adoption on the web. In *26th USENIX Security Symposium (USENIX Security 17)* (pp. 1323–1338).
28. Felt, A. P., Reeder, R. W., Ainslie, A., Harris, H., Walker, M., Thompson, C., Acer, M. E., Morant, E., & Consolvo, S. (2016). Rethinking connection security indicators. In *Twelfth Symposium on Usable Privacy and Security (SOUPS 2016)* (pp. 1–14). USENIX Association.
29. Fiesler, C., & Hallinan, B. (2018). "We are the product": Public Reactions to online data sharing and privacy controversies in the media. In *Proceedings of the 2018 CHI Conference on Human Factors in Computing Systems* (pp. 53:1–53:13). ACM.
30. Fiesler, C., & Proferes, N. (2018). "Participant" perceptions of Twitter research ethics. *Social Media + Society, 4*(1), 205630511876336.

31. Freed, D., Palmer, J., Minchala, D., Levy, K., Ristenpart, T., & Dell, N. (2018). "a stalker's paradise": How intimate partner abusers exploit technology. In *Proceedings of the 2018 CHI Conference on Human Factors in Computing Systems*, CHI '18 (pp. 1–13). Association for Computing Machinery.
32. Furnham, A., & Boo, H. C. (2011). A literature review of the anchoring effect. *The Journal of Socio-Economics, 40*(1), 35–42.
33. Gallagher, K., Patil, S., Dolan-Gavitt, B., McCoy, D., & Memon, N. (2018). Peeling the onion's user experience layer: Examining naturalistic use of the Tor browser. In *Proceedings of the 2018 ACM SIGSAC Conference on Computer and Communications Security* (pp. 1290–1305). ACM.
34. Garfinkel, S., & Lipford, H. R. (2014). *Usable security : History, themes, and challenges*. Morgan & Claypool Publishers.
35. Gerber, A. S., & Green, D. P. (2012). *Field experiments: Design, analysis, and interpretation*. W. W. Norton & Company.
36. Goodman, E., Kuniavsky, M., & Moed, A. (2012). *Observing the user experience* (2nd ed.). Morgan Kaufmann.
37. Habib, H., Colnago, J., Gopalakrishnan, V., Pearman, S., Thomas, J., Acquisti, A., Christin, N., & Cranor, L. F. (2018). Away from prying eyes: Analyzing usage and understanding of private browsing. In *Fourteenth Symposium on Usable Privacy and Security (SOUPS 2018)* (pp. 159–175). USENIX Association.
38. Habib, H., Colnago, J., Gopalakrishnan, V., Pearman, S., Thomas, J., Acquisti, A., Christin, N., & Cranor, L. F. (2018). Away from prying eyes: Analyzing usage and understanding of private browsing. In *Fourteenth Symposium on Usable Privacy and Security (SOUPS 2018)* (pp. 159–175).
39. Habib, H., Naeini, P. E., Devlin, S., Oates, M., Swoopes, C., Bauer, L., Christin, N., & Cranor, L. F. (2018). User behaviors and attitudes under password expiration policies. In *Fourteenth Symposium on Usable Privacy and Security (SOUPS 2018)* (pp. 13–30). USENIX Association.
40. Halskov, K., & Hansen, N. B. (2015). The diversity of participatory design research practice at PDC 2002–2012. *International Journal of Human-Computer Studies, 74*, 81–92.
41. Haney, J. M., Theofanos, M., Acar, Y., & Prettyman, S. S. (2018). "we make it a big deal in the company": Security mindsets in organizations that develop cryptographic products. In *Fourteenth Symposium on Usable Privacy and Security (SOUPS 2018)* (pp. 357–373). USENIX Association.
42. Hara, K., Adams, A., Milland, K., Savage, S., Callison-Burch, C., & Bigham, J. P. (2018). A data-driven analysis of workers' earnings on Amazon Mechanical Turk. In *Proceedings of the 2018 CHI Conference on Human Factors in Computing Systems* (pp. 1–14). Montreal, QC: ACM.
43. Heckman, J. J. (1990). Selection bias and self-selection. In J. Eatwell, M. Milgate, & P. Newman (Eds.), *Econometrics* (pp. 201–224). Palgrave Macmillan.
44. Hormuth, S. E. (1986). The sampling of experiences in situ. *Journal of Personality, 54*(1), 262–293.
45. Hu, H., & Wang, G. (2018). End-to-end measurements of email spoofing attacks. In *27th USENIX Security Symposium (USENIX Security 18)* (pp. 1095–1112).
46. Ion, I., Reeder, R., & Consolvo, S. (2015). "...no one can hack my mind": Comparing expert and non-expert security practices. In *Eleventh Symposium On Usable Privacy and Security (SOUPS 2015)* (pp. 327–346). USENIX Association.
47. Karunakaran, S., Thomas, K., Bursztein, E., & Comanescu, O. (2018). Data breaches: User comprehension, expectations, and concerns with handling exposed data. In *Fourteenth Symposium on Usable Privacy and Security (SOUPS 2018)* (pp. 217–234). USENIX Association.
48. Kenneally, E., & Dittrich, D. (2012). The Menlo Report: Ethical principles guiding information and communication technology research. *SSRN Electronic Journal*, 14. https://www.scinapse.io/papers/2292723020

49. Komanduri, S., Shay, R., Cranor, L. F., Herley, C., & Schechter, S. (2014). Telepathwords: Preventing weak passwords by reading users' minds. In *23rd USENIX Security Symposium (USENIX Security 14)* (pp. 591–606).
50. Kvale, S. (1996). *Interviews: An introduction to qualitative research interviewing*. Sage Publications.
51. Larson, R., & Csikszentmihalyi, M. (2014). *The experience sampling method* (pp. 21–34). Springer Netherlands.
52. Lastdrager, E., Gallardo, I. C., Hartel, P., & Junger, M. (2017). How effective is anti-phishing training for children? In *Thirteenth Symposium on Usable Privacy and Security (SOUPS 2017)* (pp. 229–239). USENIX Association.
53. Lazar, J., Feng, J. H., & Hochheiser, H. (2017). *Research methods in human computer interaction* (2nd ed.). Burlington: Morgan Kaufmann.
54. Lee, J.-J., Jaatinen, M., Salmi, A., Mattelmäki, T., Smeds, R., & Holopainen, M. (2018). Design choices framework for co-creation projects. *International Journal of Design, 12*(2), 17.
55. Lyastani, S. G., Schilling, M., Fahl, S., Backes, M., & Bugiel, S. (2018). Better managed than memorized? Studying the impact of managers on password strength and reuse. In *27th USENIX Security Symposium (USENIX Security 18)* (pp. 203–220).
56. MacCoun, R. J. (1998). Biases in the interpretation and use of research results. *Annual Review of Psychology, 49*(1), 259–287.
57. Marforio, C., Masti, R. J., Soriente, C., Kostiainen, K., & Čapkun, S. (2016). Evaluation of personalized security indicators as an anti-phishing mechanism for smartphone applications. In *Proceedings of the 2016 CHI Conference on Human Factors in Computing Systems* (pp 540–551). ACM.
58. McCambridge, J., Witton, J., & Elbourne, D. R. (2014). Systematic review of the Hawthorne effect: New concepts are needed to study research participation effects. *Journal of Clinical Epidemiology, 67*(3), 267–277.
59. McReynolds, E., Hubbard, S., Lau, T., Saraf, A., Cakmak, M., & Roesner, F. (2017). Toys that listen: A study of parents, children, and internet-connected toys. In *Proceedings of the 2017 CHI Conference on Human Factors in Computing Systems*, CHI '17 (pp. 5197–5207). ACM.
60. Melicher, W., Kurilova, D., Segreti, S. M., Kalvani, P., Shay, R., Ur, B., Bauer, L., Christin, N., Cranor, L. F., & Mazurek, M. L. (2016). Usability and security of text passwords on mobile devices. In *Proceedings of the 2016 CHI Conference on Human Factors in Computing Systems* (pp. 527–539).
61. Müller, H., Sedley, A., & Ferrall-Nunge, E. (2014). Survey research in HCI. *Ways of knowing in HCI* (pp. 229–266). Springer.
62. Murillo, A., Kramm, A., Schnorf, S., & De Luca, A. (2018). "if i press delete, it's gone": User understanding of online data deletion and expiration. In *Proceedings of the Fourteenth USENIX Conference on Usable Privacy and Security*, SOUPS '18 (pp. 329–339). USENIX Association.
63. Naiakshina, A., Danilova, A., Tiefenau, C., Herzog, M., Dechand, S., & Smith, M. (2017). Why do developers get password storage wrong?: A qualitative usability study. In *Proceedings of the 2017 ACM SIGSAC Conference on Computer and Communications Security* (pp. 311–328). ACM.
64. Newcomer, K. E., Hatry, H. P., & Wholey, J. S. (2015). Culturally responsive evaluation. In *Handbook of practical program evaluation* (p. 281). https://www-wiley-com.proxy.bnl.lu/en-us/Handbook+of+Practical+Program+Evaluation%2C+4th+Edition-p-9781118893609
65. Nissenbaum, H. (2004). Privacy as contextual integrity. *Washington Law Review, 79*(1), 119–157.
66. Nthala, N., & Flechais, I. (2018). Informal support networks: An investigation into home data security practices. In *Fourteenth Symposium on Usable Privacy and Security (SOUPS 2018)* (pp. 63–82). USENIX Association.

67. Oest, A., Safaei, Y., Zhang, P., Wardman, B., Tyers, K., Shoshitaishvili, Y., & Doupé, A. (2020). PhishTime: Continuous longitudinal measurement of the effectiveness of anti-phishing blacklists. In *29th USENIX Security Symposium (USENIX Security 20)* (pp. 379–396).
68. Oest, A., Zhang, P., Wardman, B., Nunes, E., Burgis, J., Zand, A. Thomas, K., Doupé, A., & Ahn, G.-J. (2020). Sunrise to sunset: Analyzing the end-to-end life cycle and effectiveness of phishing attacks at scale. In *29th USENIX Security Symposium (USENIX Security 20).*
69. Olejnik, K., Dacosta, I., Machado, J. S., Huguenin, K., Khan, M. E., & Hubaux, J.-P. (2017). SmarPer: Context-aware and automatic runtime-permissions for mobile devices. In *2017 IEEE Symposium on Security and Privacy (SP)* (pp. 1058–1076). IEEE.
70. Olejnik, L., Castelluccia, C., & Janc, A. (2012). Why Johnny can't browse in peace: On the uniqueness of web browsing history patterns. In *5th Workshop on Hot Topics in Privacy Enhancing Technologies (HotPETs 2012).*
71. Pearman, S., Thomas, J., Naeini, P. E., Habib, H., Bauer, L., Christin, N., Cranor, L. F., Egelman, S., & Forget, A. (2017). Let's go in for a closer look: Observing passwords in their natural habitat. In *Proceedings of the 2017 ACM SIGSAC Conference on Computer and Communications Security* (pp. 295–310).
72. Plane, A. C., Redmiles, E. M., Mazurek, M. L., & Tschantz, M. C. (2017). Exploring user perceptions of discrimination in online targeted advertising. In *26th USENIX Security Symposium (USENIX Security 17)* (pp. 935–951).
73. Prasad, S., Bouma-Sims, E., Mylappan, A. K., & Reaves, B. (2020). Who's calling? Characterizing robocalls through audio and metadata analysis. In *29th USENIX Security Symposium (USENIX Security 20)* (pp. 397–414).
74. Proferes, N., Jones, N., Gilbert, S., Fiesler, C., & Zimmer, M. (2021). Studying reddit: A systematic overview of disciplines, approaches, methods, and ethics. *Social Media + Society, 7*(2), 205630512110190.
75. Prolific (2022). Representative samples. https://researcher-help.prolific.co/hc/en-gb/articles/360019236753-Representative-samples
76. Rashidi, Y., Ahmed, T., Patel, F., Fath, E., Kapadia, A., Nippert-Eng, C., & Su, N. M. (2018). "you don't want to be the next meme": College students' workarounds to manage privacy in the era of pervasive photography. In *Fourteenth Symposium on Usable Privacy and Security (SOUPS 2018)* (pp. 143–157). USENIX Association.
77. Rea, L. M., & Parker, R. A. (2014). *Designing and conducting survey research: A comprehensive guide*. Wiley.
78. Redmiles, E. M., Kross, S., & Mazurek, M. L. (2016). How i learned to be secure: A census-representative survey of security advice sources and behavior. In *Proceedings of the 2016 ACM SIGSAC Conference on Computer and Communications Security*, CCS '16 (pp. 666–677). ACM.
79. Reeder, R. W., Felt, A. P., Consolvo, S., Malkin, N., Thompson, C., & Egelman, S. (2018). An experience sampling study of user reactions to browser warnings in the field. In *Proceedings of the 2018 CHI Conference on Human Factors in Computing Systems* (pp. 1–13). ACM.
80. Reynolds, J., Samarin, N., Barnes, J., Judd, T., Mason, J., Bailey, M., & Egelman, S. (2020). Empirical measurement of systemic 2FA usability. In *29th USENIX Security Symposium (USENIX Security 20)* (pp. 127–143).
81. Ruoti, S., Andersen, J., Heidbrink, S., O'Neill, M., Vaziripour, E., Wu, J., Zappala, D., & Seamons, K. (2016). "We're on the same page": A usability study of secure email using pairs of novice users. In *Proceedings of the 2016 CHI Conference on Human Factors in Computing Systems* (pp. 4298–4308). ACM Press.
82. Ruoti, S., O'Neill, M., Zappala, D., & Seamons, K. (2016). User attitudes toward the inspection of encrypted traffic. In *Twelfth Symposium on Usable Privacy and Security (SOUPS 2016)* (pp. 131–146). USENIX Association.
83. Russell, D. M., & Chi, E. H. (2014). Looking back: Retrospective study methods for HCI. In *Ways of knowing in HCI* (pp. 373–393). Springer.

84. Samat, S., & Acquisti, A. (2017). Format vs. content: The impact of risk and presentation on disclosure decisions. In *Thirteenth Symposium on Usable Privacy and Security (SOUPS 2017)* (pp. 377–384). USENIX Association.
85. Samat, S., Acquisti, A., & Babcock, L. (2017). Raise the curtains: The effect of awareness about targeting on consumer attitudes and purchase intentions. In *Thirteenth Symposium on Usable Privacy and Security (SOUPS 2017)* (pp. 299–319). USENIX Association.
86. Sambasivan, N., Checkley, G., Batool, A., Ahmed, N., Nemer, N., Gaytán-Lugo, L. S., Matthews, T., Consolvo, S., & Churchill, E. (2018). "privacy is not for me, it's for those rich women": Performative privacy practices on mobile phones by women in South Asia. In *Fourteenth Symposium on Usable Privacy and Security (SOUPS 2018)* (pp. 127–142). USENIX Association.
87. Sanders, E. B.-N., & Stappers, P. J. (2008). Co-creation and the new landscapes of design. *CoDesign, 4*(1), 5–18.
88. Sanders, L. (2008). An evolving map of design practice and design research. *Human Factors, 15*, 7.
89. Schechter, S. (2013). Common pitfalls in writing about security and privacy human subjects experiments, and how to avoid them. https://www.microsoft.com/en-us/research/publication/common-pitfalls-in-writing-about-security-and-privacy-human-subjects-experiments-and-how-to-avoid-them/ Tech report nr MSR-TR-2013-5.
90. Schechter, S. E., Dhamija, R., Ozment, A., & Fischer, I. (2007). The emperor's new security indicators. In *2007 IEEE Symposium on Security and Privacy (SP'07)* (pp. 51–65). IEEE.
91. Sikder, A. K., Aksu, H., & Uluagac, A. S. (2017). 6thSense: A context-aware sensor-based attack detector for smart devices. In *26th USENIX Security Symposium (USENIX Security 17)* (pp. 397–414).
92. Silberman, M. S., Tomlinson, B., LaPlante, R., Ross, J., Irani, L., & Zaldivar, A. (2018). Responsible research with crowds: Pay crowdworkers at least minimum wage. *Communications of the ACM, 61*(3), 39–41.
93. Sotirakopoulos, A., Hawkey, K., & Beznosov, K. (2011). On the challenges in usable security lab studies: Lessons learned from replicating a study on SSL warnings. In *Proceedings of the Seventh Symposium on Usable Privacy and Security* (p. 3). ACM.
94. Tan, J., Bauer, L., Bonneau, J., Cranor, L. F., Thomas, J., & Ur, B. (2017). Can unicorns help users compare crypto key fingerprints? In *Proceedings of the 2017 CHI Conference on Human Factors in Computing Systems* (pp. 3787–3798).
95. The National Commission for the Protection of Human Subjects of Biomedical and Behavioral Research. The Belmont Report. Technical report, 1979.
96. Torabi, S., & Beznosov, K. (2016). Sharing health information on Facebook: Practices, preferences, and risk perceptions of North American users. In *Twelfth Symposium on Usable Privacy and Security (SOUPS 2016)* (pp. 301–320). USENIX Association.
97. Tourangeau, R., Rips, L. J., & Rasinski, K. (2000). *The psychology of survey response*. Cambridge University Press
98. Tranfield, D., Denyer, D., & Smart, P. (2003). Towards a methodology for developing evidence-informed management knowledge by means of systematic review. *British Journal of Management, 14*(3), 207–222.
99. Vaniea, K. E., Rader, E., & Wash, R. (2014). Betrayed by updates: How negative experiences affect future security. In *Proceedings of the SIGCHI Conference on Human Factors in Computing Systems* (pp. 2671–2674). ACM.
100. Vaziripour, E., Wu, J., O'Neill, M., Whitehead, J., Heidbrink, S., Seamons, K., & Zappala, D. (2017). Is that you, Alice? A usability study of the authentication ceremony of secure messaging applications. In *Thirteenth Symposium on Usable Privacy and Security (SOUPS 2017)* (pp. 29–47). USENIX Association.
101. Vitak, J., Proferes, N., Shilton, K., & Ashktorab, Z. (2017). Ethics regulation in social computing research: Examining the role of institutional review boards. *Journal of Empirical Research on Human Research Ethics, 12*(5), 372–382.

102. Warshaw, J., Taft, N., & Woodruff, A. (2016). Intuitions, analytics, and killing ants: Inference literacy of high school-educated adults in the US. In *Twelfth Symposium on Usable Privacy and Security (SOUPS 2016)* (pp. 271–285). USENIX Association.
103. Wash, R., & Cooper, M. M. (2018). Who provides phishing training?: Facts, stories, and people like me. In *Proceedings of the 2018 CHI Conference on Human Factors in Computing Systems - CHI '18* (pp. 1–12). ACM Press.
104. Wash, R., Rader, E., Berman, R., & Wellmer, Z. (2016). Understanding password choices: How frequently entered passwords are re-used across websites. In *Twelfth Symposium on Usable Privacy and Security (SOUPS 2016)* (pp. 175–188).
105. Weber, S., Harbach, M., & Smith, M. (2015). Participatory design for security-related user interfaces. In *Proceedings 2015 Workshop on Usable Security*. Internet Society.
106. Wetzel, E., Böhnke, J. R., & Brown, A. (2016). Response biases. In F. T. L. Leong, D. Bartram, F. M. Cheung, K. F. Geisinger, & D. Iliescu (Eds.), *The ITC international handbook of testing and assessment* (pp. 349–363). Oxford University Press.
107. Yang, Y., Clark, G. D., Lindqvist, J., & Oulasvirta, A. (2016). Free-form gesture authentication in the wild. In *Proceedings of the 2016 CHI Conference on Human Factors in Computing Systems* (pp. 3722–3735). ACM.
108. Zimmer, M. (2018). Addressing conceptual gaps in big data research ethics: An application of contextual integrity. *Social Media + Society, 4*(2), 205630511876830.
109. Zimmerman, J., Forlizzi, J., & Evenson, S. (2007). Research through design as a method for interaction design research in HCI. In *Proceedings of the SIGCHI Conference on Human Factors in Computing Systems* (pp. 493–502). ACM.

Toward Valid and Reliable Privacy Concern Scales: The Example of IUIPC-8

Thomas Groß

1 Introduction

Valid and reliable measurement instruments are vital for human factors in privacy research [23]. Validity means that an instrument measures what it purports to measure. Reliability means that the instrument measures this consistently.

In this chapter, we focus on the validity and reliability of privacy concern scales. While there is a range of privacy concern and behavior measurement instruments available [8, 10, 12, 22, 33, 41, 45, 50], also discussed in studies on the privacy paradox [14, 26], we will focus on the scale Internet Users' Information Privacy Concerns (IUIPC) [33]. IUIPC has roots in the earlier scale Concerns for Information Privacy (CFIP) [45], itself a popular scale measuring organizational information privacy concern and validated in independent studies [18, 46].

IUPIC has been appraised by researchers as part of other studies [36, 43] and undergone an independent empirical evaluation of the scale itself [16] and of the applicability of the full nomology in other cultures [39]. Even though the scale was originally created in a diligent, evolutionary fashion and founded on a sound underpinning for its content validity, construct validity and internal consistency reliability were not always found up to par for the purpose of human factors in privacy research.

In this chapter, we will discuss a brief form of the Internet Users' Information Privacy Concern scale (IUIPC) [33] as a running example. The brief form, called

Supplementary Information The online version contains supplementary material available at (https://doi.org/10.1007/978-3-031-28643-8_4).

T. Groß (✉)
Newcastle University, School of Computing, Newcastle upon Tyne, UK
e-mail: thomas.gross@newcastle.ac.uk

N. Gerber et al. (eds.), *Human Factors in Privacy Research*,
https://doi.org/10.1007/978-3-031-28643-8_4

IUIPC-8, only uses eight of the original ten items and was determined to yield stronger construct validity and internal consistency reliability [16].

Our aim for this chapter is not only to present the IUIPC-8 scale itself, but also to shed light on methods for the evaluation of valid and reliable measurement instruments. To that end, we will employ confirmatory factory analysis (CFA) as the tool of choice. We will use CFA to model the ordinal non-normal data of the questionnaire, to confirm the three-dimensionality is a fixed term, typically written with hyphen of IUIPC-8, to establish global and local fit, and finally to estimate construct validity and internal consistency reliability metrics.

Chapter Overview We begin this chapter with an overview of information privacy concern and predominant measurement instruments in Sect. 2. We give a brief introduction of validity and reliability notions in Sect. 3 and lay the foundations for the use of confirmatory factor analysis as tool to evaluate measurement-instrument properties in Sect. 4. We discuss the abstract approach used for the evaluation of IUIPC-8 in Sect. 5 and include the empirical results in the validation of the instrument in Sect. 6. Section 7 highlights aspects of the scale properties and considerations for its use in practice in a general discussion. Definitions used throughout the chapter are summarized in the definition box below.

Definitions

- **Validity:** Capacity of an instrument to measure what it purports to measure [6, 35].
- **Reliability:** Extent to which a variable is consistent in what is being measured [17, p. 123].
- **Construct Validity:** Whether the measure accurately reflects the construct intended to measure [23, 35].
- **Factorial Validity:** Factor composition and dimensionality are sound.
- **Confirmatory Factor Analysis:** Factor analysis in a restricted measurement model: Each indicator is to depend only on the factors specified [25, p. 191].
- **Nested Model:** A model that can be derived from another by restricting free parameters.
- **Accept-support test:** A statistical inference, in which the acceptance of the null hypothesis supports the model, e.g., the close-fit test [25, p. 265].
- **Reject-support test:** A statistical inference, in which the rejection of the null hypothesis supports the model, e.g., the not-close-fit test [25, p. 265].
- **Fit Statistic:** A summary measure of the average discrepancy between the sample and model covariances.
- **Goodness of fit χ^2:** Measures the exact fit of a model and gives rise to the accept-support exact-fit test against null hypothesis $H_{\chi^2,0}$.

- **RMSEA:** Root Mean Square Estimate of Approximation, an absolute badness-of-fit measure estimated as $\hat{\varepsilon}$ with its 90% confidence interval, yielding a range of fit tests: close fit, not-close fit, and poor fit [25, pp. 274].
- **Bentler Comparative Fit Index (CFI):** An incremental fit index based on the non-centrality measure comparing selected against the null model.
- **Standardized Root Mean Square Residual (SRMR):** A standardized version of the mean absolute covariance residual, for which zero indicates excellent fit.
- **Standardized Factor Loading** β: Z-transformed factor score.
- **Variance Extracted** R^2: The factor variance accounted for, computed as squared standardized loading β^2.
- **Average Variance Extracted (AVE):** The average of the squared standardized loadings β^2 of indicators belonging to the same factor [25, pp. 313].
- **Heterotrait–Monotrait (HTMT) Ratio:** A metric of discriminant validity, the ratio of the avg. correlations of indicators across constructs measuring different phenomena to the avg. correlations of indicators within the same construct [20].
- **Cronbach's** α: Internal consistency based on the average inter-item correlations.
- **Congeneric Reliability** ω: The amount of general factor saturation (also called *composite reliability* [25, pp. 313] or construct reliability (CR) [17, p. 676] depending on the source).

2 Information Privacy Concern

2.1 What Is Information Privacy Concern?

Malhotra et al. [33, p. 337] ground their definition of *information privacy concern* in Westin's definition of information privacy as a foundation of their understanding of privacy concern: "the claim of individuals, groups, or institutions to determine for themselves when, how, and to what extent information about them is communicated to others." They define information privacy concern as "an individual's subjective views of fairness within the context of information privacy."

This framing of information privacy concern resonates with the interdisciplinary review of privacy studies by Smith et al. [44]. Therein, privacy concern is shown as the central antecedent of related behavior in the privacy macro-model. At the same time, the causal impact of privacy concern on behavior has been under considerable scrutiny. The observed phenomenon, the privacy attitude–behavior dichotomy, is commonly called the *privacy paradox* [14]. Investigating the privacy paradox has been a mainstay topic of the human aspects of privacy community. This investigation calls for instruments to measure information privacy concern accurately and reliably because measurement errors and correlation attenuation of invalid or unreliable privacy concern instruments could confound the research on the privacy paradox.

2.2 Information Privacy Concern Instruments

Information privacy concern can be measured by a range of related and distinct instruments [8, 10, 12, 33, 45, 50]. As a comprehensive comparison would be beyond the scope of this chapter, we refer to Preibusch's comprehensive guide to measuring privacy concern [41] for an overview of the field. We will consider the scales most closely related to IUIPC. Table 1 offers a brief overview of these instruments and their dimensions. While IUIPC is one of the most used Internet privacy concern scales, its dimensions also influenced further scales, such as Hong and Thong's Internet Privacy Concern (IPC) [22]. Still, it remains a relatively concise scale.

Table 1 Overview of selected privacy concern instruments

Instrument	Year	Dimensions	Development	Appraisals
Concern for information privacy (CFIP)	1996	Collection$_a$	[45]	[18, 46]
		Unauthorized secondary use$_b$		
		Improper access$_c$		
		Errors$_d$		
Internet users' information privacy concern (IUIPC)	2004	Control$_f$	[33]	[16, 36, 39, 43]
		Awareness$_e$		
		Collection$_a$		
Internet privacy concerns (IPC)	2004	Control	[12]	
		Vulnerability		
		Abuse		
		Finding		
Internet privacy concerns (IPC)	2013	Collection$_a$	[22]	[21, 47]
		Secondary usage$_{b'}$		
		Errors$_{d'}$		
		Improper access$_{c'}$		
		Control$_f$		
		Awareness$_e$		
Online privacy concern and protection for use on the internet (OPC)	2007	General caution	[10]	
		Technical protection		
		Privacy attitude		

Note: Dimensions with the same subscript bear relations to another. A prime indicates that the items of the dimension were considerably reformulated

First point of call is the scale *Concern for information privacy* (CFIP) [45] as a major influence on the development of IUIPC. CFIP consists of four dimensions—Collection, Unauthorized Secondary Use, Improper Access, and Errors. While both questionnaires share questions, CFIP focuses on individuals' concerns about organizational privacy practices and the organization's responsibilities. CFIP received independent empirical confirmations of its factor structure, by Stewart and Segars [46] and by Harborth and Pape [18] on its German translation.

The scale *Internet Users' Information Privacy Concern* (IUIPC) was developed by Malhotra et al. [33], by predominately adapting questions of the earlier 15-item-scale Concern for Information Privacy (CFIP) by Smith et al. [45] and by framing the questionnaire for Internet users as consumers. IUIPC is measuring their perception of fairness and justice in the context of information privacy and online companies. IUIPC-10 was established as a second-order reflective scale of *information privacy concern*, with the dimensions Control, Awareness, and Collection. The authors considered the "act of collection, whether it is legal or illegal," as the starting point of information privacy concerns. The sub-scale Control is founded on the conviction that "individuals view procedures as fair when they are vested with control of the procedures." The authors considered being "informed about data collection and other issues" as central concept of the sub-scale Awareness.

Initial appraisals of IUIPC-10 [36, 43] yielded concerns for the validity and reliability of the scale largely tied to two items on awareness and control. These validity and reliability problems were confirmed in an independent empirical evaluation of the scale [16]. Pape et al. [39] independently evaluated the full nomology of IUIPC-10 in Japan.

Internet Privacy Concerns (IPC) [12] considered Internet privacy concerns with antecedents of perceived vulnerability and control, antecedents familiar from the Protection Motivation Theory (PMT). IPC differs from IUIPC in its focus on misuse rather than just collection of information and of concerns of surveillance. In terms of the core scale of privacy concern, Dinev and Hart identified two factors:

(i) Abuse (concern about misuse of information submitted on the Internet)
(ii) Finding (concern about being observed and specific private information being found out)

It considered the two antecedents Control and Vulnerability. The IPC scale was subsequently expanded on and integrated with other scales by Hong and Thong [22] and further investigated with respect to four driver and inhibitor dimensions by Hong et al. [21]. Herein, Hong and Thong reformulated questions to more consistently express concern.

Buchanan et al.'s *Online Privacy Concern and Protection for Use on the Internet* (OPC) [10] measure considered three sub-scales—General Caution, Technical Protection (both on behaviors), and Privacy Attitude. Compared to IUIPC, OPC sports a strong focus on item stems eliciting being concerned and on measures regarding a range of concrete privacy risks.

3 Validity and Reliability

When evaluating a privacy concern instrument such as IUIPC-8, the dual vital questions for research in human factors of privacy and the privacy paradox are:

(i) Are we measuring the hidden latent construct privacy concern accurately? (validity)
(ii) Are we measuring privacy concern consistently and with an adequate signal-to-noise ratio? (reliability)

Without sufficient reliability, a measurement instrument cannot be valid [23].

Validity refers to whether an instrument measures what it purports to measure. Messick offered an early well-regarded definition of validity as the "integrated evaluative judgment of the degree to which empirical evidence and theoretical rationales support the adequacy and appropriateness of inferences and actions based on test scores" [35]. Validity is inferred—judged in degrees—not measured. In this chapter, we put our attention on the validation procedure and underlying evidence for validity and reliability. In that, we largely take content validity of IUIPC for granted. *Content validity* refers to the relevance and representativeness of the content of the instrument, typically assessed by expert judgment.

3.1 Construct Validity

Messick [34] defines *construct validity* [11], the interpretive meaningfulness, as the extent to which an instrument accurately represents a construct. This definition has also been used in more recent papers on measurement [23] as a primary kind of validity. Construct validity is typically established by the evaluation of the instrument through multiple lenses, where we will go into factorial, convergent, and discriminant validity.

Factorial Validity First, we seek evidence of *factorial validity*, that is, evidence that that factor composition and dimensionality are sound. While IUIPC is a *multidimensional* scale with three correlated designated dimensions, we require *unidimensionality* of each sub-scale, a requirement discussed at length by Gerbing and Anderson [15].

Unidimensional measurement models for sub-scales correspond to expecting *congeneric* measures, that is, the scores on an item are the expression of a true score weighted by the item's loading plus some measurement error, where in the congeneric case neither the loadings nor error variances across items are required to be equal. This property entails that the items of each sub-scale must be conceptually homogeneous.

We find empirical evidence for factorial validity of a scale's measurement model in the closeness of fit to the sample's covariance structure. Specifically, we gain supporting evidence by passing fit hypotheses of a confirmatory factor analysis

Table 2 Exact and approximate fit hypotheses

Type	Null hypothesis	Classification	RMSEA 90% CI
Exact fit	$H_0 : \varepsilon_0 = 0$	Accept-support	$\hat{\varepsilon}_L = 0$
Close fit	$H_0 : \varepsilon_0 \leq .05$	Accept-support	$\hat{\varepsilon}_L \leq .05$
Not-close fit	$H_0 : \varepsilon_0 \geq .05$	Reject-support	$\hat{\varepsilon}_U < .05$
Poor fit	$H_0 : \varepsilon_0 \geq .10$	Reject-support	$\hat{\varepsilon}_U < .10$

Note: ε_0 = RMSEA under the null hypothesis
$\hat{\varepsilon}$ = point estimate of RMSEA; $[\hat{\varepsilon}_L, \hat{\varepsilon}_U]$ = 90% Confidence Interval on $\hat{\varepsilon}$

for the designated factor structure [3, 15, 25], where we prioritize fit metrics and hypotheses based on the RMSEA included in Table 2.

Convergent and Discriminant Validity *Convergent validity* [17, pp. 675] (convergent coherence) on an item-construct level means that items belonging together, that is, to the same construct, should be observed as related to each other. Similarly, *discriminant validity* [17, pp. 676] (discriminant distinctiveness) means that items not belonging together, that is, not belonging to the same construct, should be observed as not related to each other. On a sub-scale level, we expect factors of the same higher-order construct to relate to each other, and on hierarchical factor level, we expect all 1st-order factors to load strongly on the 2nd-order factor.

In the first instance, a poor local fit and tell-tale residual patterns yield disconfirming evidence for convergent and discriminant validity. We can further inspect inter-item correlation matrices: we expect items belonging to the same sub-scale to be highly correlated and, thereby, to converge on the same construct. Correlation to items of other sub-scales should be low, especially lower than the in-construct correlations [25, pp. 196].

These principles give rise to criteria based on the average variance extracted (AVE), the Fornell–Larcker criterion [13], and the Heterotrait–Monotrait Ratio (HTMT) [1, 20]. We summarize these terms in the definition box of this chapter.

3.2 *Reliability*

Reliability is the extent to which a variable is consistent in what is being measured [17, p. 123]. It can further be understood as the capacity of "separating signal from noise" [23, 42, p. 709], quantified by the ratio of true score to observed score variance [25, pp. 90]. We evaluate *internal consistency* as a means to estimate reliability from a single test application. Internal consistency entails that items that purport to measure the same construct produce similar scores [25, p. 91]. We will use the internal consistency measures Cronbach's α, congeneric reliability ω, and AVE, defined in the definition box of this chapter. While Cronbach's α is well-known in the community, average variance extracted (AVE) offers a simple intuitive measure, and congeneric reliability provides a robust approach.

Thresholds for reliability estimates such as Cronbach's α or composite reliability ω are debated in the field, where many recommendations are based on Nunnally's original treatment of the subject, but equally often misstated [28]. The often quoted $\alpha \geq 0.70$ was described by Nunnally only to "save time and energy," whereas a greater threshold of 0.80 was endorsed for basic research [28].

When we designate a priori thresholds as criteria for internal consistency reliability, this approach needs to be put into a broader context. As for validity, reliability is judged in degrees. John and Benet-Martínez [23] discuss the arbitrariness of one-size-fits-all fixed reliability thresholds. Internal consistency reliability needs to be considered in relation to inter-item correlations and the length of a scale and, further, how these aspects fit the nature of the construct in question. Ultimately, the choice of thresholds gives rise to a bandwidth–fidelity trade-off [23]. Whether we call an instrument "reliable enough" depends on the proportion of error variance we are willing to tolerate and on the attenuation of the correlation to other variables as a consequence of that.

4 Factor Analysis as Tool to Establish Measurement Instruments

Factor analysis is a powerful tool for evaluating the construct validity and reliability of privacy concern instruments. Thereby, it constitutes validation procedure for the measurement instruments [23]. *Factor analysis* refers to a set of statistical methods that are meant to determine the number and nature of *latent variables* (LVs) or *factors* that account for the variation and covariation among a set of observed measures commonly referred to as *indicators* [9].

Confirmatory factor analysis (CFA) is a factor analysis in a restricted measurement model, that is, in which each indicator depends only on the factors specified [25, pp. 191]. CFA is commonly used to evaluate psychometric instruments. It is based on the *common factor model* (CFM), which holds that each indicator variable contributes to the variance of one or more common factors and one unique factor. Thereby, *common variance* of related observed measures is attributed to the corresponding latent factor, and *unique variance* (uniqueness) seen either as variance associated with the item or as error variance. We call the proportion of variance associated with factors *communality* and the proportion of variance not associated with factors *uniqueness*. We depict the common factor model in Fig. 1.

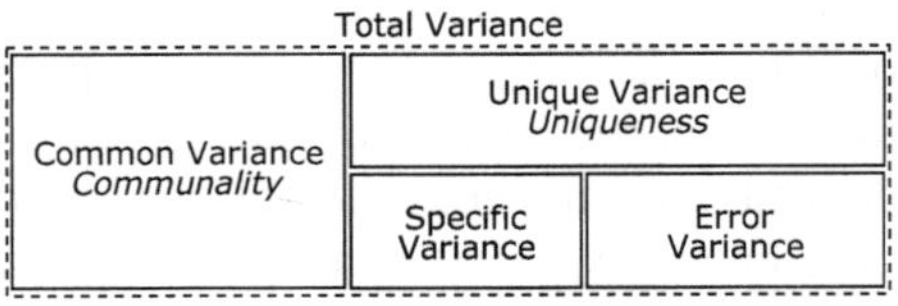

Fig. 1 Variance attribution in the common factor model (CFM)

IUIPC is based on a *reflective measurement*, that is, the observed measure of an indicator variable is seen as *caused* by some latent factor. Indicators are thereby *endogenous* variables, and latent variables *exogenous* variables. Reflective measurement requires that all items of the sub-scale are interchangeable [25, pp. 196]. In this chapter, we focus on *covariance-based confirmatory factor analysis* (CB-CFA). Therein, the statistical tools aim at estimating coefficients for parameters of the measurement model that best fit the covariance matrix of the observed data. The difference between an observed covariance of the sample and an implied covariance of the model is called a *residual*.

4.1 Estimation Methods for Ordinal Non-normal Data

The purpose of a factor analysis is to estimate free parameters of the model (such as loadings or error variance), which is facilitated by *estimators*. The choice of estimator matters because each comes with different strengths and weaknesses, requirements, and assumptions that need to be fulfilled for the validity of their use.

While *maximum likelihood* (ML) estimation is the model commonly used estimation method for CFA, it is based on *assumptions* [25, pp. 71] that are not satisfied by IUIPC:

(i) A *continuous measurement level*
(ii) *Multi-variate normal distribution* (entailing the absence of extreme *skewness*) [25, pp. 74]

The distribution requirements are placed on the endogenous variables: the indicators.

First, the Likert items used in IUIPC are *ordinal* [17, p. 11], that is, ordered categories in which the distance between categories is not constant. Lei and Wu [30] held based on a number of empirical studies that the fit indices of approximately normal ordinal variables with at least five categories are not greatly misleading. However, when ordinal and non-normal is treated as continuous and normal, the fit is underestimated, and there is a more pronounced negative bias in estimates and standard errors. While Bovaird and Kozoil [7] acknowledge robustness of the ML estimator with normally distributed ordinal data, they stress that increasingly skewed and kurtotic ordinal data inflate the Type I error rate and, hence, require another approach [25, pp. 323]. In the same vein, Kline [24, p. 122] holds the normality assumption for endogenous variables—the indicators—to be critical.

4.2 Comparing Nested Models

Nested models [25, p. 280] are models that can be derived from each other by restricting free parameters. They can be compared with a *likelihood ratio* χ^2

Difference Test (LRT) [25, p. 270]. This technique comes into play when we compare multiple models that are based on the same indicator variables, e.g., to establish which factor structure most suits the covariance matrix. We use this technique in comparing one-factor solutions with solutions with multiple factors.

4.3 Global and Local Fit

The *closeness of fit* of a factor model to an observed sample is evaluated globally with fit indices as well as locally by inspecting the residuals. We shall focus on the ones Kline [25, p. 269] required as minimal reporting.

Statistical Inference The χ^2 and RMSEA indices offer us *statistical inferences of global fit*. Such tests can either be *accept-support*, that is, accepting the null hypothesis supports the selected model, or *reject-support*, that is, rejecting the null hypothesis supports the selected model. We present them in Table 2.

Local Fit Even with excellent global fit indices, the inspection of the local fit—evidenced by the residuals—must not be neglected. Kline [25, p. 269] emphasizes "Any report of the results without information about the residuals is incomplete." Large absolute residuals indicate covariations that the model does not approximate well and that may, thereby, lead to spurious results.

5 Approach

In this section, we are weaving a general approach for creating a valid and reliable measurement instrument with specific design decisions taken for the brief information privacy concerns scale IUIPC-8 [16]. General approaches for systematic constructions of measurements [23], measurement models for survey research [5], or their reliability and validity [2] are well-documented in the literature. Here, we introduce specific considerations for IUIPC-8. The following aspects inform this evaluation:

- The scale IUIPC-8 is derived from the long-standing scale IUIPC-10. Hence, a comparison of both scales is in order.
- We will conduct confirmatory factor analyses to establish the dimensionality and construct validity of the scale.
- The IUIPC data will be from ordinal Likert items, with a skewed non-normal distribution.
- We will need sufficient sample sizes for the statistical power on RMSEA-based statistical inferences.

- We aim at a scale that yields low attenuation of the correlations of its latent variable in the relation to other variables, requiring good internal consistency reliability.

5.1 Analysis Methodology

Our analysis of IUIPC-10 and the brief variant IUIPC-8 will be supported by confirmatory factor analyses on two independent samples, one used for specification and refinement and the other used for validation. The factor analyses yield the evidence for unidimensionality of the sub-scales and the overall dimensionality of the instrument. While the creation of a new measurement instrument would often start with an exploratory factor analysis on a candidate item pool and another independent sample, here we shall focus only on the confirmatory factor analyses setting the 8-item and 10-item variants apart. The corresponding analysis process is depicted in Fig. 2.

Because IUIPC yields ordinal, non-normal data, the distribution of the data asks for careful analysis as part of the data preparation. The assumptions of a standard maximum likelihood estimation will be violated, by which we are preparing for a robust diagonally weighted least square (DWLS) estimation method as tool of choice. The specific method employed is called *WLSMV*, a robust diagonally weighted least square (DWLS) estimation with robust standard errors and mean- and variance-adjusted test statistics using a scale shift. The choice of estimation method will also impact the sample size we need to obtain: Apart from cases of small samples ($N < 200$), WLSMV was found to be less biased and more accurate than robust ML estimation (MLR) [31]. For smaller sample sizes, we would recommend MLR.

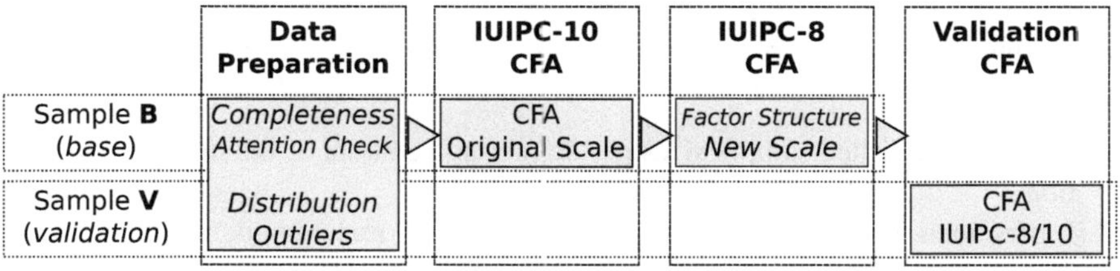

Fig. 2 Which steps were taken on what sample (adapted from [16])

5.2 Sample

The quality of the sampling method, that is, how participants are drawn from a population, has a considerable impact on the sampling and non-sampling biases introduced early in a study. In an ideal case, the target and survey population are clearly specified and the sampling frame explicitly defined [19, 48]. In terms of sampling method, random sampling, possibly stratified to be representative of the population, carries the least bias.

The sample size is determined based on three requirements:

(i) The size needed to reach a diversity approximately representative of the UK population ($N > 300$)
(ii) The minimum sample size to make DWLS estimation viable ($N > 200$)
(iii) The sample size required to reach adequate statistical power

For confirmatory factor analyses considered in this chapter, the key statistical inferences are on the χ^2 significance test for the exact fit and the RMSEA-based significance tests for approximate fit. Hence, we determined the sample size for an RMSEA close-fit test in an a priori power analysis [27, 32, 49]. We used the R package SEMPower and an IUIPC-10 model with $n_{\mathsf{par}} = 23$ free parameters and $df = 32$ degrees of freedom as benchmark. To reach $1 - \beta = 80\%$ statistical power in this constellation, we would need a target sample size of $N_{1-\beta=0.80} = 317$.

For the analysis of IUIPC-8, we employed two independent samples, B and V. Base sample B and validation sample V were designated with a sample size of 420 each, allowing for some sample size and power loss in the sample refinement and analysis.

The samples used here were used for an earlier study [16] establishing IUIPC-8 and, hence, serve for illustration and not as an independent validation of the questionnaire. The samples were recruited on Prolific Academic [38] to be representative of the UK census by age, gender, and ethnicity. The sampling frame was Prolific users who were registered on the platform as residents of the UK, consisting of 48,454 users at sampling time (August 2019). The sampling process was as follows:

1. Prolific established sample sizes per demographic strata of the intended population.
2. It presented our studies to the registered users with matching demographics.
3. The users could choose themselves whether they would participate or not.

We enforced sample independence by uniqueness of the participants' Prolific ID.

We note that the sampling method is not random, it is a crowdsourcing sample with demographics screening [29], yet Prolific has been found to obtain samples from a diverse population and with high data quality and reliability [40].

5.3 Validity and Reliability Criteria

The first consideration for construct validity is in the factorial validity of the model, where we compare multiple possible factor structures to confirm the dimensionality. Based on the overall selected model structure, we then turn to an analysis of the global and local fit in a comparison between IUIPC-10 and IUIPC-8 on samples B and V. For the fit of the models, we consider the RMSEA-based fit hypotheses shown in Table 2 as important part of our investigation. Here we are interested in getting support from the close-fit hypothesis, being aware that the CFAs will not have enough statistical power to offer a tight enough confidence interval on the RMSEA estimate to reject the not-close-fit hypothesis.

For convergent and discriminant validity, we turn to empirical criteria, especially relying on the average variance extracted (AVE) and Heterotrait–Monotrait Ratio (HTMT) in the definition box of this chapter. We gain empirical evidence in favor of convergent validity [17, pp. 675]:

(i) If the variance extracted by an item $R^2 > 0.50$ entailing that the standardized factor loading are significant and $\beta > 0.70$.
(ii) If the internal consistency (defined in Sect. 3.2) is sufficient ($AVE > 0.50$, $\omega > AVE$, and $\omega > 0.70$).

The analysis yields empirical evidence of discriminant validity [17, pp. 676]:

(i) If the square root of *AVE* of a latent variable is greater than the max correlation with any other latent variable (Fornell–Larcker criterion [13])
(ii) If the Heterotrait–Monotrait Ratio (HTMT) is less than 0.85 [1, 20]

While that would be beneficial for privacy research as well, we shall adopt reliability metrics $\alpha, \omega \geq 0.70$ as suggested by Hair et al. [17, p. 676].

6 The Validation of IUIPC-8

In this section, we are examining the model of IUIPC-8 in a diagonally weighted least square (DWLS) CFA estimation with robust standard errors and a scale-shifted mean- and variance-adjusted test statistic (WLSMV). We begin our inquiry with the characteristics of the underlying sample (Sect. 6.1) and distribution (Sect. 6.2), considering a base Sample B and an independent validation Sample V.

6.1 Sample

The demographics of both samples B and V are included in Table 3. While these samples were meant to be drawn to be UK representative, we observe an under-

Table 3 Demographics, table taken from Groß [16] licensed under CC BY-NC-ND 4.0

(a) Sample B		(b) Sample V	
	Overall		Overall
N_B	379	N_V	433
Gender (%)		Gender (%)	
Female	197 (52.0)	Female	217 (50.1)
Male	179 (47.2)	Male	212 (49.0)
Rather not say	3 (0.8)	Rather not say	4 (0.9)
Age (%)		Age (%)	
18–24	41 (10.9)	18–24	92 (21.2)
25–34	72 (19.0)	25–34	143 (33.0)
35–44	84 (22.2)	35–44	83 (19.2)
45–54	57 (15.0)	45–54	58 (13.4)
55–64	97 (25.6)	55–64	44 (10.2)
65+	28 (7.4)	65+	13 (3.0)

Table 4 Sample refinement, table taken from Groß [16] licensed under CC BY-NC-ND 4.0

	B		V	
Phase	Excl.	Size	Excl.	Size
Starting sample		473		467
Incomplete	58	415	34	433
Duplicate	25	390	0	433
FailedAC> 1	11	379	0	433
MV outlier	9	370	14	419
Final sample	$N'_B = 370$		$N'_V = 419$	

Note: $N_B = 379$, $N_V = 433$ are after attention checks

representation of elderly participants compared to the UK census age distribution. Still, the sample offers us sufficient diversity for the evaluation of the scale.

The two samples have undergone a sample refinement in stages, which Table 4 accounts for. The refinement included:

(i) Removing incomplete cases without replacement
(ii) Removing duplicates across samples by the participants' Prolific ID, to guarantee independence
(iii) Removing cases in which participants failed more than one attention check (FailedAC > 1)

The named attention checks were instructional manipulation checks [37] distributed over the wider questionnaire.

Overall, of the $N_C = 848$ complete cases, only 4.2% were removed due to duplicates or failed attention checks. After this refinement, a small number of multivariate outliers were removed.

6.2 Descriptives

Evaluating the sample distribution, we found the indicator variables to be negatively skewed. The distributions tail off to the left. The Control and Awareness indicators suffer from positive kurtosis. We found that the indicator distributions as well as the IUIPC sub-scale distributions exhibited substantial non-normality. We illustrate these aspects in Table 5 and Fig. 3.

We observed that the two samples displayed approximately equal distributions by sub-scales. Controlling for the difference between Samples B and V, we found that none of their sub-scale means was statistically significantly different, the maximal absolute standardized mean difference being 0.13—a small magnitude.

Our IUIPC-10 samples yielded 6% univariate outliers by the robust outlier labeling rule and 3% multi-variate outliers with a Mahalanobis distance of 12 or greater [25, pp. 72]. We removed these MV outliers as indicated in Table 4.

These observations on the distribution of the samples are relevant for the choice of estimator for the confirmatory factor analysis to come. A maximum likelihood (ML) estimation would require continuous measurement with multi-variate normality. These assumptions are clearly not fulfilled. While a robust maximum likelihood estimation (MLM) can also be considered, we opted for a diagonally weighted least square (DWLS) estimation with robust standard errors

Table 5 Means (SDs) of the summarized sub-scales of IUIPC-8 and the original IUIPC-10 (adapted from [16])

	Sample B	Sample V	Malhotra et al. [33]
ctrl	5.92 (0.92)	5.86 (0.99)	5.67 (1.06)
aware	6.64 (0.53)	6.56 (0.66)	6.21 (0.87)
collect	5.58 (1.12)	5.60 (1.04)	5.63 (1.09)
iuipc	6.05 (0.60)	6.01 (0.63)	5.84 (1.01)

Note: iuipc is the flat mean of all items of the scale

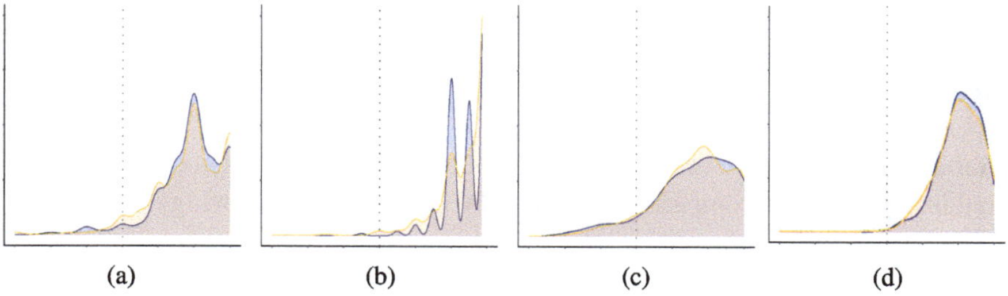

Fig. 3 Density of IUIPC-8 sub-scale responses across samples (B: violet, V: orange). *Note:* All graphs are on the same scale (adapted from [16]). (**a**) Control. (**b**) Awareness. (**c**) Collection. (**d**) IUIPC-8 overall

Table 6 Comparison of different model structures of IUIPC-8 on Sample B with WLSMV estimation

	One factor	Two factors	Three factors (1st order)	Three factors (2nd order)
$\chi^2(df)$	732.43 (20)	88.98 (19)	22.77 (17)	22.77 (17)
χ^2/df	36.62	4.68	1.34	1.34
CFI	.93	.99	1.00	1.00
RMSEA	.32 [.30, .34]	.11 [.09, .13]	.07 [.05, .09]	.07 [.05, .09]
SRMR	.22	.09	.04	.04
Scaled $\chi^2(df)$	790.34 (20)	105.24 (19)	46.76 (17)	46.76 (17)

and scale-shifted mean- and variance-adjusted test statistics (WLSMV), typically considered preferred for ordinal/non-normal data.[1]

6.3 Construct Validity

Factorial Validity

First, we investigate the three dimensionality of IUIPC-8. To that end, we computed confirmatory factor analyses on one-factor, two-factor, and the hypothesized three-dimensional second-order model displayed in Table 6. The two-factor solution was statistically significantly better than the one-factor solution, $\chi^2(1) = 215.065$, $p < .001$. In turn, the three-factor solutions were statistically significantly better than the two-factor solution, $\chi^2(2) = 30.165$, $p < .001$. Hence, given the results of the likelihood ratio tests (LRTs) on these nested models, we choose the three-dimensional second-order model. This is the model also shown in the path plot of Fig. 4.

Model Fit

Global Fit Second, we evaluate the global fit as a measure of factorial validity. We do this in a two-way comparison of WLSMV CFAs on the following dimensions:

(i) IUIPC-10 vs. IUIPC-8
(ii) Base sample B and validation sample V

Table 7 reports on the fit statistics of the four CFA models.

[1] Groß [16] also employed a DWLS estimation in the evaluation of IUIPC-8. The models of that work, however, were computed with WLSMVS, an estimation method using Satterthwaite-style test statistic scaling.

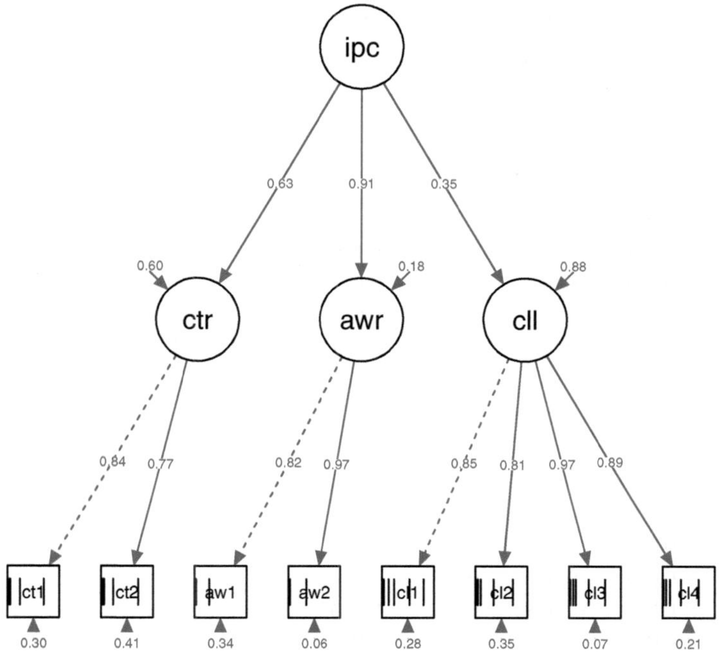

Fig. 4 CFA paths plot with standardized estimates of IUIPC-8 on Sample B. *Note:* The dashed lines signify that the raw factor loading was fixed to 1 (cf. Table 9, figure was adapted from [16])

Table 7 Fit statistic comparison of IUIPC-10 and IUIPC-8 (adapted from [16])

	Sample	
Instrument	Base B	Validation V
IUIPC-10	$\chi^2(32) = 275.087$, $p < .001$	$\chi^2(32) = 220.96$, $p < .001$
	CFI[a] = .96	CFI = .96
	RMSEA[a] = .14 [.13, .16], $p_{\epsilon_0 \leq .05} < .001$	RMSEA[a] = .12 [.10, .13], $p_{\epsilon_0 \leq .05} < .001$
	SRMR = .10	SRMR = .07
IUIPC-8	$\chi^2(17) = 46.764$, $p < .001$	$\chi^2(17) = 36.673$, $p = .004$
	CFI[a] = 1.00	CFI[a] = 1.00
	RMSEA[a] = .07 [.05, .09], $p_{\epsilon_0 \leq .05} = .086$	RMSEA[a] = .05 [.03, .08], $p_{\epsilon_0 \leq .05} = .394$
	SRMR = .04	SRMR = .03

Note: [a]Robust estimation with scaled test statistic. RMSEA reported with 90% CI

Table 8 Residuals of the WLSMV CFA of IUIPC-8 on Sample B

	(a) Correlation residuals							
	1	2	3	4	5	6	7	8
1. ctrl1	–							
2. ctrl2	0	–						
3. awa1	−0.017	0.003	–					
4. awa2	−0.017	0.032	0	–				
5. coll1	−0.071	−0.068	**−0.132**	**−0.115**	–			
6. coll2	−0.005	−0.037	0.071	0.065	0.034	–		
7. coll3	0.04	−0.009	−0.019	0.003	−0.012	0.003	–	
8. coll4	0.082	0.015	0.078	0.034	0.008	−0.036	0.004	–

Note: Correlation residuals in absolute >0.1 are marked

	Covariance residuals							
	1	2	3	4	5	6	7	8
1. ctrl1	–							
2. ctrl2	0	–						
3. awa1	−0.017	0.003	–					
4. awa2	−0.017	0.032	0	–				
5. coll1	−0.071	−0.068	−0.132	−0.115	–			
6. coll2	−0.005	−0.037	0.071	0.065	0.034	–		
7. coll3	0.04	−0.009	−0.019	0.003	−0.012	0.003	–	
8. coll4	0.082	0.015	0.078	0.034	0.008	−0.036	0.004	–

Note: Standardized residuals are not available for estimator WLSMV

Because IUIPC-10 and IUIPC-8 models are non-nested, we cannot use likelihood ratio test (LRT) to evaluate their difference. In the WLSMV estimation, we are left with comparing fit measures.[2]

Regarding the global fit reported in Table 7, we notice that all CFA models fail the exact-fit test on the χ^2 test statistic. To evaluate approximate fit, we draw attention to the root mean square estimate of approximation (RMSEA), its confidence interval, and the close-fit hypothesis $H_{\varepsilon_0 \leq .05,0}$. We observe that the IUIPC-10 models are not supported by the close-fit test, and the IUIPC-8 models are Sample B: $p_{\epsilon_0 \leq .05} = .086$ and Sample V: $p_{\epsilon_0 \leq .05} = .394$. Hence, we conclude that the IUIPC-8 model shows a sufficient approximate close fit, even if not an exact fit.

Local Fit The good global fit for IUIPC-8 shown in Table 7 alone is not sufficient to vouch for the overall fit of the model. For this purpose, we inspect the correlation and raw residuals in Table 8. Therein, we observe slightly reduced correlation residuals

[2] On the maximum likelihood (ML) estimation used by Groß [16], a Vuong non-nested LRT was available as a test of difference. That work also considered a Consistent Akaike Information Criterion (CAIC) directly derived from the χ^2 metric typically used for ML estimations. Banks and Joyner [4] offer detailed analysis of AICs for different estimations.

between coll1 and the awareness indicator variables. These negative correlation residuals mean that the CFA model overestimates the correlation between the indicator variables in question. The correlation residuals in the validation model (included in the online supplementary materials) show lower deviations. Hence, we believe both models to hold acceptable local fit.

CFA Model, Convergent, and Discriminant Validity

We illustrate the selected second-order CFA model for IUIPC-8 in Fig. 4. Table 9 contains the corresponding factor loadings with their standardized solutions. The standardized loadings of the model give us confidence in the convergent validity of the model: the average variance extracted (AVE) was greater than 50% for all first-level factors. This observation holds equally for the standardized factor loadings of the validation CFA, summarized in the online supplementary materials.

In terms of discriminant validity, we first consider the Fornell–Larcker criterion in Table 10. As required, we find that the square root of the AVE displayed on the diagonal of the matrices is greater than the inter-factor correlations in the rest of the matrix. This holds for both the base Sample B and the validation Sample V.

Further, we evaluate the HTMT criterion in Table 11. We are satisfied with this criterion for Samples B and V at a threshold of 0.85. Hence, we conclude that the scale offers sufficient discriminant validity.

6.4 Reliability: Internal Consistency

Table 9 also includes reliability metrics derived from the WLSMV CFA model of IUIPC-8. For both the base Sample B and the validation Sample V, we observe that the congeneric reliability ω is consistently greater than .70. By that, the reliability criteria established in Sect. 5.3 are fulfilled, and we can expect a good signal-to-noise ratio for the scale.

7 Discussion

We have seen that IUIPC-8 offers good construct validity and reliability. The outcomes of our analysis are summarized in Table 12. It can serve as a useful measurement instrument for information privacy concern. As any measurement instrument, IUIPC-8 is a working solution, which may be proven wrong eventually and superseded by more refined scales [23].

In terms of bandwidth–fidelity trade-off [23], IUIPC-8 offers a greater fidelity than the original scale IUIPC-10 [33], at the expense of bandwidth. Because of the greater congeneric reliability, we expect less attenuation in the correlations to other

Table 9 Factor loadings and their standardized solution of the WLSMV CFA of IUIPC-8 on Sample B

Factor	Indicator	Factor loading				Standardized solution				Reliability				
		λ	SE_λ	Z_λ	p_λ	β	SE_β	Z_β	p_β	R^2	AVE	α	ω	S/N_ω
ctrl	ctrl1	1.00^+				0.84	0.06	14.21	<.001	0.70	0.65	0.72	0.72	2.57
	ctrl2	0.91	0.13	7.22	<.001	0.77	0.06	13.18	<.001	0.59				
aware	awa1	1.00^+				0.82	0.06	14.07	<.001	0.66	0.80	0.76	0.79	3.67
	awa2	1.19	0.15	7.91	<.001	0.97	0.06	16.41	<.001	0.94				
collect	coll1	1.00^+				0.85	0.02	50.30	<.001	0.72	0.77	0.91	0.91	10.18
	coll2	0.95	0.03	37.56	<.001	0.81	0.02	40.59	<.001	0.65				
	coll3	1.14	0.02	49.25	<.001	0.97	0.01	102.10	<.001	0.93				
	coll4	1.05	0.02	44.24	<.001	0.89	0.01	60.34	<.001	0.79				
iuipc	collect	0.30	0.05	5.56	<.001	0.35	0.06	5.57	<.001	0.12				
	ctrl	0.53	0.08	6.62	<.001	0.63	0.10	6.54	<.001	0.40				
	aware	0.74	0.12	6.00	<.001	0.91	0.12	7.27	<.001	0.82				

Note: $^+$ fixed parameter; the standardized solution is STDALL

Table 10 First-level correlations and $\sqrt{AVE}$ as evidence for the Fornell–Larcker criterion for discriminant validity on IUIPC-8

(a) Sample B

	1	2	3
1. ctrl	0.646		
2. aware	0.573	0.802	
3. collect	0.221	0.316	0.772

Note: The diagonal contains the $\sqrt{AVE}$

(b) Sample V

	1	2	3
1. ctrl	0.663		
2. aware	0.63	0.718	
3. collect	0.268	0.415	0.69

Table 11 Heterotrait–Monotrait ratios as criterion for discriminant validity on IUIPC-8

	Sample B		
	1	2	3
1. ctrl	–		
2. aware	0.47	–	
3. collect	0.14	0.25	–

	Sample V		
	1	2	3
1. ctrl	–		
2. aware	0.55	–	
3. collect	0.23	0.34	–

variables that with IUIPC-10: according to classical test theory, such correlations are bounded by the square root of its reliability. Thereby, IUIPC-8 can be useful in investigating relations to other variables, such as impact of privacy concern on privacy behavior.

The restriction to eight items also bears limitations that need to be considered carefully. First, the factors Control and Awareness are based on a narrower footing in terms of content validity, that is, in terms of coverage of relevant and representative aspects of the construct. We also need to consider CFA model identification. While IUIPC-8 as a whole is identified because of the two-indicator rule [25, Rule 9.1], the sub-scales Control and Awareness on their own will not be identified. Hence, they cannot be used as robust measurement instruments of their own.

In terms of future work, it would be preferable to refine IUIPC-8 with further items rounding out the sub-scales Control and Awareness, while maintaining a high construct validity and reliability. Ideally, each factor would have three or more indicators. While this chapter largely focused on the construct validity and reliability in the form of internal consistency of the scale itself, a more comprehensive evaluation of privacy concern scales is vital. For IUIPC-8, we considered internal consistency as reliability (that is, generalizing across items). At the same time, retest reliability (generalizability across times) and equivalence reliability (generalizability across forms) are still research areas to expand. In addition, the investigation of IUIPC in its full nomology is important, such as pursued by Pape et al. [39] in the case of the use of the scale in Japan.

Table 12 Selected evidence for construct validity and reliability criteria on Samples B and V under WLSMV estimation (adapted from [16])

		Construct validity								Reliability	
		Factorial				Convergent		Divergent		Internal consistency	
		$H_{\chi^2,0}$	$H_{\hat{\varepsilon}\leq .05,0}$	$H_{\hat{\varepsilon}>.05,0}$	$H_{\hat{\varepsilon}>.10,0}$	$\beta > .70$	$AVE > .50$	$\sqrt{AVE} > \forall \bar{r}$	$HTMT < .85$	$\alpha > .70$	$\omega > .70$
IUIPC-10	B	○	○	○	○	○	◐	○	●	○	○
	V	○	○	○	○	○	○	○	●	○	○
IUIPC-8	B	○	●	○	●	●	●	●	●	●	●
	V	○	●	○	●	●	●	●	●	●	●

Note: ● supports the model, ○ rejects the model, ◐ criterion is borderline fulfilled, β standardized loading, *AVE* average variance extracted, $\bar{r}$ correlation with other factor, *HTMT* Heterotrait–Monotrait ratio, ω composite reliability

8 Summary

This chapter considers the validity and reliability of privacy concern scales, with IUIPC-8 as an example of a brief information privacy concern scale:

- We introduced validity and reliability concepts, focusing on construct validity and congeneric reliability.
- We discussed confirmatory factory analysis as a tool to establish the properties of measurement instruments.
- We discussed CFA estimation methods for ordinal and non-normal data as found with the IUIPC scale.
- We included an empirical analysis of IUIPC-8 on both a base sample and an independent validation sample.
- We evaluated validity and reliability criteria in the comparison of IUIPC-10 and the brief form IUIPC-8.

Acknowledgments This work was supported by ERC Starting Grant CASCAde (GA no 716980).

Appendix

Materials and Sample

We included the used IUIPC-10 questionnaire in Table 13.

For the reproducibility of the maximum likelihood estimation, Table 14 contains the correlations, means, and standard deviations (SDs) of Sample B. The OSF supplementary materials contain more precise covariance matrices of all samples.

Thresholds

In this section, we include the WLSVM thresholds for the base and the validation models. Table 15 shows the thresholds of the main model.

Table 13 Items of the instrument Internet users' information privacy concerns (IUIPC-10) [33], adapted from [16]

Construct	Item	Question
Control (ctrl)	ctrl1	Consumer online privacy is really a matter of consumers' right to exercise control and autonomy over decisions about how their information is collected, used, and shared
	ctrl2	Consumer control of personal information lies at the heart of consumer privacy
	[ctrl3]	I believe that online privacy is invaded when control is lost or unwillingly reduced as a result of a marketing transaction
Awareness (aware)	awa1	Companies seeking information online should disclose the way the data are collected, processed, and used
	awa2	A good consumer online privacy policy should have a clear and conspicuous disclosure
	[awa3]	It is very important to me that I am aware and knowledgeable about how my personal information will be used
Collection (collect)	coll1	It usually bothers me when online companies ask me for personal information
	coll2	When online companies ask me for personal information, I sometimes think twice before providing it
	coll3	It bothers me to give personal information to so many online companies
	coll4	I am concerned that online companies are collecting too much personal information about me

Note: The questionnaire is administered with 7-point Likert items, anchored on 1 = " Strongly Disagree" to 7 = "Strongly Agree." The items in squared brackets ctrl3 and awa3 are included in IUIPC-10, but not in IUIPC-8

Table 14 Correlations, means, and standard deviations of base Sample B

	1	2	3	4	5	6	7	8	9	10
1. ctrl1										
2. ctrl2	0.56									
3. ctrl3	0.25	0.27								
4. awa1	0.25	0.23	0.25							
5. awa2	0.32	0.32	0.30	0.62						
6. awa3	0.23	0.19	0.26	0.32	0.31					
7. coll1	0.05	0.05	0.26	0.05	0.11	0.33				
8. coll2	0.10	0.06	0.28	0.22	0.24	0.30	0.66			
9. coll3	0.16	0.09	0.32	0.16	0.22	0.40	0.76	0.72		
10. coll4	0.19	0.10	0.31	0.23	0.23	0.47	0.71	0.62	0.81	
M	5.97	5.96	6.68	6.62	5.26	5.76	5.69	5.73	5.97	5.96
SD	0.93	0.93	1.00	0.55	0.56	0.82	1.36	1.11	1.24	1.22

Note: $N_B = 370$

Table 15 Thresholds table of fitted model on Sample B

	Variable	t1	t2	t3	t4	t5	t6
1	ctrl1	−2.782	−1.926	−1.583	−0.705	0.524	
2	ctrl2	−1.846	−1.537	−0.722	0.564		
3	awa1	−2.782	−1.810	−0.564			
4	awa2	−2.549	−1.885	−0.404			
5	coll1	−2.782	−1.776	−1.115	−0.731	0.075	0.842
6	coll2	−2.297	−1.632	−1.251	−0.382	0.588	
7	coll3	−1.972	−1.416	−1.128	−0.310	0.517	
8	coll4	−2.782	−2.139	−1.515	−1.066	−0.382	0.501

References

1. Ab Hamid, M., Sami, W., & Sidek, M. (2017). Discriminant validity assessment: Use of Fornell & Larcker criterion versus HTMT criterion. In *Journal of Physics: Conference Series* (Vol. 890, pp. 012163). IOP Publishing.
2. Alwin, D. F. (2010). *Handbook of survey research*, chapter How good is survey measurement—assessing the reliability and validity of survey measures (2nd ed., pp. 263–313). Emerald Group Publishing Limited.
3. Anderson, J. C., Gerbing, D. W., & Hunter, J. E. (1987). On the assessment of unidimensional measurement: Internal and external consistency, and overall consistency criteria. *Journal of Marketing Research, 24*(4), 432–437.
4. Banks, H. T., & Joyner, M. L. (2017). AIC under the framework of least squares estimation. *Applied Mathematics Letters, 74*, 33–45.
5. Bohrnstedt, G. W. (2010). *Handbook of survey research*, chapter Measurement models for survey research (2nd ed., pp. 263–313). Emerald Group Publishing Limited.
6. Borsboom, D., Mellenbergh, G. J., & Van Heerden, J. (2004). The concept of validity. *Psychological Review, 111*(4), 1061.
7. Bovaird, J. A., & Koziol, N. A. (2012). Measurement models for ordered-categorical indicators. In R. H. Hoyle (Ed.), *Handbook of structural equation modeling* (pp. 495–511). The Guilford Press.
8. Braunstein, A., Granka, L., & Staddon, J. (2011). Indirect content privacy surveys: Measuring privacy without asking about it. In *Proceedings of the Seventh Symposium on Usable Privacy and Security* (pp. 1–14).
9. Brown, T. A. (2015). *Confirmatory factor analysis for applied research*. The Guilford Press.
10. Buchanan, T., Paine, C., Joinson, A. N., & Reips, U.-D. (2007). Development of measures of online privacy concern and protection for use on the Internet. *Journal of the American Society for Information Science and Technology, 58*(2), 157–165.
11. Cronbach, L. J., & Meehl, P. E. (1955). Construct validity in psychological tests. *Psychological Bulletin, 52*(4), 281.
12. Dinev, T., & Hart, P. (2004). Internet privacy concerns and their antecedents-measurement validity and a regression model. *Behaviour & Information Technology, 23*(6), 413–422.
13. Fornell, C., & Larcker, D. F. (1981). Evaluating structural equation models with unobservable variables and measurement error. *Journal of Marketing Research, 18*(1), 39–50.
14. Gerber, N., Gerber, P., & Volkamer, M. (2018). Explaining the privacy paradox: A systematic review of literature investigating privacy attitude and behavior. *Computers & Security, 77*, 226–261.

15. Gerbing, D. W., & Anderson, J. C. (1988). An updated paradigm for scale development incorporating unidimensionality and its assessment. *Journal of Marketing Research, 25*(2), 186–192.
16. Groß, T. (2021). Validity and reliability of the scale Internet users' information privacy concerns (IUIPC). *Proceedings of the Privacy Enhancing Technology Symposium, 2021*(2), 235–258.
17. Hair, J. F., Black, W. C., Babin, B. J., & Anderson, R. E. (2019). *Multivariate data analysis* (8th ed.). Cengage Learning.
18. Harborth, D., & Pape, S. (2018). German translation of the concerns for information privacy (CFIP) construct. SSRN 3112207.
19. Henry, G. T. (1990). *Practical sampling* (Vol. 21). Sage.
20. Henseler, J., Ringle, C. M., & Sarstedt, M. (2015). A new criterion for assessing discriminant validity in variance-based structural equation modeling. *Journal of the Academy of Marketing Science, 43*(1), 115–135.
21. Hong, W., Chan, F. K., & Thong, J. Y. (2021). Drivers and inhibitors of Internet privacy concern: A multidimensional development theory perspective. *Journal of Business Ethics, 168*(3), 539–564.
22. Hong, W., & Thong, J. Y. (2013). Internet privacy concerns: An integrated conceptualization and four empirical studies. *MIS Quarterly, 37*, 275–298.
23. John, O. P., & Benet-Martínez, V. (2000). Measurement: Reliability, construct validation, and scale construction. In *Handbook of research methods in social and personality psychology*.
24. Kline, R. B. (2012). Assumptions in structural equation modeling. In R. H. Hoyle (ed.), *Handbook of structural equation modeling* (pp. 111–125). The Guilford Press.
25. Kline, R. B. (2015). *Principles and practice of structural equation modeling* (4th ed.). The Guilford Press.
26. Kokolakis, S. (2017). Privacy attitudes and privacy behaviour: A review of current research on the privacy paradox phenomenon. *Computers & Security, 64*, 122–134.
27. Kyriazos, T. A., et al. (2018). Applied psychometrics: Sample size and sample power considerations in factor analysis (EFA, CFA) and SEM in general. *Psychology, 9*(8), 2207.
28. Lance, C. E., Butts, M. M., & Michels, L. C. (2006). The sources of four commonly reported cutoff criteria: What did they really say? *Organizational Research Methods, 9*(2), 202–220.
29. Landers, R. N., & Behrend, T. S. (2015). An inconvenient truth: Arbitrary distinctions between organizational, Mechanical Turk, and other convenience samples. *Industrial and Organizational Psychology, 8*(2), 142–164.
30. Lei, P.-W., & Wu, Q. (2012). Estimation in structural equation modeling. In R. H. Hoyle (ed.), *Handbook of structural equation modeling* (pp. 164–179). The Guilford Press.
31. Li, C.-H. (2016). Confirmatory factor analysis with ordinal data: Comparing robust maximum likelihood and diagonally weighted least squares. *Behavior Research Methods, 48*(3), 936–949.
32. MacCallum, R. C., Browne, M. W., & Sugawara, H. M. (1996). Power analysis and determination of sample size for covariance structure modeling. *Psychological Methods, 1*(2), 130.
33. Malhotra, N. K., Kim, S. S., & Agarwal, J. (2004). Internet users' information privacy concerns (IUIPC): The construct, the scale, and a causal model. *Information Systems Research, 15*(4), 336–355.
34. Messick, S. (1980). Test validity and the ethics of assessment. *American Psychologist, 35*(11), 1012.
35. Messick, S. (1987). Validity. *ETS Research Report Series, 1987*(2), i-208.
36. Morton, A. (2013). Measuring inherent privacy concern and desire for privacy—a pilot survey study of an instrument to measure dispositional privacy concern. In *2013 International Conference on Social Computing* (pp. 468–477). IEEE.
37. Oppenheimer, D. M., Meyvis, T., & Davidenko, N. (2009). Instructional manipulation checks: Detecting satisficing to increase statistical power. *Journal of Experimental Social Psychology, 45*(4), 867–872.
38. Palan, S., & Schitter, C. (2018). Prolific.ac—a subject pool for online experiments. *Journal of Behavioral and Experimental Finance, 17*, 22–27.

39. Pape, S., Ivan, A., Harborth, D., Nakamura, T., Kiyomoto, S., Takasaki, H., & Rannenberg, K. (2020). Re-evaluating Internet users' information privacy concerns: The case in Japan. *AIS Transactions on Replication Research, 6*(1), 18.
40. Peer, E., Brandimarte, L., Samat, S., & Acquisti, A. (2017). Beyond the Turk: Alternative platforms for crowdsourcing behavioral research. *Journal of Experimental Social Psychology, 70*, 153–163.
41. Preibusch, S. (2013). Guide to measuring privacy concern: Review of survey and observational instruments. *International Journal of Human-Computer Studies, 71*(12), 1133–1143.
42. Revelle, W., & Condon, D. M. (2018). Reliability. In *The Wiley handbook of psychometric testing: A Multidisciplinary reference on survey, scale and test development* (1st ed., pp. 709–749). Wiley.
43. Sipior, J. C., Ward, B. T., & Connolly, R. (2013). Empirically assessing the continued applicability of the IUIPC construct. *Journal of Enterprise Information Management, 26*(6), 661–678.
44. Smith, H. J., Dinev, T., & Xu, H. (2011). Information privacy research: An interdisciplinary review. *MIS Quarterly, 35*(4), 989–1015.
45. Smith, H. J., Milberg, S. J., & Burke, S. J. (1996). Information privacy: Measuring individuals' concerns about organizational practices. *MIS Quarterly, 20*(2), 167–196.
46. Stewart, K. A., & Segars, A. H. (2002). An empirical examination of the concern for information privacy instrument. *Information Systems Research, 13*(1), 36–49.
47. Terlizzi, M. A., Brandimarte, L., & Sanchez, O. (2019). Replication of Internet privacy concerns in the mobile banking context. *AIS Transactions on Replication Research, 5*(1), 8.
48. Williams, B. (1978). *A sampler on sampling*. Wiley.
49. Wolf, E. J., Harrington, K. M., Clark, S. L., & Miller, M. W. (2013). Sample size requirements for structural equation models: An evaluation of power, bias, and solution propriety. *Educational and Psychological Measurement, 73*(6), 913–934.
50. Xu, H., Dinev, T., Smith, H. J., & Hart, P. (2008). Examining the formation of individual's privacy concerns: Toward an integrative view. In *ICIS 2008 Proceedings* (p. 6).

Achieving Usable Security and Privacy Through Human-Centered Design

Eduard C. Groen, Denis Feth, Svenja Polst, Jan Tolsdorf, Stephan Wiefling, Luigi Lo Iacono, and Hartmut Schmitt

1 Introduction

Scope and Motivation Numerous examples show that cybersecurity and data protection measures need to be designed in such a way that end users can interact safely with digital systems (e.g., [81, 100]). This user orientation is addressed by the field of *usable security and privacy* (USP). USP aims to support the design of security and data protection measures in a way that: (1) users, designers, and developers are supported in the best way possible in their security- or privacy-related projects and (2) the measures contribute to a continuously positive user experience. Because of our research background and projects, in this chapter, we will focus primarily on usable privacy. However, usable security and usable privacy usually go hand in hand, which makes it sensible to view them as a common research discipline. Thus, our recommendations should be equally applicable to security-related topics.

Problem and Idea In practice, the fields of requirements engineering (RE) and user experience (UX) design are tasked with translating data protection regulations into a system's implementation through requirements and design concepts. Although both disciplines have decades-long expertise with security and data protection requirements, the changes in international data protection regulations

E. C. Groen (✉) · D. Feth · S. Polst
Fraunhofer Institute for Experimental Software Engineering IESE, Kaiserslautern, Germany
e-mail: eduard.groen@iese.fraunhofer.de; denis.feth@iese.fraunhofer.de

J. Tolsdorf · S. Wiefling · L. Lo Iacono
Hochschule Bonn-Rhein-Sieg, Sankt Augustin, Germany
e-mail: jan.tolsdorf@h-brs.de; stephan.wiefling@h-brs.de; luigi.lo_iacono@h-brs.de

H. Schmitt
HK Business Solutions GmbH, Friedrichsthal, Germany
e-mail: hartmut.schmitt@hk-bs.de

N. Gerber et al. (eds.), *Human Factors in Privacy Research*,
https://doi.org/10.1007/978-3-031-28643-8_5

and the digital transformation impose new challenges on these disciplines. A particularly challenging question is how to equip end users of systems—both data processors and data subjects[1]—with the appropriate decision support, transparency, empowerment, and other resources.

Proper design of USP requires new questions to be answered to make the right data protection design choices, such as: What understanding do stakeholders—particularly end users—have of privacy and data protection? What kind of privacy-specific needs do they have? And how can we categorize stakeholders into groups? The answers to these questions strongly affect the way a system is designed and its security and privacy properties are made usable, thereby achieving *usable security and privacy* (USP; see Sect. 2.2). Functionally, they influence what the system should do; quality-wise, they affect how well it is adapted to the stakeholders' characteristics and context. Unfortunately, existing security frameworks (e.g., BSIMM, SAMM, Common Criteria), models (e.g., MS SDL), and best practices (e.g., the OWASP guides) barely consider usability and user experience.

Our idea is to firmly cement security and privacy into the *human-centered design* (HCD) process. The HCD process includes the human perspective in a software system's design process and ensures that it is developed in a way that its interaction helps human actors make the "right" security and data protection decisions intuitively in the corresponding use cases, thereby minimizing potential errors by misconfiguration. Techniques for RE and UX that guide practitioners to obtain the correct privacy-related answers are still emerging. Including these techniques in the HCD process helps to ensure that the stakeholders are properly considered during the design of these security and privacy aspects.

Contribution and Structure In this chapter, we present three complimentary USP-oriented methods developed on the basis of good practices, which can be used in HCD, RE, and UX design processes. Together, these methods provide practitioners tasked with designing the USP for a system with a practical toolkit, helping them to assure that USP aspects are sufficiently considered in the HCD process:

1. Eliciting and modeling the **mental models** of end users with respect to security and privacy in order to understand the stakeholders' assumptions and expectations (Sect. 3)
2. Eliciting and analyzing the **privacy needs** of data subjects and the **data usage needs** of data processors in order to understand the stakeholders' privacy-related requirements (Sect. 4)
3. Collecting and structuring user characteristics along dimensions into **user group profiles** and **privacy personas** in order to understand typical stakeholder perspectives on the protection of their data (Sect. 5)

[1] According to the GDPR [27], *data subjects* are natural persons whose personal data are processed; *data processors* are legal entities or individuals that process personal data of others; *processing* includes gathering, storing, using, transferring, and deleting personal data.

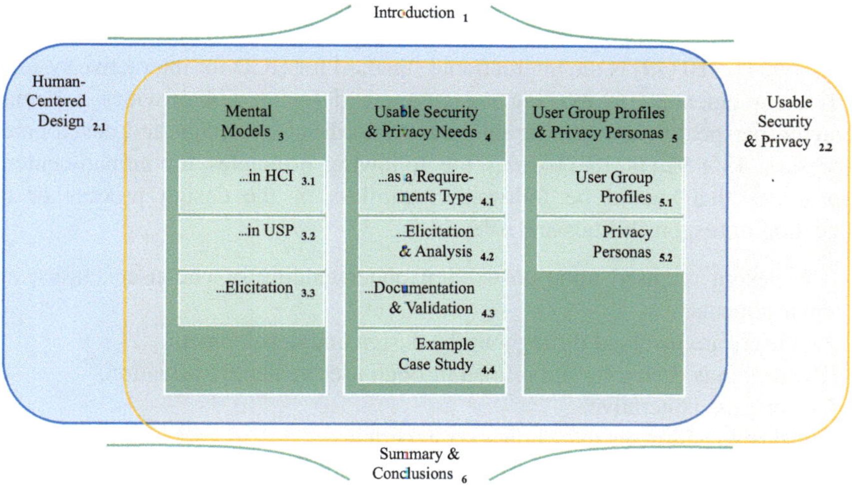

Fig. 1 Overview of this chapter's structure

Figure 1 presents the structure of this chapter. In Sect. 2, we first describe the two concepts underlying this chapter: HCD and USP. We then present the three aforementioned methods in Sects. 3–5. In Sect. 6, we conclude by outlining implications of the USP elicitation techniques. This structure enables readers interested in applying a particular technique to consult only its corresponding section, while we recommend that casual readers follow this chapter's sequential order.

2 Background

2.1 Human-Centered Design

HCD reflects the consideration of end users in the design of systems. The goal and main argument for using an HCD process is to increase the fit of the product to the requirements of end users by involving the stakeholders themselves in the design process. Stakeholder involvement is also intended to minimize the risk of erroneous design decisions. So far, however, the various requirements for security and user experience have mostly been elicited as an incidental by-product—if at all—in HCD. With regard to usable security and privacy (USP) requirements, no best practices exist for many application domains yet, let alone verified research experience. Feth, Maier and Polst [30] proposed a model for USP, using smart homes as an application example. They mapped different parts of their model to the activities in the HCD process. However, mental models and personas were not considered in the model, and USP needs were not addressed as detailed as in this chapter.

ISO 9241-210 [49] is the international standard for HCD for interactive systems. It is complementary to existing design methodologies and provides a human-centered perspective that can be integrated into different design and development processes. ISO 9241-210 provides the following principles for human-centered approaches that should be followed, regardless of the design process or the allocation of responsibilities and roles:

- The design is based upon an explicit understanding of end users, tasks, and environments.
- End users are involved throughout the design and development.
- The design is driven by and refined through user-centered evaluation.
- The process is iterative.
- The design addresses the whole user experience.
- The design team performing HCD includes multidisciplinary skills and perspectives; this does not require a team to be large, but it should be sufficiently diverse to collaboratively make trade-off decisions regarding design and implementation at appropriate times.

The HCD process consists of four activities for designing an interactive system, which in this chapter we relate to USP:

Activity 1: Understanding and Specifying the Context of Use According to ISO 9241-210, the context-of-use description shall include the following:

- *The end users and other stakeholder groups:* There can be a range of different user groups as well as other stakeholder groups whose needs are important, also regarding security and privacy.
- *The characteristics of the end users or groups of users:* End users have different needs and characteristics regarding privacy and security. Eliciting the end users' mental models promotes a better understanding of their subjective conception of technical processes (e.g., data processing) and tasks. The identified user groups can be described in the form of user group profiles or personas. Their relation to personal data helps determine whether they have privacy needs as data subjects and/or data usage needs as data processors.
- *The goals and tasks of the end users:* The types and frequency of tasks that end users typically perform can be part of the persona descriptions, while USP needs are intertwined with particular goals regarding the use or protection of personal data.
- *The environment(s) of the system:* Relevant questions concerning the technical, physical, or socio-cultural environment can be, e.g., Do the end users need to interact with the system when they are preoccupied with other activities? Are there presumably many bystanders? Can someone watch over a user's shoulder and read the screen? Is the data transferred via public or private Wi-Fi?

Activity 2: Specifying the User Requirements Identifying user needs and specifying the functional and other requirements for the system being designed are crucial activities. A specific subset of needs are USP needs, which can, among other things, be analyzed in order to derive further privacy requirements. The intended context of use includes the (personal) data used in and transferred by the system. A quality model can ensure that the goals of different stakeholders are taken into account and that all relevant quality aspects are considered: data quality, product quality, quality in use, process quality, and structural quality [82].

Activity 3: Producing Design Solutions Established best practices facilitate the design of solution prototypes and final designs. In the USP context, three different levels of best practices can be distinguished [92]: (1) *principles* as general fundamentals that should be considered during the development process; (2) *guidelines* as descriptions for adopting these principles, and (3) *patterns* as reusable and proven solutions to commonly occurring problems appearing in system development.

Activity 4: Evaluating the Design User-centered evaluation is an essential element of HCD. Because user tests are usually time-consuming and costly, it is advisable to first conduct an expert-based heuristic evaluation for USP [29]. Another part of the evaluation is the assessment of compliance with legal standards, for which the USP needs provide a helpful basis. In the European Union, the GDPR has the greatest regulatory impact, especially as it defines the rights of the data subject, which include the right of access, the right to rectification, and the right to erasure.

2.2 *Usable Security and Privacy*

As mentioned earlier, USP refers to inter- and transdisciplinary methods for designing security- and privacy-enhancing measures in such a way that: (1) users and security engineers (e.g., designers and developers) are supported in the best way possible in their security- or privacy-related goals and projects and (2) the measures contribute to a continuously positive user experience (e.g., promoting intuitive decision-making on choices regarding data privacy) [39, 81].

USP gained attention and relevance in the mid-1990s when computers entered every household and the Internet became widespread. In 1996, Zurko and Simon [105] proposed three categories for a user-friendly security agenda: (1) usability testing for security systems; (2) security models and mechanisms for user-friendly systems, and (3) consideration of the end users' needs as the primary goal(s) for secure system development. This was a radically new perspective, as end users were often still regarded as a security threat at the time. Other works such as those by Whitten and Tygar [99], Adams and Sasse [3], and Blythe, Koppel and Smith [8] built upon this work.

In the 2000s, USP gained momentum in research. Several standard works dedicated to this topic were published—such as [7, 23, 46, 85]—and many studies were conducted [20, 35–37, 87]. 34 of those earliest works were collected in an anthology [21] focusing on realignment of usability and security, authentication mechanisms, secure systems, privacy and anonymity systems, and commercialization of usability. Garfinkel and Lipford [39] provide a good summary of the field up to 2014, while the work of Fischer-Hübner et al. [31] (in German) is also still up-to-date in many areas.

In recent years, the field of USP expanded to cover topics including ubiquitous computing, smart home, and online privacy. Current research trends also reflect trends in social challenges and themes. Two examples are inclusiveness and diversity, which are both increasingly getting attention in the USP community [59, 70]. Moreover, the surge in employees working from home during and after the COVID-19 pandemic has increased the need for USP [26]. For a more detailed summary of USP research, please refer to the chapter "Empirical Research Methods in Usable Privacy and Security".

3 Mental Models in Security and Privacy

Mental models are personal internal representations of the *external reality* that help people understand their surroundings and guide their actions [52]. On a more abstract level, the external reality can represent any kind of *target system* or problem space that people have to deal with: It can be simple and concrete, like finding our way to the kitchen, or complex and abstract, like dealing with climate change. Mental models essentially convey an individual's perception, imagination, knowledge, and comprehension of a particular target system. When people deal with security and privacy issues in cyberspace, their actions are inevitably the result of their concepts regarding technology, tools, or threats contained in their mental models. In case of misconceptions, people may bypass security measures or avoid using privacy settings because they do not understand how they work or what benefit they bring. It is therefore important to consider end users' mental models in the design of a system, as this helps designers and developers of security and privacy mechanisms to align those mechanisms with the end users' understanding and expectations. This can help increase end users' acceptance and enables them to make informed decisions regarding security and privacy. In Sect. 3.1, we will first define key properties of mental models that are specific to their application in human–computer interaction (HCI). In Sect. 3.2, we will detail for which purposes mental models are suitable in usable security and privacy (USP) and provide examples. In Sect. 3.3, we will conclude this section by outlining how mental models can be elicited in practice.

3.1 Mental Models in Human–Computer Interaction

In the field of HCI, mental models are commonly used to capture the various elements of an individual's awareness and perception of theoretical concepts or the specific information of systems they use [74, 93]. Human beings employ (predominantly simplistic) mental models to grasp complicated processes and systems in their daily lives, rather than spending a lot of time studying them in depth [19]. Nevertheless, irrespective of their accuracy, mental models guide people's decision-making process in both familiar and unfamiliar situations [19, 51]. An end user's mental model is created through interactions with the target system (e.g., the Internet), respectively its system image (e.g., an Internet browser) [71]. Individuals construct mental models of unknown systems by attempting to explain their observations and experiences using analogies from concepts they are familiar with [15]. Thus, the model is affected by an end user's experience and understanding. However, a mental model does not have to be technically correct; it only needs to be practical.

The elicitation of mental models can provide insights into the perceptions and sensations of individuals, which in turn helps to better understand the reasons for and the factors influencing their behavior [19]. If one then tries to elicit an end user's mental model, a conceptualization of this model emerges (i.e., a model of a model). The insights gained from this model can be used to align the target system with the end user's mental model by either supporting them in their understanding or adapting the design of the target system or system image. For example, conceptualized models can be used to design a system in such a way that the cognitive effort required to use it is minimized.

Mental models are generally considered to be vague and highly contextual representations [71]. Based on observations, the use of mental models is subject to the following restrictions [71], which have also been confirmed in related studies [10, 32, 58, 62, 77, 84, 88]: (1) Mental models are incomplete, unstable, and simplified. (2) Mental models have no sharp boundaries. (3) Mental models are "unscientific" and tend to be incorrect. (4) The ability of end users to use mental models is limited. Consequently, there cannot be one unambiguous mental model for a target system; rather—due to subjectivity—several models must always be considered. If the complexity of a target system exceeds the cognitive abilities of a human being, they depend on using a more or less suitable mental model to plan the actions they assume to be "correct" for achieving a goal.

3.2 Mental Models in Usable Security and Privacy

In the field of USP research, mental models are often studied regarding particular tools and technologies (e.g., password managers [12], Wi-Fi [55]), abstract systems

Table 1 Overview of mental model studies in USP

Topic	Context	Stakeholder	Publications
Risk communication	Privacy and security	Computer users	[10, 68]
Smart home	Privacy and security	End users	[103, 104]
Internet use, attacker models, threats, protection strategies	Online privacy and security	Online users, computer users, security experts	[24, 66, 67, 77, 80]
Computer security warnings	Computer security	Lay users vs. experts	[9]
Firewalls	Computer security	Computer users	[78]
Computer security threats	Computer security	Computer users	[54, 96]
Phishing	Online security	Online users	[25]
Influence of mass media	Online security	Online users	[34]
HTTPS, connection security	Online security	Lay users, experts	[33, 56]
Internet	Online security	Lay users, experts	[53]
(End-to-end) Secure communication	Secure communication	Online users	[1, 69, 79, 80]
Encryption mechanisms	Security	Online users	[101]
Passwords and password managers	Security	Online users	[12, 91, 98]
Mobile apps	Mobile privacy	Mobile users	[62]
Online behavioral advertising	Online privacy	Online users	[102]
Internet use and online privacy literacy	Online privacy	Online users, children	[16, 40, 58, 65]
Wi-Fi	Online privacy	Online users	[55]
K-anonymity, anonymous credentials	Privacy-enhancing tech.	Online users	[84, 94]
TOR network	Privacy-enhancing tech.	Lay users vs. experts	[38]
Privacy in employment	Privacy perceptions	Employees	[88]
Folk definitions of privacy	Privacy perceptions	Online users	[60, 72]
Home network maintenance	Technology	Computer users	[76]

(e.g., the Internet [53], smart home [103, 104]), or other abstract concepts (e.g., privacy perceptions [60, 72, 88]). Table 1 presents a non-exhaustive overview of the body of literature on mental model studies in USP and maps these to the stakeholders they address. Numerous studies have sought to understand how end users—and laypeople in particular—envision networks and communication channels, what entities they assume are involved in them, and what threats they believe these entities pose to security and privacy. For example, lay users tend to underestimate the complexity and multi-layered nature of Internet communication, meaning that the actual (personal) data flow remains obscure to them [38, 53, 67].

At the same time, end users also underestimate the capabilities of secure protocols because their complexity exceeds the end users' knowledge and understanding [1, 56]. From previous studies, it is evident that the nature of security and privacy does not permit a mental model that is universally true. Instead, individuals use highly simplified models [2] and rely on various incomplete and poorly formed sub-models [77]. Because the complexity of information systems is often high, simplified mental models can cause end users to behave in unexpected ways, such as unintentionally disclosing private information [2]. Surveying mental models in USP can help mitigate such effects because they help researchers and developers understand why end users may or may not use certain tools or security and privacy mechanisms. Comprehensive summaries of the contents and applications of mental models in security and privacy can be found in [17, 93]. We can distinguish between three main purposes of using mental models in USP:

Purpose 1: Developing Systems in Which Cognitive Effort Is Optimized for Usability [9, 61, 84, 93] Mental models are frequently used to address the common difficulty in USP in order to ensure that the end users of a system accurately perceive the presented information [5, 9, 11] and to facilitate security- and privacy-preserving behavior [12, 84, 100]. For example, a study on security warnings revealed that novice end users and experienced end users seek out different cues when confronted with a warning and also perform different actions [9]. Likely due to their more limited knowledge and experience, novice end users tend to ignore potential security risks because they, for example, do not understand what "SSL certificate" means or because they believe that "saving" and "opening" a file is equivalent. As a result, novice end users may lower their device's overall security level or run unknown software and just wait to see what happens. So, instead of presenting end users with warnings that require them to engage in manual and complex security checks, better wording and automated security checks are ways to increase both usability and security. Many other mental model studies on information systems and security or privacy mechanisms in use contexts highlight similar issues [12, 56, 69, 104]. However, few researchers have used mental models in the development process to inform the design of metaphors and to ensure that security or privacy information is conveyed as intended [5].

Purpose 2: Effective Communication Between Researchers, Experts, Developers, and Laypeople [56, 78, 94, 96] Studies in USP have repeatedly found that lay users and developers or researchers do not speak the same language; for example, lay users talk about "encryption" but actually refer to concepts related to "authentication" [1, 80]. Also, mental models of actual encryption are limited to concepts of symmetric encryption because asymmetric encryption is beyond laypersons' understanding [101]. In such cases, the laypersons' mental models can serve as a tool or template onto which the knowledge of experts can be mapped, thereby making expert knowledge accessible to non-experts. At the same time, the aspects that laypersons consider to be pivotal to their decision-making can be identified and mapped to experts' mental models [11]. Effective communication may also refer to certain UI designs or support tools. These can be designed to correct misconceptions in laypersons' mental models, thereby facilitating security- and privacy-enhancing decisions [32]. This is especially relevant because people have been found to behave rationally in their decision-making, acting congruently with the framework of their mental models [11]. USP research has also suggested approaches that stimulate "false" mental models of end users, which can still lead to a secure use of tools [95].

Purpose 3: Capturing and Exploring Concerns, Expectations, and Understanding [34, 40, 53, 65, 79, 103] End users' mental models can be used, e.g., to get an overview of the variety of mental models that an information system should support [66, 75]. In particular, mental models can serve as an additional basis for dividing the potential target user group into smaller and more homogeneous subgroups of end users that share certain key characteristics related to privacy or security (see also Sect. 5.1 on user group profiles). In doing so, the designers in an HCD process can decide which subgroups to prioritize or give more attention to based on their mental models, for example, user groups with mental models that lead to potentially undesirable, non-privacy-compliant, or insecure behavior [75]. Figure 2 shows three mental models that were identified in a study on employees' understanding of their right to privacy in employment [88]. Each mental model is characterized by different objectives, desires for self-determination and transparency, and acceptance of restrictions. Naming the different mental models makes them more "tangible" and allows them to be used by researchers or developers when implementing privacy controls to take specific sets of properties into consideration. For example, based on the three models, when employers promote a new information system as privacy-friendly, although all employees expect greater control over their personal information, only "Privacy Doctrinairists" would also expect greater transparency. In contrast, a system that only provides transparency may not even be perceived as privacy-friendly by employees with the other two mental models.

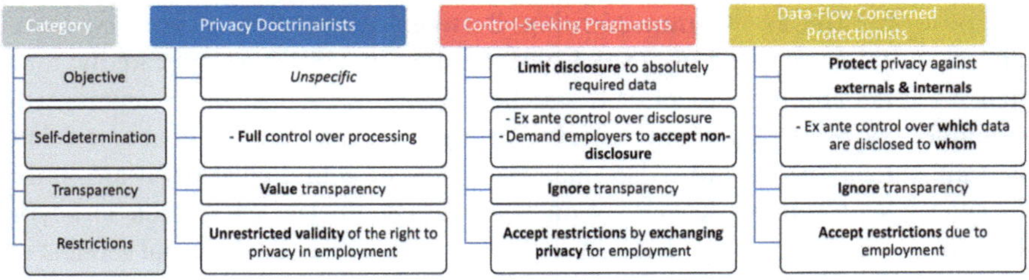

Fig. 2 Example mental models of the right to privacy in employment (adapted from [88])

3.3 Mental Model Elicitation

Eliciting mental models involves extracting an individual's internal representation of a target system. Common elicitation techniques can be divided into two categories according to their methodology [73]: (1) *direct methods*, which rely upon a stakeholder's ability to articulate and partially structure their knowledge and train of thoughts, and (2) *indirect methods*, which employ analyses of written or verbal recordings when stakeholders might be unaware of how they perceive the target system. In the latter case, an evaluator interprets the results of previously processed tasks and structures them, e.g., by frequency.

Open-ended semi-structured interviews are a frequently used instrument for eliciting mental models because the stakeholders can express themselves freely while allowing the interviewer to explore relevant aspects in greater depth by asking targeted follow-up questions [93]. Different support methodologies can be used during these interviews, such as card-sorting tasks or verbal and graphical methods. To ensure the mental models are elicited as completely as possible, purely verbal elicitation can be supplemented, e.g., by presenting the interviewees with illustrations depicting typical elements of the contextual topic and asking them to sort these according to relevance, draw them for themselves, or verbally explain and define certain terms [44, 68]. The interviewees can also be asked to solve practical tasks. During all activities, participants should be encouraged to describe their thought process aloud, which allows inferences to be made about their mental model [68].

Some researchers also use a combination of *focus groups* and individual interviews [84]. Focus groups are a special type of workshop that allows a larger number of subjects to be interviewed simultaneously [57]. Moreover, they help to uncover previously unidentified aspects through discussions between the participants when their opinions diverge. However, researchers should be aware that participants may adapt or even withhold their personal opinions due to group dynamics [57]. Other elicitation techniques are based on hypothetical scenarios that put stakeholders in a situation where they must make decisions according to their mental model [9, 25, 84, 95]. For example, participants may role-play a hypothetical end user who has to manage their privacy and security in everyday tasks [25]. All of these methodologies have their respective advantages and limitations [6]. In order to overcome these limitations, it is a common practice to employ two or more elicitation techniques [55, 78, 84].

When eliciting mental models by means of *surveys*, covering all topics of interest poses a challenge. A sound understanding of the target system is usually required. For this purpose, it can be helpful to first model the target system completely and then derive the survey from it. This is also referred to as an *expert model approach* [10, 68]. First, a model of the target system is created, which ideally contains a complete overview of influencing factors and their relationships. The modeling process may include literature reviews and expert involvement. The model may be revised and fine-tuned over several iterations [88]. Subsequently, questions are derived from the created model, which can be used in the context of an interview study, among other things. In this way, each aspect of the previously created model can be examined specifically. Initial conceptualizations of mental models can be based upon these results. The results can then be verified or validated using a survey for which questions are formulated that test the key points of the previously found conceptualizations. To assure that the outcomes have statistical power, measures must be taken to conduct such a survey with a sufficiently large number of participants. If the survey confirms the initial conceptualizations, the mental model can be further tested with experts, for example, through a practical evaluation in which the mental model is tested against several application scenarios.

In most cases, the interviews are analyzed by means of inductive coding. Here, an initial code list can be created either by coding a few transcripts [53, 58] or by using the available literature and expert knowledge of the research group [77]. To make the mental models more tangible, some researchers create word clouds of the codes to identify their relevance by frequency [84]. Other researchers use graphical approaches [16]. For example, they split the interviewees' responses into short phrases and identify connections between two objects within a statement. These are then represented by nodes in a diagram. The relationships between the nodes are visualized by paths, which are also taken from the analyzed statement (action, relationship). If the elicitation is based on an expert model, this can also serve as a code book for deductive coding. If a statement cannot be assigned to a code, the expert model is expanded to include it. During the evaluation, frequently occurring nodes or paths can be highlighted to visualize the frequency of keywords in the expert model. This visualization can furthermore be used to evaluate the accuracy of the statements [68].

4 Usable Security and Privacy Needs

In Sect. 3, we learned how end users think about usable security and privacy (USP) in the context of (software) systems. With this understanding, we can elicit the end-user needs that flow from these perceptions. Because we found that there is no widely used process for integrating USP into the design process of a secure system, we developed an approach based on the human-centered design (HCD) process that proves very helpful. In this section, we present the specific aspects of our approach to inspire design processes in other organizations. Although the focus here is on USP, keep in mind that aspects other than USP also need to be considered that pertain solely to user experience (UX; e.g., usage needs) and security (e.g., risk analysis).

The USP needs play well into the phases of the HCD process (see Sect. 2.1). Regarding *context of use*, stakeholders are assessed in terms of their role regarding personal data, which helps determine whether they have privacy needs and/or data usage needs. When *specifying the user requirements*, these needs are elicited and documented, and other types of requirements can be derived from them through analysis. The needs can be used as principles that guide the activities toward *producing design solutions*. Finally, they play a crucial role in uncovering and negotiating requirements conflicts when *evaluating the design*.

We will begin with an overview of the five types of requirements that are central to this approach (Sect. 4.1). We will then detail the activities for eliciting and analyzing them (Sect. 4.2) and then those for documenting and validating (Sect. 4.3) USP needs. Finally, we will apply them in a real-world example (Sect. 4.4).

4.1 USP Needs as a Requirements Type

In essence, a *user requirement* describes "a need perceived by a stakeholder" [43]. Typically, these needs are aimed at *what* a (software) system should do (functional requirement) and *how well* that system should do this (quality requirement) [42]. The requirements regarding data protection are somewhat different because the end users' concern is not so much with the system, but with what is (potentially) being done with personal data. These *needs* are therefore often more abstract than functional and quality requirements and cannot be translated directly into organizational measures or software properties. In this chapter, we consider a need to be a goal expressed by a data subject or data processor regarding the processing of personal data.

Various models and methods have been developed to transform abstract legal requirements into suitable measures and to evaluate these measures. The Standard Data Protection Model (SDM) [4] by the technology working group of the German Data Protection Conference standardizes the implementation of the GDPR requirements in concrete technical and organizational measures. One approach to defining privacy requirements is to consider "data privacy" a software product quality characteristic—much like "security"—and taking the seven protection goals from the SDM as subcharacteristics [82]. This helps to define and organize security and privacy requirements. To promote USP, the Usable Privacy Cube model [50] considers both objective and perceived usability criteria when evaluating data protection.

Still, documenting privacy aspects in the form of a traditional user requirement might cause them to either be too unspecific (e.g., "The system shall maintain the users' privacy.") or too much in the solution space (e.g., "When the user logs in, the system shall perform the following actions: ..."), while other notations such as soft goal models do not differentiate between privacy aspects. Something appeared to be missing in between: a different type of need that must be elicited in order to understand what drives the stakeholders. This led us to propose *USP needs*, which we introduce as a novel concept in this chapter: a set of five needs organized into two logical groups as shown in Fig. 3. Each of these needs represents a desire expressed by a stakeholder regarding how personal data are handled [83]. It is paramount to distinguish the stakeholders into the two main user group profiles regarding data

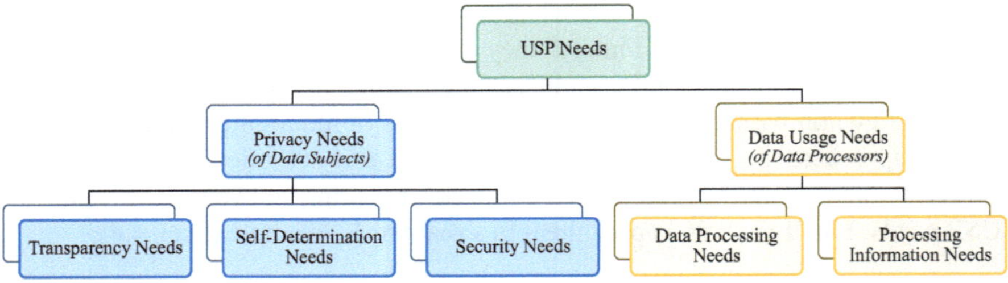

Fig. 3 The five usable security and privacy needs, grouped by type of end user

privacy—data subjects and data processors—because they have divergent needs that, as we will see, might contradict each other.

What sets USP needs apart from typical user requirements is that they are general purpose, i.e., not oriented toward a particular software implementation. They more strongly relate to the legal aspects of personal data. This way, when the needs of both user groups are fulfilled optimally, the effectiveness and the legal compliance of the developed system are inherently maximized, which helps to argue against needs that data protection regulations do not tolerate.

Let us start with *data subjects*. They are concerned with knowing how their data are protected and what their data are being used for, and they want to exert control over (what happens with) their data. This translates into three *privacy needs*:

- **Transparency need:** The data subjects' need or desire for understandable information and openness about the processing of their personal data.
- **Self-determination need:** The data subjects' need or desire for autonomous control over the processing of their personal data.
- **Security need:** The data subjects' desire for their personal data to be protected, particularly with regard to the privacy violations that should be prevented. These will often be phrased negatively, i.e., a need for something *not* to occur.

Data processors, on the other hand, want to use personal data for particular purposes and understand what is allowed. This translates into two *data usage needs*:

- **Data processing need:** The data processors' need to process certain personal data for a specific purpose, including the ability to access such data.
- **Processing information need:** The data processors' need for information about regulations regarding the processing of personal data in order to be legally compliant.

4.2 USP Needs Elicitation and Analysis

In this section, we will describe our recommended approach for embedding the needs into typical requirements engineering (RE) activities, with suggestions on how needs are elicited in workshops and interviews. The elicitation of needs should begin very early in the RE phase; this activity can be initiated as soon as the most important stakeholders have been identified during the initial stakeholder analysis. This is possible because USP needs describe a personal perception or desire that is largely independent of the system being developed. The needs analysis is performed in various stages depending on the status of other requirements artifacts and illustrates the useful contribution of USP needs within the RE activities:

- **Open needs analysis:** The basic needs are determined for the key stakeholders through workshops or interviews. This results in an initial set of basic needs that provide input for deriving user requirements, akin to soft-goal analysis.
- **Scenario-based needs analysis:** Once the project's topic, scope, and goals have crystallized and high-level scenarios have been formulated, the second type of needs analysis can be performed. This analysis associates basic needs with scenarios, while further needs are uncovered as the domain is understood better. In workshops and interviews, scenarios can be used to trigger the stakeholders to express previously uncovered needs. As this activity aims to enrich the understanding of a scenario with the needs that apply to it, scenarios do not have to be associated with all the needs nor vice versa.
- **Detailed needs analysis:** When the to-be (process) situation has been described and use cases have been formulated, it becomes possible to triangulate the use cases with the user requirements and needs associated with them. This ensures that the stakeholders' privacy-related needs have been considered and that the system will deliberately promote, ignore, or actively prevent these needs from getting fulfilled. At this stage, workshops and interviews are usually no longer performed; this is only recommended if the analysis reveals gaps in the elicitation.

The first two analyses, in particular, require active participation of the stakeholders. We recommend eliciting the needs through workshops, but if the project context demands it, semi-structured interviews can also be used. Below, we provide a general-purpose template for both elicitation techniques, which can be tailored to specific project contexts (e.g., a company for which a privacy solution is being designed, the analysis of a particular website, or the definition of a process in which personal data are processed). We also suggest holding at least two different workshops: one with stakeholders who are primarily data subjects to elicit their privacy needs and one with stakeholders who are predominantly data processors to elicit their data usage needs.

The *workshop for data subjects* should be held with at least six participants so that at least two groups can be formed. Ideally, a much larger workshop with a diverse sample of stakeholders is best, but with more than fifteen participants, the workshop becomes harder to manage. Moderation cards should be prepared in three colors, such as yellow, green, and blue. To write down their needs, the participants should use the following template, which is derived from user stories [14]: *"As a <data subject>, I would like* <**need**>, *so that* <rationale>"—see Sect. 4.3 for several examples. The workshop can be structured as follows:

- Form groups of three.
- Each person in the group selects one data class that is likely to contain personal data about them.
- For each data class, the group repeats the following steps:
  - Discuss what a company, individuals, or third parties do with this data class, and for what purpose. Optionally: describe typical security problems with protecting this data class.
  - As a data subject, which needs do you have (accordingly) regarding the protection of this data? Write each need on a yellow card using the sentence template. (This question elicits security needs.)
  - As a data subject, what would you like to know regarding the collection, processing, or use of this data? Write each need on a green card using the sentence template. (This question elicits transparency needs.)
  - As a data subject, what need do you have regarding self-determination? Write each need on a blue card using the sentence template. (This question elicits self-determination needs.)
- Each person in the group picks one data protection, transparency, and self-determination need that they find most important. If another person in the group already picked their most important need, they should choose their second most important one. It is also possible for the workshop organizers to use a prioritization technique (e.g., "buy-a-feature" [41]).
- Discuss as a group what use cases/overall features of a system should be available to fulfill the selected needs.

The *workshop for data processors* can be performed with fewer participants than the workshop with data subjects because this is a smaller stakeholder group whose goals are more homogeneous. Six or nine participants will therefore suffice. Prepare moderation cards in two colors, such as purple and orange. Their needs are also documented as a user story, but using this template instead: *"As a <data processor>, I would like* <**need**>, *so that* <rationale>." The workshop can be structured as follows:

- Form groups of three.
- Each person in the group selects one class of personal data that they are likely to process.
- For each data class, the group repeats the following steps:
  - Discuss what you, your company, other individuals, or third parties do with this data class, and for what purpose. Optionally: describe typical problems regarding the processing of this data class.
  - As a data processor, which needs do you have (accordingly) regarding the processing of this data? Write each need on a purple card using the sentence template. (This question elicits data processing needs.)
  - As a data processor, what would you like to know regarding the collection, processing, or use of these data? Write each need on an orange card using the sentence template. (This question elicits processing information needs.)

- Each person in the group picks one data processing need and one processing information need that they find most important. If another person in the group already picked their most important need, they should choose their second most important one.
- Discuss as a group what use cases/overall features of a system should be available to fulfill the selected needs.

In case the above two workshops cannot be organized, or whenever there are key stakeholders who cannot participate in the workshops, we recommend including the following questions in an *elicitation interview* with the stakeholders:

1. Which of your personal data do you consider worth protecting in the context of the system under development?
2. What (potential) problems do you see in protecting your privacy?
3. In what way should a tool that allows you to set and monitor your privacy settings improve the protection of your privacy? (This question implicitly probes for self-determination needs.)
4. For a specific data category: Which actor or role has or should have access to these data, and what do they use it for?
5. Which need do you have regarding the protection (or, for data processors: processing) of these data? (This question elicits security or data processing needs.)
6. What would you want to know regarding the collection, processing, or use of these data? (This question elicits transparency or processing information needs.)
7. Which of each discussed need is most important to you, and why?

4.3 USP Needs Documentation and Validation

When needs have been elicited during the open or scenario-based needs analysis, they should be documented accordingly. In addition to guidelines on how to document them, in this section, we present a procedure for validating the needs by examining their legal basis.

Typically, documenting the needs begins with typing up the needs from the workshop's moderation cards. We recommend including every elicited need from the workshops or interviews, but differentiating between the needs that the participants identified as important as opposed to those they did not, for example, by prioritizing them according to the MoSCoW method into must-, should-, could-, and will-not-have needs [13]. At this stage, it is important to assure the quality of the contents by checking for wrongly attributed needs (written on the wrong color card) and verifying that the expressed need and the rationale make sense and are self-explanatory. The need should also be given a name, which can often be derived

Table 2 Examples of a security need and a data processing need, adapted from [89]

Attribute	Content
Name	**Business email communication**
Description	As a *social partner*, I would like *my email communication to not be disclosed to others*, so that *I can protect both the content and my contacts.*
Priority	Nice-to-have
Name	**View email contents of employees**
Description	As an *employer*, I would like *to be able to view the email content of my employees*, so that *I can detect misconduct in internal or external communication.*
Priority	Should-have

quite simply from the keywords in the *need* section of the template. Table 2 shows examples of a documented security need and a data processing need.[2]

The key difference in validating needs compared to typical requirements is that instead of verifying with the stakeholders that the documented needs are correct,[3] they are instead analyzed for their legal merit based on applicable regulations, legislature, and case law. The resulting *legal interpretations* form an important basis for assessing to what degree a particular need can be met in the intended context: Should it be allowed, limited, or forbidden? We recommend storing the legal interpretations as separate entities that are subsequently linked to one or more needs. For example, in the German-language documentation of the TrUSD project [89], the needs shown in Table 2 are linked to the two legal interpretations "Processing for purposes of the employment relationship" and "Processing of business emails/determination of private email use," while the data processing need is additionally linked to the legal interpretation "Communication control." These descriptions explain that the processing is permissible pursuant to the German Federal Data Protection Act (BDSG) under specific conditions (e.g., that business and private emails are clearly separated) and for specific purposes detailed in the BDSG and the GDPR.

There is a constant tension between whether limiting one's personal privacy is justified for a specific processing purpose. The use of personal data to optimize work processes often sparks concerns regarding monitoring and performance evaluation,

[2] In chapters 4–8 in [89], we provide a catalog of 139 USP needs for organizational settings in German. It provides 46 transparency needs, 11 self-determination needs, 38 security needs, 39 data processing needs, and 5 processing information needs.

[3] Needs are subjective and do not describe an implementable aspect of a system, so there is no real need to validate them with stakeholders.

which shows that especially *data processing needs* and *security needs* may clash.[4] A key activity within RE is to analyze requirements during the requirements validation phase in order to identify conflicting requirements that need to be reconciled. This is important because failing to identify such conflicts might lead to an implemented system that does not satisfy the needs of at least one stakeholder. By specifying these needs, requirements reconciliation can be performed, and the decisions made to address conflicting needs are made explicit. In many cases, this will involve communicating the legal interpretation to the stakeholders involved (e.g., that privacy legislation does not permit a need to be fulfilled, or that higher value is assigned to a different need). The best solution approach depends on the context. For example, if potentially many data subjects have a concern, the explanation of why the system helps fulfill a particular data processing need could take the form of an information campaign explaining the necessity and benefits of processing the data and what security measures are being taken. Similarly, if a project reveals there is a great demand for transparency, this can be met through the development of solutions such as privacy dashboards in organizations [90] or privacy cockpits in digital ecosystems [22].

4.4 Example Case Study

In 2021, the regional public broadcaster *L1* of the Dutch province of Limburg got international media coverage due to a serious privacy-related incident. A quickly escalating dispute caused their newly appointed director to be suspended after nine months; a court ruling particularly blamed a disorganized works council.[5] Problems had arisen even before the director took up his post, with staff disputing the Supervisory Board's appointment procedure.[6] Dissatisfaction over his communication and leadership style caused employees to respond in a way the director described as a guerilla war waged against him.[7] But things really culminated when he presented a draft of the new privacy regulations that would infringe on the workers' privacy through the use of hidden cameras in the office, and—if there were compelling reasons—access to browsing histories and email accounts, including

[4] For example, an employee may not wish for their employer to know that they are ill (security need), but an employer has the right to know this. However, the employer may only use this knowledge for specific purposes such as resource planning and aggregated analyses (data processing needs). Using this information to send a collective get-well card is only allowed with the data subject's consent, and individual assessments based on this information are strictly prohibited.

[5] https://amp.nos.nl/artikel/2387272-bestuurder-peter-elbers-van-regionale-omroep-l1-op-non-actief.htm.

[6] https://www.limburger.nl/cnt/dmf20200922_00176921.

[7] https://www.limburger.nl/cnt/dmf20201105_93947605.

those of journalists, the company physician, and members of the works council.[8] The director had the sole power to determine what he considered compelling, and he would also be responsible for handling any complaints. The outrage among staff, Dutch journalists and lawyers, and in society as a whole resulted in the draft being retracted just three days later.

This case study shows that the director had several *data processing needs*, such as "View email contents of employees" (shown in Table 2) and "View browsing history of employees" for the purpose of assessing individual employees. Typically, professional correspondence may be reviewed if it is clearly distinct from private communications, but this assessment should then be performed by a superior, not by the director. This specific situation, however, uncovers a domain-specific type of *processing information need*: "Privacy rights of journalists." The fundamental principle that safeguards freedom of the press limits the ability to put journalists under surveillance to ensure that they can exert their duty of protecting their sources.[9] The director should have been aware that this kind of data processing contradicts the special rights of journalists that safeguard the *security needs* of "Business email communication" (shown in Table 2) and "Protect the identity of news sources from others" (including their employer), to which they are legally and ethically entitled. This demonstrates that these kinds of needs are not general purpose; while work emails may normally be monitored under certain conditions, these particular *security needs* are prioritized as "must-have" for journalists.

The director also had the *data processing need* "Video surveillance of employee activities." Under strict conditions, data protection legislation allows video surveillance for specific purposes (e.g., preventing illegal activities or industry espionage; improving work floor safety). However, monitoring employee activities in non-public spaces using CCTV cameras is only allowed if it is the mildest and most suitable measure, and should in that case be openly announced instead of through the use of hidden cameras. For the same reasons as above, this conflicts with and is overruled by journalists' *security needs*.

5 User Group Profiles and Privacy Personas

Section 4 described what needs the end users of a software system have with regard to usable security and privacy (USP). But who are these end users, and how can we typify them? The ISO 9241-210 standard [49] names two artifacts for describing user characteristics: user group profiles and personas. Although they are introduced as part of the *context of use*, they can also be used in other activities of the human-centered design (HCD) process, for example to specify the usage requirements of specific groups. These artifacts can accompany the development

[8] https://www.volkskrant.nl/cultuur-media/directeur-limburgse-omroep-11-wil-eigen-personeel-kunnen-volgen-met-camera-s~bba48bd7/.

[9] In this context, *sources* are professional contacts who provide journalists with newsworthy information.

team throughout the development process up to the evaluation, which can be carried out as a walkthrough from the perspective of a specific persona. We will discuss the concept of user group profiles in Sect. 5.1 and that of (privacy) personas in Sect. 5.2.

User characteristics strongly influence the context in which a system is used. It is therefore useful to gather and analyze relevant information about them in order to understand the current context and to specify the context for the future system. User group profiles summarize typical characteristics of end users, while personas are concrete examples of typical end users [86]. As examples of the characteristics of different user types, ISO 9241-210 cites end users with different levels of experience or physical capability.

5.1 User Group Profiles

Concerning data protection, the most essential user group profiles are the two main types of end users distinguished in the GDPR [27]: *data subjects*, whose personal data are processed, and *data processors*, who process personal data. Section 4.1 demonstrates one practical use of these profiles.

The consumer study "DsiN-Sicherheitsindex 2022" [63] distinguishes five different groups of end users of Internet services by their knowledge and behavior and provides suggestions on how to address security deficits for each (percentages according to DsiN):

1. **Fatalistic users** (17.7%) see dangers lurking everywhere but question the effectiveness of security measures. They often do not realize that their own behavior is an important component in the security concept.
2. **Outsiders** (5.3%) often feel overwhelmed by new digital offers but consider themselves to be primarily responsible for protecting their personal data.
3. **Thoughtless users** (37.1%) have a very high level of security knowledge but apply it too rarely. They are the least concerned about being at risk and have little interest in risk reduction measures.
4. **Driving users** (22.2%) are open to new things and try out more new digital services and offers than other end users. Due to their open-mindedness and curiosity, they are particularly suitable as multipliers to raise awareness.
5. **Considerate users** (17.8%) have the highest security knowledge and are also forerunners in the implementation processes. They are the most cautious and privacy-aware users when it comes to new digital offerings.

A similar approach is taken by Dupree et al. [97]. They divide end users of privacy and security tools into five categories according to their attitudes, beliefs, and behaviors: *marginally aware*, *fundamentalist*, *struggling amateur*, *technician*, and *lazy expert*. Some of these categories are compatible with the DsiN classification; for example, the lazy expert resembles the thoughtless users. Based on the rather abstract user group profiles, Dupree et al. also created personas (see Sect. 5.2) that cover the user space of privacy and security tools (e.g., "Henry—

The Lazy Expert"). With respect to end users' attitude and motivation toward giving feedback, Groen et al. [45] identified seven categories: *privacy-tolerant and socially ostentatious*, *privacy-fanatical but generous*, *passive and stingy*, *loyal & passionate*, *incentive seekers*, *perfectionists & complainers*, and *impact seekers*. Due to cultural differences, corresponding categorizations often only apply to the inhabitants of the country examined. For example, in a recent study, 65% of the participants in Cyprus were open to sharing their facial images with public administration for identity purposes, compared to 9% of the participants in Germany, Poland, and Romania [28]. For user group profiles that describe end users according to their use of particular security measures, it must be noted that corresponding security measures often become outdated after a few years, which may cause these user classifications to also become outdated over time.

User group profiles are a helpful means of painting a much more accurate picture of the key stakeholders that directly interact with the system. For example, the stakeholder group of *end users* can correspond to the five DsiN groups. By categorizing them accordingly, it is possible to analyze and document the needs and requirements of this stakeholder group in a much more differentiated way.

5.2 *Privacy Personas*

Personas are fictitious individuals representing typical user groups as archetypes [18]. Creating personas is not the same as defining user groups or creating user group profiles. Personas are descriptions of stereotypical individual end users that are derived from the identified user groups in order to emphasize the most important characteristics and details of the respective user group [47]. Usually, as many personas are created as are needed to cover all relevant user groups [64].

The intention behind creating personas is to get a more vivid description of the end users than with the more abstract user group profiles. The basis for the creation of personas can be quantitative or qualitative data collections, online surveys, interviews, or participatory observations of potential end users. Personas for the USP domain, or *privacy personas*, should emphasize the different ways in which personal data are handled and the different security needs of end users, among other things. Importantly, no discriminatory aspects should be highlighted nor associations made with real people [48]. Cooper, Reimann and Cronin [18] recommend that after the research with end users is complete, the distinct aspects of user behavior be listed as a set of behavioral variables. While demographic variables such as age or geographic location influence behavior, behavioral variables are much more useful in developing effective personas. The most important variables for distinguishing behavioral patterns according to Cooper et al. are *activities*, *attitudes*, *aptitudes* (e.g., education, training), *motivations*, and *skills* (related to the product domain and technology). In enterprise applications, behavioral variables are often closely related to job roles. Therefore, they recommend listing the variables separately for each role, i.e., creating a separate persona for each role.

Table 3 Example description of a privacy persona [89]

Attribute	Content
Name	**Ian Frederick** (sales employee)
Who am I?	37 years, male, single
Attitude toward digital work	Ian is aware of the importance of data protection in digitized work processes, especially as he handles customer data in sales work
Reasons for using the system	To see which consents have been given to the employer
Reasons for not using the system	Complicated handling; missing help options
Personality classification	Ian is extroverted, partly analytical and partly creative, neither particularly chaotic nor organized, is team-oriented, and has partial freedom in terms of time
Interests, motives, and goals	Well-established, simple processes for all sales and marketing activities; fast, centralized access to all required data
Problems and challenges	Both customer data and employee data must be kept up-to-date; this only works if all colleagues play their part
Personal environment and self-perception	Ian is appreciated by all colleagues as a team player and finds very good ways to approach different customer personalities
Typical working day	Everyday exposure to technology in the work environment, both as an end user (e.g., CRM and ERP system) and in sales (product demonstrations) for data protection
Qualifications and skills	IT specialist; Ian is involved in many of the company's projects

Personas can be used by a system's design and development team to imagine themselves in the role of current and future end users and better emphasize with them. This enables them to better understand their needs and play through different usage scenarios from the end users' point of view. By understanding the way the end user thinks and acts, it is easier to make the right design decisions—in overall product development, but also during the design of security features and data protection mechanisms to ensure they become as user-friendly as possible for specific user groups [64].

Various projects in the area of USP [22, 90] have developed templates and examples that support the creation of personas. Templates make it possible to evaluate research data and summarize the collected findings in a structured and clear way so that they can be referred to in the further course of development. Table 3 shows one of eight personas developed for a privacy dashboard that caters to the goals and needs of employees. Figure 4 shows a persona template for representing different user groups of digital ecosystems used to design and develop privacy cockpits. In both examples, some variables of "conventional" templates were adapted or further specified in order to collect and analyze USP needs in a structured way.

In addition to persona templates, workshop concepts for supporting the creation of personas have been proposed. Workshops for creating personas are particularly

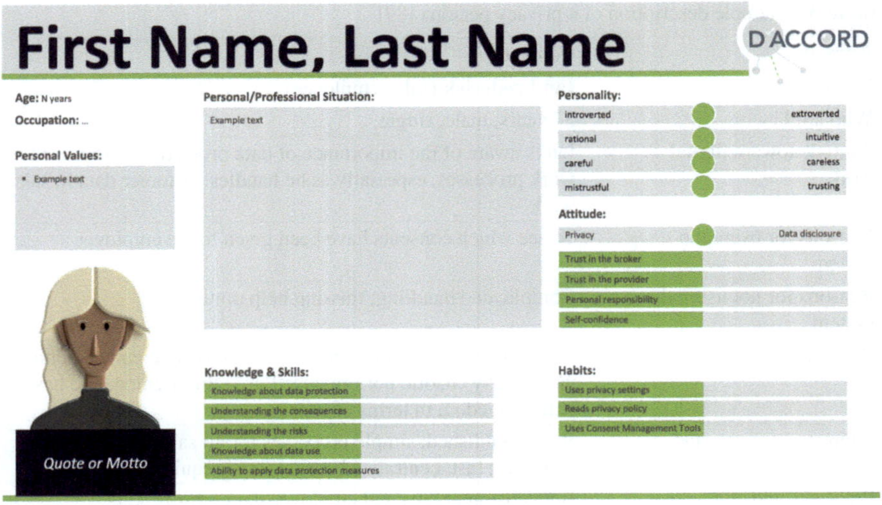

Fig. 4 Persona template for end users of digital ecosystems

useful if no comprehensive research material is available. For example, a workshop concept for elaborating personas in companies or organizations was developed that involves around 18 participants (including the moderator) and has a duration of two hours [48]. This workshop's schedule is as follows:

- Welcome and round of introductions.
- Presentation of the method.
- Formation of small groups.
- Each group works out an organization-specific persona using the persona template (e.g., on a Metaplan wall).
- Each group presents its persona in a group discussion.
- Summarization of the results in a feedback round.

6 Summary and Conclusion

In this chapter, we presented three methods regarding the interface between *human-centered design* (HCD) and *usable security and privacy* (USP): (1) mental models in security and privacy, (2) USP needs, and (3) stakeholder descriptions using user group profiles and privacy personas. These methods are complimentary in that they elicit or collect different types of information, with their own documentation formats and contributions to the design of a digital system.

The methods can play a constructive role throughout the HCD process. They can all be used to specify and understand the stakeholders' *characteristics* regarding USP: Mental models enable this by conceptually exploring their subjective perception and assumptions (implicit expectations); USP needs by inventorying their desires and requirements (explicit expectations), and profiles/personas by organizing them into logical groups. In other process steps, *goals and tasks* can be identified from the USP needs and included in the persona descriptions. Through analysis, *user requirements* can be derived from analyzing the USP needs. For the *design solutions*, mental models can inform patterns on end users' preconceptions, while USP needs provide possible principles. Finally, all methods help to *evaluate the design*: in terms of how well the system plays into the mental models, in terms of assuring that the USP needs are being fulfilled or overruled by other USP needs and comply with legal standards, and in terms of considering the usage scenarios of the system from the perspective of each persona.

Together, the three methods augment the HCD process with practical approaches to analyzing and assuring that USP is correctly implemented in a system by ensuring that the stakeholders are known, understood, and validated in the system's design. The additional work involved in applying these methods is manageable and can be justified by their contribution of employing good requirements engineering (RE) and user experience (UX) design practices, with which they integrate perfectly in our experience. Their use makes a positive contribution to a system's overall quality, not only in terms of constraints (e.g., improved assurance of compliance with data protection regulations), but also in terms of system quality (e.g., because security aspects have been analyzed in more depth) and quality in use (e.g., greater trust in the system). Specifically, we argue that these techniques will help the system to better achieve two core principles stipulated in Article 25 of the GDPR: (1) *Data Protection by Design* or *Security by Design*, which postulates the consideration of technical and organizational measures in the system design and development from the very beginning to ensure the best possible privacy and security as well as smooth human–machine interaction, and (2) *Data Protection by Default* or *Security by Default*, which postulates that the privacy and security of a system should not rely on end users making good settings, but rather that the default settings should already be as user-friendly, privacy-promoting, and secure as possible.

By presenting these methods, we hope to support the reader with practical knowledge and skills to help them achieve better USP in their systems. We do not claim that these are the only USP-related techniques that can be used in the HCD process, but in our context, we found these methods to be sufficient supplements to the tried and tested RE and UX techniques for achieving our goals. We do encourage the reader to try these approaches for themselves.

Acknowledgments This work is funded by the German Federal Ministry of Education and Research (BMBF) (grant numbers 16KIS1506K, 16KIS1507, and 16KIS1508). We thank Sonnhild Namingha for proofreading this chapter and Jannis von Albedyll for providing insightful comments.

References

1. Abu-Salma, R., Sasse, M. A., Bonneau, J., Danilova, A., Naiakshina, A., & Smith, M. (2017). Obstacles to the adoption of secure communication tools. In *Proc. of IEEE Symposium on Security and Privacy (SP)* (pp. 137–153). IEEE.
2. Acquisti, A., & Grossklags, J. (2005). Privacy and rationality in individual decision making. *IEEE Security & Privacy, 3*(1), 26–33.
3. Adams, A., & Sasse, M. A. (1999). Users are not the enemy. *Communications of the ACM, 42*(12), 40–46.
4. AK Technik of the Independent Data Protection Supervisory Authorities of the Federation and the Länder. (2020). The standard data protection model. Technical report, UAG Standard Data Protection Model of the AK Technik of the Independent Data Protection Supervisory Authorities of the Federation and the Länder.
5. Angulo, J., Fischer-Hübner, S., Pulls, T., & Wästlund, E. (2015). Usable transparency with the data track: A tool for visualizing data disclosures. In *Proc. of the 33rd Annual ACM Conference Extended Abstracts on Human Factors in Computing Systems (CHI EA)* (pp. 1803–1808). ACM Press.
6. Asgharpour, F., Liu, D., & Camp, L. J. (2007). Mental models of security risks. In S. Dietrich, & R. Dhamija (Eds.), *Financial cryptography and data security*. Lecture notes in computer science (pp. 367–377). Springer.
7. Balfanz, D., Durfee, G., Smetters, D. K., & Grinter, R. E. (2004). In search of usable security: Five lessons from the field. *IEEE Security & Privacy, 2*(5), 19–24.
8. Blythe, J., Koppel, R., & Smith, S. W. (2013). Circumvention of security: Good users do bad things. *IEEE Security & Privacy, 11*(5), 80–83.
9. Bravo-Lillo, C., Cranor, L. F., Downs, J., & Komanduri, S. (2011). Bridging the gap in computer security warnings: A mental model approach. *IEEE Security & Privacy Magazine, 9*(2), 18–26.
10. Camp, L. J. (2009). Mental models of privacy and security. *IEEE Technology and Society Magazine, 28*(3), 37–46.
11. Cassaigne, N. (2002). The dashboard: A knowledge conversion tool. In *IEEE International Engineering Management Conference* (Vol. 1, pp. 292–297). IEEE.
12. Chiasson, S., van Oorschot, P. C., & Biddle, R. (2006). A usability study and critique of two password managers. In *Proc. of the 15th Conference on USENIX Security Symposium* (pp. 1–16). USENIX Association.
13. Clegg, D., & Barker, R. (1994). *Case method fast-track: A RAD approach*. Addison-Wesley.
14. Cohn, M. (2004). *User stories applied: For agile software development*. Addison-Wesley Longman Publishing.
15. Collins, A., & Gentner, D. (1987). How people construct mental models. In *Cultural models in language and thought* (pp. 243–265). Cambridge University Press.
16. Coopamootoo, K. P., & Groß, T. (2014). Mental models of online privacy: Structural properties with cognitive maps. In *Proc. of the 28th International BCS Human Computer Interaction Conference (BCS-HCI)*, BCS-HCI '14 (pp. 287–292). BCS.
17. Coopamootoo, K. P. L., & Groß, T. (2014). Mental models for usable privacy: A position paper. In D. Hutchison, T. Kanade, J. Kittler, J. M. Kleinberg, A. Kobsa, F. Mattern, J. C. Mitchell, M. Naor, O. Nierstrasz, C. Pandu Rangan, B. Steffen, D. Terzopoulos, D. Tygar, G. Weikum, T. Tryfonas, & I. Askoxylakis (Eds.), *Human aspects of information security, privacy, and trust* (Vol. 8533, pp. 410–421). Springer International Publishing.
18. Cooper, A., Reimann, R., & Cronin, D. (2012). *About Face 3: The essentials of interaction design*. Wiley.
19. Craik, K. J. W. (1943). *The nature of explanation*. University Press, Macmillan.
20. Cranor, L. F., & Garfinkel, S. (2004). Guest editors' introduction: Secure or usable? *IEEE Security & Privacy, 2*(5), 16–18.

21. Cranor, L. F., & Garfinkel, S. (2005). *Security and usability: Designing secure systems that people can use*. O'Reilly Media.
22. D'accord-Konsortium. (2022). D'accord—Adaptive Datenschutz-Cockpits in digitalen Ökosystemen (2022). https://daccord-projekt.de/
23. DeWitt, A. J., & Kuljis, J. (2006). Aligning usability and security: A usability study of Polaris. In *Proc. of the 2nd Symposium on Usable Privacy and Security* (pp. 1–7). ACM.
24. Dourish, P., de la Flor, J. D., & Joseph, M. (2003). Security as a practical problem: Some preliminary observations of everyday mental models. In *Proc. of CHI 2003 Workshop on HCI and Security Systems* (p. 3). ACM.
25. Downs, J. S., Holbrook, M. B., & Cranor, L. F. (2006). Decision strategies and susceptibility to phishing. In *Proc. of the 2nd Symposium on Usable Privacy and Security*, SOUPS '06 (pp. 79–90). ACM.
26. Emami-Naeini, P., Francisco, T., Kohno, T., & Roesner, F. (2021). Understanding privacy attitudes and concerns towards remote communications during the COVID-19 pandemic. In *Proc. of the 17th Symposium on Usable Privacy and Security,* SOUPS'21 (pp. 695–714). USENIX Association.
27. European Union. (2016). General Data Protection Regulation. (2016). https://eur-lex.europa.eu/eli/reg/2016/679/2016-05-04. Regulation (EU) 2016/679.
28. European Union Agency for Fundamental Rights. (2020). *Your rights matter: Data protection and privacy: Fundamental rights survey*. Publications Office of the European Union.
29. Feth, D., & Polst, S. (2022). Benutzerfreundliche Umsetzung von Datensouveränität in Digitalen Ökosystemen. Whitepaper, Fraunhofer IESE.
30. Feth, D., Maier, A., & Polst, S. (2017). A user-centered model for usable security and privacy. In T. Tryfonas (Ed.), *Human aspects of information security, privacy and trust* (pp. 74–89). Springer.
31. Fischer-Hübner, S., Grimm, R., Lo Iacono, L., Möller, S., Müller, G., & Volkamer, M. (2011). Gebrauchstaugliche Informationssicherheit. *Die Zeitschrift für Informationssicherheit Jg. 4*, 14–19.
32. Fischer-Hübner, S., Pettersson, J. S., & Angulo, J. (2015). HCI requirements for transparency and accountability tools for cloud service chains. In M. Felici & C. Fernández-Gago (Eds.), *Accountability and security in the cloud: First summer school, cloud accountability project, A4Cloud, Malaga, Spain, June 2–6, 2014, Revised Selected Papers and Lectures* (pp. 81–113). Springer.
33. Friedman, B., Hurley, D., Howe, D. C., Felten, E., & Nissenbaum, H. (2002). Users' conceptions of web security: A comparative study. In *CHI '02 Extended Abstracts on Human Factors in Computing Systems*, CHI EA '02 (pp. 746–747). ACM.
34. Fulton, K. R., Gelles, R., McKay, A., Roberts, R., Abdi, Y., & Mazurek, M. L. (2019). The effect of entertainment media on mental models of computer security. In *Proc. of the 15th USENIX Conference on Usable Privacy and Security*, SOUPS'19 (pp. 79–95). USENIX Association.
35. Furnell, S. M., Jusoh, A., & Katsabas, D. (2006). The challenges of understanding and using security: A survey of end-users. *Computers & Security, 25*(1), 27–35.
36. Furnell, S. (2005). Why users cannot use security. *Computers & Security, 24*(4), 274–279.
37. Furnell, S. (2007). Making security usable: Are things improving? *Computers & Security, 26*(6), 434–443.
38. Gallagher, K., Patil, S., & Memon, N. (2017). New me: Understanding expert and non-expert perceptions and usage of the Tor anonymity network. In *Proc. of the 13th USENIX Conference on Usable Privacy and Security*, SOUPS '17 (pp. 385–398). USENIX Association.
39. Garfinkel, S., & Lipford, H. R. (2014). Usable security: History, themes, and challenges. *Synthesis Lectures on Information Security, Privacy, and Trust, 5*(2), 1–124.
40. Gerber, N., Zimmermann, V., & Volkamer, M. (2019). Why Johnny fails to protect his privacy. In *2019 European Symposium on Security and Privacy Workshops (EuroS PW)* (pp. 109–118). IEEE.

41. Gkatzidou, V., Giacomin, J., & Skrypchuk, L. (2021). *Automotive human centred design methods*. De Gruyter.
42. Glinz, M. (2007). On non-functional requirements. In *Proc. of the 15th IEEE International Requirements Engineering Conference,* RE'07 (pp. 21–26). IEEE.
43. Glinz, M. (2017). A glossary of requirements engineering terminology. https://www.ireb.org/en/cpre/cpre-glossary/
44. Grenier, R. S., & Dudzinska-Przesmitzki, D. (2015). A conceptual model for eliciting mental models using a composite methodology. *Human Resource Development Review, 14*(2), 163–184.
45. Groen, E. C., Seyff, N., Ali, R., Dalpiaz, F., Doerr, J., Guzmán, E., Hosseini, M., Marco, J., Oriol, M., Perini, A., & Stade, M. (2017). The crowd in requirements engineering: The landscape and challenges. *IEEE Software, 34*(2), 44–52.
46. Gutmann, P., & Grigg, I. (2005). Security usability. *IEEE Security & Privacy, 3*(4), 56–58.
47. Harley, A. (2015). Personas make users memorable for product team members. https://www.nngroup.com/articles/persona/
48. Institut für Technologie und Arbeit (ITA). (2021). Entwicklung eines Privacy Dashboard-Modellierungsrahmenwerks: D2.3 Dokumentation des Vorgehensmodells. Version 6. https://www.trusd-projekt.de/wp/wp-content/uploads/2022/06/TrUSD-D2.3-Partizipatives-Vorgehensmodell.pdf
49. ISO. (2019). *Ergonomics of human-system interaction—part 210: Human-centred design for interactive systems*. Standard.
50. Johansen, J., & Fischer-Hübner, S. (2020). Making GDPR usable: A model to support usability evaluations of privacy. In M. Friedewald, M. Önen, E. Lievens, S. Krenn, & S. Fricker (Eds.), *Privacy and identity management. Data for better living: AI and privacy: 14th IFIP WG 9.2, 9.6/11.7, 11.6/SIG 9.2.2 International Summer School, Windisch, Switzerland, August 19–23, 2019, Revised Selected Papers* (pp. 275–291). Springer International Publishing.
51. Johnson-Laird, P. N. (1986). *Mental models: Towards a cognitive science of language, inference, and consciousness*. Cognitive Science Series. Harvard University Press.
52. Jones, N., Ross, H., Lynam, T., Perez, P., & Leitch, A. (2011). Mental models: An interdisciplinary synthesis of theory and methods. *Ecology and Society, 16*(1), article 46.
53. Kang, R., Dabbish, L., Fruchter, N., & Kiesler, S. (2015). "My data just goes everywhere:" User mental models of the Internet and implications for privacy and security. In *Proc. of the 11th Symposium on Usable Privacy and Security*, SOUPS'15 (pp. 39–52). USENIX Association.
54. Kauer, M., Günther, S., Storck, D., & Volkamer, M. (2013). A comparison of American and German folk models of home computer security. In L. Marinos & I. Askoxylakis (Eds.), *Human aspects of information security, privacy, and trust.* Lecture Notes in Computer Science 8030 (pp. 100–109). Springer.
55. Klasnja, P., Consolvo, S., Jung, J., Greenstein, B. M., LeGrand, L., Powledge, P., & Wetherall, D. (2009). "When I am on Wi-Fi, I am fearless": Privacy concerns & practices in everyday Wi-Fi use. In *Proc. of the SIGCHI Conference on Human Factors in Computing Systems*, CHI '09 (pp. 1993–2002). ACM.
56. Krombholz, K., Busse, K., Pfeffer, K., Smith, M., & von Zezschwitz, E. (2019). "If HTTPS were secure, I wouldn't need 2FA" - end user and administrator mental models of TTPS. In *Proc. of the 2019 Symposium on Security and Privacy (SP)* (pp. 246–263). IEEE.
57. Krueger, R. A., & Casey, M. A. (2015). *Focus groups: A practical guide for applied research* (5th ed.). SAGE.
58. Kumar, P., Naik, S. M., Devkar, U. R., Chetty, M., Clegg, T. L., & Vitak, J. (2017). 'No telling passcodes out because they're private': Understanding children's mental models of privacy and security online. In *Proc. of the ACM on Human-Computer Interaction, 1*(CSCW), 64:1–64:21.
59. Kumar, D., Kelley, P. G., Consolvo, S., Mason, J., Bursztein, E., Durumeric, Z., Thomas, K., & Bailey, M. (2021). Designing toxic content classification for a diversity of perspectives.

In *Proc. of the 17th Symposium on Usable Privacy and Security*, SOUPS'21 (pp. 299–318). USENIX Association.
60. Kwasny, M., Caine, K., Rogers, W. A., & Fisk, A. D. (2008). Privacy and technology: Folk definitions and perspectives. In *Extended abstracts on human factors in computing systems*, CHI EA '08 (pp. 3291–3296). ACM.
61. Lederer, S., Hong, J. I., Dey, A. K., & Landay, J. A. (2004). Personal privacy through understanding and action: Five pitfalls for designers. *Personal and Ubiquitous Computing, 8*(6), 440–454.
62. Lin, J., Sadeh, N., Amini, S., Lindqvist, J., Hong, J. I., & Zhang, J. (2012). Expectation and purpose: Understanding users' mental models of mobile app privacy through crowdsourcing. In *Proc. of the 2012 ACM Conference on Ubiquitous Computing (UbiComp)* (pp. 501–510). ACM Press.
63. Littger, M. (2022). Studie von Deutschland sicher im Netz e.V. zur digitalen Sicherheitslage von Verbraucher:innen in Deutschland. https://www.sicher-im-netz.de/dsin-sicherheitsindex-2022
64. Lo Iacono, L., Schmitt, H., Feth, D., Jakobi, T., Gorski, P. L., Dölle, M., Nehren, P., Kropp, E., Hausmann, S., Hofmeister, A., Frydyada de Piotrowski, A., & Balthasar, M. (2019). Arbeitskreis Usable Security & Privacy: Nutzerzentrierter Schutz sensibler Daten. Fachschrift. 3., aktualisierte Ausgabe. Technical report, German UPA e.V.
65. Maceli, M. (2019). Librarians' mental models and use of privacy-protection technologies. *Journal of Intellectual Freedom & Privacy, 4*(1), 18–32.
66. Maier, J., Padmos, A., S. Bargh, M., & Wörndl, W. (2017). Influence of mental models on the design of cyber security dashboards. In *Proc. of the 12th International Joint Conference on Computer Vision, Imaging and Computer Graphics Theory and Applications* (pp. 128–139). SCITEPRESS - Science and Technology Publications.
67. Mbewe, E. S., & Chavula, J. (2022). Security mental models and personal security practices of Internet users in Africa. In Y. H. Sheikh, I. A. Rai, & A. D. Bakar (Eds.), *E-infrastructure and e-services for developing countries*. Lecture Notes of the Institute for Computer Sciences. Social Informatics and Telecommunications Engineering (pp. 47–68). Springer International Publishing.
68. Morgan, M. G., Fischhoff, B., Bostrom, A., & Atman, C. J. (Eds.) (2002). *Risk communication: A mental models approach*. Cambridge University Press.
69. Naiakshina, A., Danilova, A., Dechand, S., Krol, K., Sasse, M. A., & Smith, M. (2016). Poster: Mental models – User understanding of messaging and encryption. In *Proc. of the 1st IEEE European Symposium on Security and Privacy* (article 18). IEEE.
70. Napoli, D., Baig, K., Maqsood, S., & Chiasson, S. (2021). "I'm literally just hoping this will work:" Obstacles blocking the online security and privacy of users with visual disabilities. In *Proc. of the 17th Symposium on Usable Privacy and Security*, SOUPS'21 (pp. 263–280). USENIX Association.
71. Norman, D. A. (1983). Some observations on mental models. In D. Gentner & A. L. Stevens (Eds.), *Mental models* (pp. 7–14). Lawrence Erlbaum Associates.
72. Oates, M., Ahmadullah, Y., Marsh, A., Swoopes, C., Zhang, S., Balebako, R., & Cranor, L. F. (2018). Turtles, locks, and bathrooms: Understanding Mental models of privacy through illustration. *Proceedings on Privacy Enhancing Technologies, 2018*(4), 5–32.
73. Olson, J. R., & Rueter, H. H. (1987). Extracting expertise from experts: Methods for knowledge acquisition. *Expert Systems, 4*(3), 152–168.
74. Payne, S. J. (2007). Mental models in human-computer interaction. In A. Sears & J. A. Jacko (Eds.), *The human-computer interaction handbook: Fundamentals, evolving technologies and emerging applications*, Human Factors and Ergonomics Ser. (2nd ed., p. 14). CRC Press.
75. Piekarska, M., Zhou, Y., Strohmeier, D., & Raake, A. (2015). Because we care: Privacy dashboard on Firefox OS. In *Proc. of the 9th Workshop on Web 2.0 Security and Privacy (W2SP)* (article 1). IEEE.

76. Poole, E. S., Chetty, M., Grinter, R. E., & Edwards, W. K. (2008). More than meets the eye: Transforming the user experience of home network management. In *Proc. of the 7th ACM Conference on Designing Interactive Systems*, DIS '08 (pp. 455–464). ACM.
77. Prettyman, S. S., Furman, S., Theofanos, M., & Stanton, B. (2015). Privacy and security in the brave new world: The use of multiple mental models. In T. Tryfonas & I. Askoxylakis (Eds.), *Human aspects of information security, privacy, and trust*. Lecture Notes in Computer Science 8030 (pp. 260–270). Springer International Publishing.
78. Raja, F., Hawkey, K., & Beznosov, K. (2009). Revealing hidden context: Improving mental models of personal firewall users. In *Proc. of the 5th Symposium on Usable Privacy and Security*, SOUPS'09 (article 1). ACM Press.
79. Renaud, K., Volkamer, M., & Renkema-Padmos, A. (2014). Why doesn't Jane protect her privacy? In E. De Cristofaro & S. J. Murdoch (Eds.), *14th International Symposium on Privacy Enhancing Technologies (PETS)*. Lecture Notes in Computer Science 8555 (pp. 244–262). Springer International Publishing.
80. Ruoti, S., Monson, T., Wu, J., Zappala, D., & Seamons, K. E. (2017). Weighing context and trade-offs: How suburban adults selected their online security posture. In *Proc. of the 13th Symposium On Usable Privacy and Security*, SOUPS'17 (pp. 211–228). USENIX Association.
81. Sasse, M. A., Smith, M., Herley, C., Lipford, H., & Vaniea, K. (2016). Debunking security–usability tradeoff myths. *IEEE Security & Privacy, 14*(5), 33–39.
82. Schmitt, H., & Groen, E. C. (2021). Qualitätsmodell zur Förderung des Beschäftigtendatenschutzes. *Datenschutz und Datensicherheit - DuD, 45*(1), 28–32.
83. Schmitt, H., & Polst, S. (2020). Anforderungen und Rahmenwerk für den betrieblichen Datenschutz. *Softwaretechnik-Trends, 40*(1), 9–10.
84. Schomakers, E.-M., Lidynia, C., & Ziefle, M. (2018). Hidden within a group of people – mental models of privacy protection. In *Proc. of the 3rd International Conference on Internet of Things, Big Data and Security* (pp. 85–94). SCITEPRESS - Science and Technology Publications.
85. Schultz, E. E., Proctor, R. W., Lien, M.-C., & Salvendy, G. (2001). Usability and security an appraisal of usability issues in information security methods. *Computers & Security, 20*(7), 620–634.
86. Shirogane, J. (2014). Support method to elicit accessibility requirements. In D. Zowghi & Z. Jin (Eds.), *Requirements Engineering* (pp. 210–223). Springer.
87. Stanton, J. M., Stam, K. R., Mastrangelo, P., & Jolton, J. (2005). Analysis of end user security behaviors. *Computers & Security, 24*(2), 124–133.
88. Tolsdorf, J., Dehling, F., Reinhardt, D., & Lo Iacono, L. (2021). Exploring mental models of the right to informational self-determination of office workers in Germany. *Proc. on Privacy Enhancing Technologies (PoPETs), 2021*(3), 5–27.
89. TrUSD-Konsortium. (2021). Deliverables 1.1 & 1.2: Anforderungen und Anwendungsszenarien (Version 6.0). https://www.trusd-projekt.de/wp/wp-content/uploads/2021/09/TrUSD-D1.1_1.2-Anforderungen.pdf
90. TrUSD-Konsortium. (2022). TrUSD – Transparente und selbstbestimmte Ausgestaltung der Datennutzung im Unternehmen. https://www.trusd-projekt.de/
91. Ur, B., Bees, J., Segreti, S. M., Bauer, L., Christin, N., & Cranor, L. F. (2016). Do users' perceptions of password security match reality? In *Proc. of the 2016 CHI Conference on Human Factors in Computing Systems*, CHI '16 (pp. 3748–3760). ACM.
92. USecureD-Konsortium. (2022). USecureD Tools – Werkzeuge für Usable Security. https://das.h-brs.de/usecured
93. Volkamer, M., & Renaud, K. (2013). Mental models: General introduction and review of their application to human-centred security. In M. Fischlin & S. Katzenbeisser (Eds.), *Number theory and cryptography: Papers in honor of Johannes Buchmann on the occasion of his 60th birthday*. Lecture Notes in Computer Science 8260 (pp. 255–280). Springer.

94. Wästlund, E., Angulo, J., & Fischer-Hübner, S. (2011). Evoking comprehensive mental models of anonymous credentials. In *Proc. of the 2011 IFIP WG 11.4 International Conference on Open Problems in Network Security*, iNetSec'11 (pp. 1–14). Springer.
95. Wash, R., & Rader, E. (2011). Influencing mental models of security: A research agenda. In *Proc. of the 2011 New Security Paradigms Workshop*, NSPW '11 (pp. 57–66). ACM.
96. Wash, R. (2010). Folk models of home computer security. In *Proc. of the 6th Symposium on Usable Privacy and Security*, SOUPS'10 (article 11). ACM Press.
97. (Weber) Dupree, J.-L., Lank, E., & Berry, D. M. (2018). A case study of using grounded analysis as a requirement engineering method. *Science of Computer Programming, 152*(C), 1–37.
98. Weirich, D., & Sasse, M. A. (2001). Pretty good persuasion: A first step towards effective password security in the real world. In *Proc. of the 2001 Workshop on New Security Paradigms*, NSPW '01 (pp. 137–143). Association for Computing Machinery.
99. Whitten, A., & Tygar, J. D. (1998). Usability of security: A case study. Technical report, Carnegie-Mellon Univ Pittsburgh, PA, Dept of Computer Science.
100. Whitten, A., & Tygar, J. D. (1999). Why Johnny can't encrypt: A usability evaluation of PGP 5.0. In *USENIX security symposium* (Vol. 348, pp. 679–702). USENIX Association.
101. Wu, J., & Zappala, D. (2018). When is a tree really a truck? exploring mental models of encryption. In *Proc. of the Fourteenth USENIX Conference on Usable Privacy and Security*, SOUPS '18 (pp. 395–409). USENIX Association.
102. Yao, Y., Lo Re, D., & Wang, Y. (2017). Folk models of online behavioral advertising. In *Proc. of the 2017 ACM Conference on Computer Supported Cooperative Work and Social Computing*, CSCW '17 (pp. 1957–1969). ACM.
103. Zeng, E., Mare, S., & Roesner, F. (2017). End user security and privacy concerns with smart homes. In *Proc. of the 13th Symposium on Usable Privacy and Security*, SOUPS'17 (pp. 65–80). USENIX Association.
104. Zimmermann, V., Bennighof, M., Edel, M., Hofmann, O., Jung, J., & von Wick, M. (2018). 'Home, smart home'—exploring end users' mental models of smart homes. In R. Dachselt & G. Weber (Eds.), *Mensch und Computer 2018—workshopband* (article 122). Gesellschaft Für Informatik e.V.
105. Zurko, M. E., & Simon, R. T. (1996). User-centered security. In *Proc. of the 1996 Workshop on New Security Paradigms* (pp. 27–33). ACM.

What HCI Can Do for (Data Protection) Law—Beyond Design

Timo Jakobi and Maximilian von Grafenstein

1 Introduction

In the digital context, there are increasingly points at which users encounter data protection topics, especially to provide consent. At the same time, companies are heavily involved in handling data in a legally compliant manner and obtaining permission to process data. Entire industries have built up around "managing" user consent. However, a significant part of the added value of such solutions is to promise a high consent rate: That is, to apply designs that are lawful, but still nudge users to disclose data as much as possible. This does not have to happen through deceptive design [18, 37, 62] (see also the chapter "The Hows and Whys of Dark Patterns: Categorizations and Privacy") but can also work by achieving transparency in terms of privacy and security safeguards (low risks) and presentation of added value (large benefit) [24, 51]. Nevertheless, it is usually done in the interest of the data processor that in turn can be to the detriment of a free decision by the customer aka data subject. The data-driven economy fosters exactly such unbalanced relations between the entities that gather and process personal information and the individuals who are often unaware of the extent and the significance of the processing [20]. By large, there are three ways of influencing said imbalance: First, by the increasing business incentive for companies to collect and make use of (especially: personal) data, actors become more likely to engage in more excessive data collection practices. Second, by the increasing complexity and opaqueness of algorithms used, it is becoming more difficult to explain data processing, especially to non-tech-

T. Jakobi (✉)
Technical University of Applied Sciences Nuremberg Georg Simon Ohm, Nuremberg, Germany
e-mail: timo.jakobi@th-nuernberg.de

M. von Grafenstein
Universität der Künste Berlin, Berlin, Germany
e-mail: m.von-grafenstein@udk-berlin.de

N. Gerber et al. (eds.), *Human Factors in Privacy Research*,
https://doi.org/10.1007/978-3-031-28643-8_6

savvy people. Third, the complexity of organizational and market structures likewise makes it more difficult to communicate what data are used, in which ways, and to whom it is transferred to or obtained by.

In this area of tension, the HCI sub-community called "Usable Privacy" is researching, among other things, the optimization of the user-friendliness of privacy management in terms of both awareness [9, 11, 32, 45, 74, 82] and control [3, 14, 15, 23, 31, 34, 58]. Often times, such research quite naturally intersects with data protection law: A classic example of this is the extensive study of privacy notices, which, in essence, illustrates a problem also present for other applications of law: the content is written by lawyers for lawyers and is hardly understandable for laypersons [64, 65]. Privacy communication tends to be too long, legalistic, and as such not graspable for consumers. In this context, various research contributions apply methods of information visualization, for example, to make texts easier to consume and clearer [49, 55, 56, 61, 78, 81].

The HCI community has been researching solutions to this problem in a variety of ways for a long time. However, so far existing concepts and structures of law are largely taken as "given" and designed around them to lay out user-friendly interfaces. Lawyers' formulations and contents themselves usually remain untouched—at most, there is a discussion on prioritizing information on different layers or extension with icons or graphics. There are two flaws of researching data protection law in HCI, and both of which sadly make up for the majority of work: The first kind of studies plays along with legal requirements undisputed. Such studies are, for example, very popular in the domain of privacy policies, where the legal text receives a decent polishing with design measures, but the actual content remains unchanged and even unchallenged. The second kind, on the contrary, researches alternatives from a user perspective but ends up neglecting factual legal necessities altogether or does not have the ambition to include them into their reasoning. Both kinds of studies cannot and should not claim to engage in actual research in the domain of data protection law—let alone an interdisciplinary one. A deeper examination of legal concepts is typically left out, such that the actual potential of truly interdisciplinary work is often not exploited. Yet HCI, more than any other research field, has appropriate methods to inform jurisprudence and policymaking in the context of privacy in the digital space. On the side of HCI, however, a strategic and thorough engagement with legal concepts likewise is largely missing as of today.

In this chapter, therefore, we present the extensive impulses that are also coming from legal sciences themselves that motivate a more thorough engagement of HCI and legal sciences in a multi-stakeholder environment, which HCI so often claims to be sensitive for. To this end, we turn to the example of data protection legislation and discuss the legislative intentions surrounding the European General Data Protection Regulation (GDPR), which is not only forming the basis for data protection and use in all of the European Union, but moreover has become a blueprint for many international privacy legislations. GDPRs' requirement of "effectiveness" of technical and organizational protection measures (Art. 25 GDPR) is in the center of interest, since, as by dominating legal interpretation, these include "data protection by design," but also design measures, as these are ultimately reflected in technical measures and thus open the door for collaboration with HCI.

Afterward, we discuss in how far empirical research—and especially HCI-related research—has already engaged with law. We especially carve out the differences of the "Legal Design" approach and argue why HCI is more suitable for answering complex legal research questions.

We finally present practical examples to demonstrate three different levels of such cooperation between HCI and legal scholars, namely implementation, evaluation, and identification. We argue that especially the last point requires thorough collaboration and engagement with legal concepts discussed among legal scholars and has so far barely been conducted. However, we further argue that such engagement is necessary for HCI to unfold its capabilities of identifying both problems and solutions in multi-stakeholder environments such as data protection regulation (and arguably any regulation).

2 The Call for Effective Measures: A Door Opener for Empirical Sciences

But why should lawyers even care about HCI and its research methods (for an introduction to HCI research methods, please refer to the chapter "Achieving Usable Security and Privacy Through Human-Centered Design")? First off, there is the value that iterative, user-centered design promises to all: Insights about how the research artifact (here: data protection legislation) unfolds in real life and thus input to design a better version in the next iteration. But there are more formal reasons for lawmakers to include HCI into reasoning, too:

With the GDPR, new rules for the processing of personal data have been applied throughout the EU since May 25, 2018. In recognizing the general problem of uncertainty in a complex world, the GDPR lays out certain processing principles (Art. 5 GDPR [69]) such as transparency, purpose limitation, data minimization as well as a set of specific rights for data subjects to intervene, correct or delete data, which the controller has to implement into the technical and organizational design of the processing operations. The regulations also relate to the way in which data processors must inform the data subjects about the processing (for example, Art. 5 sect. 1 lit. a alt. 3 and Art. 12 et seq.) and on this basis can obtain an "informed consent" from the data subject (in particular Art. 5 sect. 1 lit. a Alt. 2, Art. 6 sect. 1 lit. a and Art. 7) [69].

Interestingly, GDPR also states requirements that should sound familiar to researchers in Usable Privacy and beyond: For example, Article 12 (1) sent. 1 GDPR stipulates that the responsible person must take appropriate measures to provide the data subject with all information "in a precise, transparent, comprehensible and easily accessible form in clear and simple language" [69]. These requirements pick up on the general principle of fairness and transparency, which are a traditional playing field for research into Usable Privacy.

Additionally, said requirements must be implemented into the technical and organizational design of the processing operations according to the approach "Data Protection by Design and by Default" under Art. 25 GDPR [69], resembling the widely known Privacy by Design approach. The same article provides further guidance on how to determine the appropriate measures, namely based on the risks to the rights and freedoms of natural persons arising from the data processing. Their evaluation entails appropriate technical and organizational measures to be taken, which are designed to implement the data protection principles (such as legality and transparency/...) to protect data subjects, effectively [29]. In their guidelines on Art. 25 Data Protection by Design and by Default, the European Data Protection Board (EDPB) also defines key performance indicators, and to use qualitative methods and to seek feedback [26]. Similarly, the Article 29 Working Group states:

> It is not only mandatory to disclose certain information about data practices, but even the comprehensibility and presentation of that information assume a central role to demonstrate compliance and its quality should even be empirically evaluated. [7]

The lawyers reading this may excuse us here, but "even" in the legal literature, it is the prevailing opinion that the formulations "appropriate measures" as defined in Article 12 (1) and "technical and organizational measures" as defined in Article 25 (1) also cover mechanisms and methods of user experience design [29, 40, 42]. It is simply too easy to draw the connection. Although Usable Privacy focuses on understanding demands in and designing usable solutions for user interactions in the digital sphere, its outcomes very much pose technical measures as designs are finally implemented in code. Especially in the digital domain, the measures to implement the principles of the GDPR, such as transparency, purpose limitation, or intervention rights, depend to a large extent on their usability and utility [13]. Thus, effectiveness hinges essentially on design issues and the evaluation of the user perspective.

Here, legal science now faces a major problem: How should such effectiveness be evaluated? The concept of effectiveness poses challenges for the legal sciences as to how this proof should be provided with the existing set of methods. Lawyers—which is not to take as a criticism—typically are not qualified for conducting appropriate empirical evaluations and lack the methodological expertise to fulfill said requirements.

HCI, on the other hand, has a strong empirical tradition of measuring "usability" [44] and therefore a natural alignment to scientifically measure effectiveness of data protection implementation and the user's satisfaction of the interaction with a system [35]. The HCI community has conducted intense research on the needs for control in privacy management [3, 23, 34], transparency [4, 4, 23], system intelligibility and accountability [1, 12, 47, 68], and privacy awareness [33, 46]. Moreover, from a legal perspective, such recourse to experts for assessments and requirements is traditionally not alien to the legal sciences, especially in IT security. For example, regulation of smart metering as critical infrastructure has frequently turned to IT experts to draft security requirements. The cooperation of the normative–deductive legal approach with the empirical–inductive one of HCI unleashes great fruitful potential for the development of data protection—both on the level of legal texts, their interpretation in court, and in practical use with users.

3 Going Beyond Designing Law: The Case for the Full Toolbox of HCI Research

While Usable Privacy research does share a connection to data protection law in some ways, a close relative to the field would be legal design. Its self-declared goal is to make legal processes more accessible to laypersons and communicating them appropriately:

> Legal design is a way of assessing and creating legal services, with a focus on how usable, useful, and engaging these services are.[1]

Typically, legal design focuses on the state of the legal system and seeks to apply design thinking methods to communicate better. The ambition of legal design—much like HCI—so far, however, has largely limited itself to taking legal framework conditions for granted. Principles of information visualization and UX design are applied to legal content to make it easier to consume and clearer [49, 55, 56, 61, 78, 81]. To put it bluntly, however, privacy notices now look prettier, include a table of contents, are interactive and easier to consume, but remain basically unreadable for laymen because they are still fundamentally constructed to be a document from and for lawyers to fence out possible disagreement.

Similarly, several labs[2] digitalize legal documents and processes to effectively make them accessible to a much wider audience. These developments are important, yet they do not go beyond a certain level of engagement with legal sources themselves. They are about problem solving in applying the law, but much less about the problems of legal standards (and their implementation) in the first place. It might be the close relation to product development that, by asking how legal design can find solutions for clients, hinders actual legal innovation. After all, building solutions quite naturally places great emphasis on practicability and legal certainty. As a result, the core of legal design focuses on what design (and its competencies) can do for the legal system [25] and remains rather uncritical of the existing interpretations of the law itself and its underlying concepts.

While this focus of designing the law into artifacts that are usable and accessible is absolutely to be applauded, it leaves out much of the capabilities of (empirical) design research. What is largely missing in legal design—if not structurally, then de facto—is the innovation component of law itself. The claim is to methodically

[1] https://lawbydesign.co/legal-design/.

[2] For example, the Legal Tech Design lab from Stanford Law School: https://www.legaltechdesign.com/.

improve law or its interpretation. Finding that privacy notices are illegible is important. Designing them to structure text, provide links, and finally apply principles of information visualization is a major step into the right direction. However, it is also within HCI's capabilities—if not even: responsibility—to explore potential for improvement in multi-stakeholder environments such as the legal system is by nature. Constructively addressing the issue would mean identifying ways to improve the communication of the legal requirements themselves. From this perspective, enriching privacy notices with interactivity and layered approaches only scratches the surface of the issue of the lack of readability of privacy notices.

What is needed is scientific data protection law research or call it a research stream for legal design. There are concepts and principles in data protection law that are abstract and especially under the imperative of effectivity, call for empirical evaluation, such as the requirement of "transparency," "fairness," or the principle of purpose limitation" in GDPR. Even if those are not new, they still need to be filled with meaning in a comparable and sound way, respecting all stakeholders, and based on scientific reasoning. Arguably, the toolbox of HCI is extremely well -suited to investigate how law unfolds in practice. Both its quantitative and qualitative methods are needed to be able to explore problems existing and then to design and test alternative solutions reliably. It also calls for the HCI community to strongly engage with the legal framework conditions, especially to identify the aspects of the law that need negotiation of interests and interpretation. Such commitment, however, calls for a deeper inspection of also normative requirements. These sometimes may be quite apparent, such as in the case of Article 12 GDPR, which is even named "transparent communication[..]," arguably triggering every HCI researcher right away. In other cases, the normative requirements analysis takes some more effort, as for example the principle of purpose limitation. While not a new principle in data protection, its provision is to strike a balance between information of data subjects on the limits of data processing and practically necessary leeway for processors [8]. With evolving technological means, this balance arguably has shifted, and the implementation of the principle needs evaluation in terms of its effectiveness to inform data subjects. Such components that open a design space need to be filled with meaning, some more obvious, others less so. Especially data protection principles such as those of transparency and purpose limitation are overarching themes in the GDPR, which desperately need scientific interpretation and evaluation for data controllers as well as lawyers to rely on. But there also are more structured aspects, such as requirements for data subject rights and their implementation and assessment of communication requirements and access. In essence, while law benefits from incorporating empirical research of HCI as an example of evidence-based regulation, HCI can benefit from a deeper engagement with legal provisions to make its research more applicable and legally certain, thus increasing connectivity of its research for law.

4 Levels of Engagement: How HCI and Law Can Make Data Protection More Effective

Research in Human–Computer Interaction typically targets improving interactive systems for human interaction. Next to this very down-to-earth perspective stemming from usability engineering and user experience design, a noticeable subgroup of researchers also follows a value-driven, normative agenda. For the example case of data protection, such an agenda manifests inter alia in bringing together law and HCI following the joint goal of providing more user-friendly, safer data protection.

What is more, HCI, like arguably no other discipline, has the methods to inform regulation in both planning and enforcement—not only of data protection. As a result, interdisciplinary collaboration also has a two more levels, beyond designing Usable Privacy experiences (see Table 1). On the backdrop of the urge for effectiveness, HCI can pose as the instance of constant benchmarking of data protection law itself. Such evaluation should in the end not only help users, but likewise reliably inform processors of their duties and possibilities and finally enable regulators to let GDPR match its aim to both protect data subjects but also enable the free flow of data. While there may have been intuitive feelings about this gap between regulatory goal and real world, scientific studies provide well-grounded and reliable insights.

Finally, on a third level, HCI can also provide innovation to regulation and its interpretation: Once a non-effective mechanism is identified, HCI can follow up with its multi-stakeholder design process methods, to craft new tradeoffs reflecting both normative, data subject, data controller, and other stakeholders. For doing so, however, an actual in-depth engagement with legal provisions is necessary. In the following, we briefly outline two of the few design spaces in which HCI already engages with data protection law more or less extensively. For doing so, we turn to the cases of Cookie Banners and Data Subject Rights, to highlight and demonstrate our argument of levels of engagement. These three different levels of engagement are loosely connected to a user-centered design lifecycle, and we prototypically name them: implementation, evaluation, and identification.

Table 1 Overview of different levels of conceptual engagement of HCI research with data protection law.

Level	Key targets of HCI
Implementation	Studies applying UX Design methods and principles of information visualization to design more easily consumable legal documents and interactions. Researching features to improve e.g., user awareness and perceived controllability
Evaluation	Studies directed on understanding the current implementation of law and its effectiveness to reach regulatory goals. Aiming at revealing best/worst practices, informing law about use and misuse and as such the effectiveness of law
Identification	Studies challenging legal concepts or helping to (re-)interpret them, by identifying user requirements from normative provisions

4.1 Case 1: Cookie Banners

Cookies and other tracking mechanisms are very important pieces for today's Web, especially for commercial websites to conduct (re-)targeting via ads or other personalization. GDPR requires that such cookies, which do not serve the sole purpose of making the website work from a technical perspective, may only be used after consent of the user. Yet, due to its relevance for maximizing commercial success and the value ascribed to profiles, entire corporate units and research departments are concerned with the optimization of Consent Management. However, this optimization always takes place from a corporate perspective.

In recent years, there were several milestone legal decisions declaring several design practices unfair and thus illegal. Especially well-known is the Planet49 ruling by the European Court of Justice, making pre-ticked boxes illegal for consent [52]. Since then, when designing the consent, care must be taken to ensure that it is designed as a genuine opt-in, i.e., that the user must actually act actively in order to agree. In addition, multiple consents for different purposes must not be handled with a single submit button, but require the user's separate active action, for example by checking boxes for the various data processing operations. The topic, remains in the regulatory to-do list, as the highly discussed ePrivacy regulation [85] also foresees adaptions to consent for cookies.

There is also a long list of HCI research diving into Web tracking (e.g., [2, 9, 14, 25, 43] and even more so on cookie banners [10, 27, 36, 38, 60, 63, 79, 83]. With regard to legal provisions, typically, studies in this domain look into dark design patterns [36, 38, 79, 83] (see also the chapter "The Hows and Whys of Dark Patterns: Categorizations and Privacy") or evaluate compliance [22, 63, 79] in terms of transparency and controllability. The bottom line of these in-part large-scale studies is that cookie banners often strategically make use of unfair practices to undermine the users' free choice to accept or deny cookies.

4.2 Case 2: Data Subject Rights

In both data protection law and research of Usable Privacy, awareness and control over the collection and use of personal data are understood to be cornerstones of digital sovereignty. For example, the European General Data Protection Regulation (GDPR) provides data subjects with the right to access data collected by organizations but remains unclear on the concrete process design.

HCI research has quickly picked up on the design space provided by GDPR. One of the many ambiguities that spark researchers' interest surrounds the articles 12–18 GDPR. These articles formulate requirements on designing data subject rights, which, as so often in the artifact-bound and context-specific world of HCI, need to be filled with meaning. The design of data subject rights is crucial when it comes to the ability of customers to exercise their right and fulfill regulatory aims such as "transparency."

According to Article 12 of the GDPR, the controller shall provide information about actions taken regarding the subject access request (SAR) without undue delay and within one month of receipt of the request. What is more, Article 15 of the GDPR requires that the process of claiming data will result in information being provided "in a concise, transparent, intelligible and easily accessible form" [69]. Regarding the design of the process of exercising data subject rights such as the right to access, Art. 12 (2) GDPR [69] highlights that the respective data "controller shall facilitate the exercise." Lawmakers thus generally see controllers as responsible for helping their users exercise their right to access.

Currently, however, there is still much to be done to reduce uncertainty about which measures will be judged as sufficiently compliant with concepts such as "understandable," "transparent," and "accessible" in court—all the more so as these terms partly overlap, depending on the content [89].

4.3 *Implementation: What Can Design Do for Law?*

This is where, by large, both most HCI research and especially legal design have their focus. However, the level of interdisciplinarity needed to conduct such studies implementing law is rather low. Especially in the case that provisions of GDPR directly resemble, albeit sometimes complex, concepts known in HCI, such as transparency, understandability, and controllability. HCI typically then applies its own methods to qualitatively understand phenomena in the context of interacting with cookie banners and then try to design for improving on those specific aspects. On the downside, these studies have a strong focus on singular aspects and typically work on a qualitative level, thus struggling to be picked up by lawmakers who look for strong evidence on where and how their normative intentions work or fail.

Cookie Banners: Optimizing UX Design Putting HCI concepts and methods to practice, usability engineering contributes to designing digital technologies such that they become usable at work, and even joyful and desirable for everyday life. With the rise of the digital economy, understanding users (aka customers) has been increasingly professionalized for commercial exploitation, too. On the one hand, knowing customers in terms of e.g., their practices, psychology, demands and behaviour has broadly improved Usability and UX of consumer technology. On the other hand, these advances have also led to the creation of user journeys that nudge users into pressing the "buy" button as seamlessly—if not: quickly—as possible; or into disclosing as muchpersonal information about themselves as possible for commercial use. Given the increase of informational power asymmetries by such "usable, useful, and joyful" technologies, it stands to reason that the same methods can be used with similar success for other, more ethical value-oriented objectives, such as enshrined in data protection law. Of course, the description of the excesses of commercial success of HCI methods is not meant to disparage the merits of HCI research in general. Regarding the legal debate, recent cookie banner design has gained a lot of attention in terms of what not to do [10, 27, 36, 38, 60, 63, 79, 83]. Few studies, however, provide best practices and bright patterns. Habib et al. compare several design options to find that fully blocking consent interfaces with in-line cookie options accompanied by a persistent button to later change consent decisions work best for fulfilling GDPRs goals [39]. Utz et al. provide a field study on privacy notices [88], as well as Kulyk et al. [59]. These studies, however, are small scale. While important for HCI to sensitize, such studies can rarely make an impact on regulation, since they do not show a prevailing scheme of legislation being (in-)effective. Rather few studies, such as Graß et al. [36], provide insights for law makers and data processors alike for best practices. In their work, Graß et al. evaluate bright design patterns on cookie banners found in the wild.

Data Subject Rights: Information Visualization for Interactive Dashboards Relatively few studies so far have explicitly targeted the implementation of provisions and data subject rights provided by the GDPR. The data subject rights in GDPR are currently being researched, such as the right to data portability [21, 93]. Closely related to the right to access data, transparency-enhancing tools have been proposed—mostly following a dashboard approach. For example, similar to the Usable Privacy dashboard by Raschke et al. [75] mentioned above, Olausson developed a dashboard specifically targeting nurses' work [67]. Tolsdorf et al. [86] qualitatively compared ten implementations of dashboards, comparing their levels of compliance. Still, dashboard implementations are scarce in practice and are often only adopted by big players on the market. Looking at manual subject access requests (SAR), Alizadeh et al. interviewed customers of German loyalty card systems, who were asked to make use of their right to access [5]. The scope of the study, however, is limited to a single organization, focusing on how data are provided and the potential to help users with their privacy practices. With a similar perspective on supporting sense-making and data literacy, Pins et al. [71] designed and tested a prototype that visualizes the interaction with voice assistants based on data of SARs from Amazon Alexa and Google Assistant. These studies, however, pose singular islands of knowledge providing insights often from a qualitative stance, which lack representative power to inform lawmakers.

4.4 Evaluation: How Well Is Law Currently Working?

The evaluation of current implementations of law is an important prerequisite to being able to innovate law. While almost all HCI research studies do have evaluative parts, their focus lies on testing their very own implementations. The scientific identification and verification of ineffective law, however, needs different kinds of evaluations. These should be rather large-scale assessments of the current state of affairs. They may cover both practical implementation of law and rather abstract concepts, motivating methods and implementations. The evaluative perspective often remains rather "destructive," in showing how things do or do not work out as meant by law. Still, a core benefit for law and business lies in the provision of worst practices, which can then be made public, avoided, or even fined if taken as a guideline for non-compliance.

Cookie Banners: The "Notice and Choice" mechanism Noticeably, evaluative studies on the issue of Cookie Banners are quite popular in HCI. The majority of the community is keenly looking at the emergence of "dark patterns" in interaction design [36, 38, 79, 83]. Often, Web scraping technologies are used to defer data sharing practices, information provided, and designs applied. Such studies often explicitly target evaluation of existing solutions. For example, Degeling et al. especially measured the impact of GDPR on cookie use and banners [22]. Matte et al. evaluated compliance of IABs consent design [63]. Leenes and Kosta report a case study on how Dutch regulation failed to meet its goals in practice [60]. Regarding data sharing practices, Okoyomon et al. examined 68,051 apps and found out that 10% of those shared personal identifiers with third-party services but did not declare such conduct in their privacy policy. What is more, only 22% of these apps explicitly named third parties, concluding that it is impossible for users to know where their data are being used [66].

On a more conceptual level, the key mechanism in Cookie Banners (and beyond) to obtain a lawful basis to collect and process data is "notice and choice." HCI and related research have long shown that this mechanism has its limits. Cranor et al. state that notice and choice mechanisms are necessary to understand where and under what conditions personal data flow, yet they also conclude that it is insufficient to properly protect privacy [19]. Fred et al. attest that it is a "poor mechanism for communicating with individuals about privacy" [16], and Warner and Sloan go as far as to say that there "is no acceptable way to rescue Notice and Choice" [92]. Still, data protection regulation such as the GDPR and ePrivacy Directive rely on consent for the processing of data and the use of tracking technologies. More concretely, the case of infamous cookie banners shows, at least to some extent, how HCI research can detect and prove the strategic exploitation of legal loopholes to the detriment of the user—beyond an individually intuitive feeling of judges. For example, a whole range of studies show how unfair practices are used to undermine existing law through nudging (see the chapter "Privacy Nudges and Informed Consent? Challenges for Privacy Nudge Design") or dark pattern design (see the chapter "The Hows and Whys of Dark Patterns: Categorizations and Privacy") or even straight out disregarding user choice [10, 22, 36, 48, 63, 79].

To sum up, there is a lot of work on the evaluation of legal provisions such as controllability and transparency of cookies banners in HCI. While in general a very positive sign of HCI research turning to concrete legal issues, the broad adoption consent as a research topic arguably also is supported by another factor: It is rather easy for the HCI community to evaluate cookie banners, as the legal parameters to test against (mainly control and transparency) fall in line with research fields that are long established in Usable Privacy research, too. So, whereas the field of application is new, the exact research interests are pre-existing in HCI research. Moreover, cookie banners were a new phenomenon at the time, which literally any Internet user was exposed to. While this is not a bad thing per se, it shows how HCI did not have to engage with legal issues a lot to identify a potentially misguided and ineffective regulation. Instead, it was brought to the researchers' personal and professional attention, and unavoidable.

Data Subject Rights: Assessing the Usability of Implementation of GDPR Rights Usability studies reflect that process design is crucial for users being able to complete an interaction [6, 43]. However, little is known about the factors facilitating or hindering subject right requests in terms of process design. Insights about factors such as user needs and capabilities are highly useful not only from a research perspective, but also for organizations to use data protection as a competitive advantage by optimizing their customer experience [50, 90, 96]. While there is some research on usability of dashboard solutions for addressing data subject rights (especially for the right to access), the market adoption of such generally desirable solutions has been low.

Most studies into the provisions of the GDPR from a consumer perspective in the field of Usable Privacy adopt a consumer perspective on the right to access and/or deletion. However, such studies largely ignore the challenge of getting data in the first place. Instead, studies focus either on the compatibility of the data provided as a result of the subject access request (SAR) with user demands in terms of supporting privacy practices and data literacy [5, 71]—or they take the provision of data for granted by building dashboards on top of the data [75]. Urban et al. [87] contacted 39 companies to check for several SAR parameters such as response time, reaction to questions, and the disclosed information in the context of online advertisement. Similar work has been done by Kröger et al. [57] for app vendors who conducted a longitudinal study on several SAR items such as response time, data provided, and security mechanisms. However, these studies do not apply a processual lens, nor do they evaluate the phases they identify in a user-centered way; instead, these studies merely check against legal provisions.

For companies and organizations in general, the still relatively new GDPR framework raises uncertainty regarding requirements for compliance with the regulation. For the case of the right to access data, organizations need to implement a process for users to claim their data, but it is still unclear how such a SAR process should be designed, how authentication should work, how data should be requested, provided, presented, and explained to customers in a compliant and customer-friendly way.

To the best of our knowledge, there is only one large-scale study on usability assessments of the implementation of SAR processes, namely Pins et al. [72]. In what we believe is kind of a best practice for evaluating legal processes from a user lens, Pins et al. defined a five-phase user experience journey regarding the right to access: finding, authentication, request, access, and data use. Second, and based on this model, Pins et al. had 59 participants exercise their right to access and evaluate the usability of each phase. Drawing on 422 data sets spanning 139 organizations, they inform both law and Usable Privacy research on the current state of affairs with a robust, empirical body. Their paper is one of the first larger scaled approaches to a structured approach for evaluating the design factors that drive or hinder users when conducting data subject rights. This information, however, is relevant both from the standpoint of research on Usable Privacy and for assessing the controller's role in facilitating the user in this process, as demanded in the GDPR. Both business and data protection agencies can now draw upon this work, to understand best and worst practices for compliance on the aforementioned abstract concepts of, e.g., transparency, processor facilitation, etc. We argue that for the effectiveness and evolution of GDPR implementation to match its regulatory goals, such studies are of great importance, and HCI is the premier field to provide such insights.

4.5 Identification: Challenging Existing Legal Interpretations and Concepts

The usable implementation of regulation is an important endeavor, and it holds potential to carve out wholly new ways of interacting. Likewise, the evaluation of interactions and its legal provisions can inform law on how regulations manifest in practice, paving the way to make improvements.

On a third level, there is the structured identification and framing of design spaces in data protection in the first place. As part of design science, defining a design space is a mapping of dimensions of a research artifact, which is an approach to guide practitioners in designing new solutions [70]. Formally, or informally, as part of gaining contextual understanding in the early phases of the design process—for HCI the identification of a design space is a core competency because outlining a design space conceptually provides researchers with information on the room to maneuver, by showing options available, and often also tools such as taxonomies and a vocabulary to compare, categorize, and communicate different implementations styles. HCI frequently uses such design space definitions. In the realm of Usable Privacy, for example, Schaub et al. [81] came up with a design space for effective privacy notices. Similarly, Feng et al. [30] come up with a design space for privacy choices. It takes deep understanding and critical assessment of the field of the matter at hand.

However, when working with data protection law issues, HCI seems to sometimes forget about this aspect of its work. Taking their task seriously would mean to either explore the field autonomously or take in the necessary competencies from data protection law scholars. Serious interdisciplinary research must also delve into and question the formulations and requirements of the legal system to be able to map the full design space and thus tap into the full potential of multi-stakeholder design processes.

Cookie Banner: The Future of Consent Next to values and attitudes, HCI can also measure actual behavior in data protection decision-making: For example, HCI has highlighted what is known as "consent fatigue" [17, 73, 77]: Users are confronted with providing consent so often—e.g., in Cookie Banners—that the central point of making an "informed" decision, which constitutes the very heart of giving consent, is at question. This is where regulation can and should arrive at better solutions that are sustainable in the long run because they fulfill the regulatory requirements for a free yet informed consent.

Agent systems represent one possible approach. They offer the possibility of articulating privacy needs without at the same time requiring continuous management from page to page, or not going beyond information. In this respect, the creation of an agent is not necessarily in conflict with a current action goal (to visit a web page) but could be part of the setup process of a browser on the one hand or continuously available for redesign on the other.

Here, however, much depends on the design and given possibility for negotiation. If both users and operators offer no leeway, then even such an approach will come to naught in that it could again amount to blocking an offer if a user does not agree to the terms. Because even if there is a ban on tying, many Web offers de facto refinance themselves with personalized advertising [28].

A bridge in this regard could be the offering of non-personalized advertising as "compensation" or monetary compensation as mediation. In pursuing this alternative, however, entirely new challenges arise [22]. Not only would the question of appropriate pricing in relation to data to be disclosed elsewhere be rekindled. More than that, such a decision could also lead to a division of society in that privacy on the Web would become a luxury good that has to be bought for money.

It is up to HCI to identify new potential ways of finding a new balance of the legal requirement to provide consent individually every time, and the limited capability and willingness of users to make informed decisions regarding a secondary goal such as privacy.

Data Subject Rights: The Implementation of the Principle of Purpose Limitation The principle of purpose limitation is, among other things, to inform the data subject about the limits of data processing and thus key to many transparency mechanisms of GDPR. They should be unambiguous and help data subjects to identify uses of data that data subjects "might find unexpected, inappropriate or otherwise objectionable" [8]. However, generally accepted patterns have crept in, so to speak, which do not effectively develop their actual protective effect because they are highly generic and to some extend even arbitrary [80]: "Analytics," "User experience enhancement," or "profiling" are often used phrases that do not delineate clear boundaries for data subjects. In legal practice, the question of how a purpose must be specified such that data subjects can recognize usage that they might find "unexpected, inappropriate or otherwise objectionable" [8] remains unsolved.

Accordingly, an essential question that arises to increase the effectiveness of data protection law is: How can purposes be formulated so that they can fulfill their actual purpose? An answer to this question is of utmost importance because most other processing principles and legal provisions depend on how the purpose is specified.

While the work on Usable Privacy policies in HCI has found that privacy policies lack readability and understandability, it does not seek to reformulate purposes to make them more meaningful, but rather to take the wording used as given (e.g.: [64, 65, 76, 84]). The very fact that purpose specifications form a design space resource where data subjects are likely to have justified interests must spark HCI interest.

In this regard, there are first steps which HCI could build upon. For example, there is growing interest in the HCI community to use privacy risks as perceived by users to inform about the potential implications of data disclosure. Several studies started looking into the perceived privacy risks as a design resource (especially in the realm of embedded and networked devices such as those of the so-called Internet of Things) [33, 41, 46, 49, 51, 53, 54, 94, 95]. However, these efforts are rather investigating potential design resources for increasing user awareness of privacy implications, when using services. They largely neglect or are unaware of the fact that such approach aligns well with the overall risk-based approach of GDPR and could help data subject to gain an understanding of what data would be used for by processors. Studies specifically do not take into account the principle of purpose limitation, or its first component of purpose specification as expressed in privacy policies.

A no exception is the work of Jakobi et al. who seek to bridge the gap between both risk approaches despite their—also conceptual—differences for the connected car context [49]. On a broader level, in an example of interdisciplinary collaboration of HCI and law research, von Grafenstein et al. [91] first identify this aspect of the legal design space and outline three possible alternatives in formulations stemming from a set of cross-technology focus groups on perceived privacy risks.

They suggest, their categories of what they call "unfavorable data uses" aka "privacy risks" could serve as a reference scheme for data controllers when specifying their processing purposes in future practice. Such an approach could ensure indication of information relevant and useful for data subjects and thus effectively manage privacy expectations. With these perceived privacy risks at least indirectly referencing to risks to the fundamental rights of the data subjects, these rights can serve as an immediate scale to further adjust the protection measures that is well-known to legal scholars, too.

The research results from von Grafenstein et al. are based, so far, on qualitative methods and thus suffer representativeness. Still, even qualitative methods also serve a certain proof of effectiveness [26]. Moreover, the mixed method set of HCI also provides roads to make these empirical results stronger, for example, via triangulation with quantitative methods. This new concept for the formulation of data processing purposes holds potential for more meaningful communication of the ins and outs of data processing for users. Further steps are now to be taken in a classic user-centered design process, which will cover both levels of collaboration previously mentioned: the development, evaluation and implementation of potential solutions, and comparison of existing and future options.

5 The Road Ahead

The requirement to implement the legal norms into the processing design in an effective manner (Art. 25 GDPR) constitutes a recent shift toward including empirical evidence into legal reasoning, which is not yet fully understood. By explicitly declaring the effectiveness of the protection measures to be the legally required result, the legislator raises the question of which methods can be used to test and assure such effectiveness. Extending the legal conformity assessment to the real effects of the required measures opens this assessment to (non-legal) methodologies that are specialized for assessing such empirical facts. This does not mean that lawyers must directly incorporate these methodologies and findings into the legal interpretation of Art. 25 GDPR. Instead, they are usually considered as an (sometimes more, sometimes less) important factor in the interpretation of the norm. However, this effect can become rather dominant in legal practice because the interpreter of the law (e.g., a data protection authority or legal court) cannot easily ignore the methodically assured findings of the other discipline, since these describe the factual situation on which the interpretation of the law is based.

In fact, this interdisciplinary opening in Article 25 fits into a larger development in the regulation discourse. Under the label of evidence-based policymaking, for example, the debate has been discussed for quite some time now not only the increased rationalization of the law by referring to non-legal disciplines, but also the possible pitfalls of this approach, such as the increase in complexity when considering the effects of regulation instruments in the legal reasoning.

Since law and its enforcement must also scale and remain effective in the digital realm, technology such as automated usability evaluation may play an important part for future compliance assessments. Automated evaluation may to some extent provide legal certainty for data controllers and likewise support data protection authorities. It is on HCI to provide tools that meet all stakeholders' needs: Data controllers want to be able to assess their future tools and products early and constantly with low efforts in terms of fulfilling data protection requirements. Users want to know about companies that champion data protection and data protection agencies want to be able to consult and oversee processors to maintain a high level of data protection in practice.

In this chapter, we showed how legal and HCI research can benefit from each other's competencies and showed how HCI research so far has (not) seriously engaged with data protection regulation on a broader scale. We argue that both fields can adapt concepts and methods to make the interdisciplinary work even more effective to reach its very "own" objectives. Beyond our specific example, the critical task of mapping the design space will be important to allow for transfer to other data protection principles and rules, especially those whose effectiveness depends on their usability. While HCI has a long history and strong methodology to jump to help here, on the HCI side, the engagement with legal concepts needs to be strengthened to be able to critically assess rooms to maneuver to make law more effective jointly.

References

1. Abdul, A., Vermeulen, J., Wang, D., Lim, B. Y., & Kankanhalli, M. (2018). Trends and trajectories for explainable, accountable and intelligible systems: An HCI research agenda. In *Proceedings of the 2018 CHI Conference on Human Factors in Computing Systems*, CHI '18 (pp. 1–18). Association for Computing Machinery.
2. Acar, G., Eubank, C., Englehardt, S., Juarez, M., Narayanan, A., & Diaz, C. (2014). The Web never forgets: Persistent tracking mechanisms in the wild. In *Proceedings of the 2014 ACM SIGSAC Conference on Computer and Communications Security*, CCS '14 (pp. 674–689). Association for Computing Machinery.
3. Acquisti, A., Adjerid, I., & Brandimarte, L. (2013). Gone in 15 seconds: The limits of privacy transparency and control. *IEEE Security & Privacy, 11*(4), 72–74.
4. Agozie, D., & Kaya, T. (2021). Discerning the effect of privacy information transparency on privacy fatigue in e-government. *Government Information Quarterly, 38*(4), 101601.
5. Alizadeh, F., Jakobi, T., Boldt, J., & Stevens, G. (2019). GDPR-reality check on the right to access data: Claiming and investigating personally identifiable data from companies. In *Proceedings of Mensch Und Computer 2019*, MuC'19 (pp. 811–814). Association for Computing Machinery.
6. Anderson, K. (2001). Internet use among college students: An exploratory study. *Journal of American College Health, 50*(1), 21–26.
7. Art. 29 Data Protection Working Party 2017. Guidelines on transparency under Regulation 2016/679. Technical Report #17/EN WP260 REV.01.
8. Art. 29 Data Protection Working Party. Opinion 03/2013 on purpose limitation. Technical report #00569/13/EN WP 203.
9. Balebako, R., Jung, J., Lu, W., Cranor, L. F., & Nguyen, C. (2013). "Little brothers watching you": Raising awareness of data leaks on smartphones. In *Proceedings of the Ninth Symposium on Usable Privacy and Security (SOUPS)* (pp. 1–11).
10. Bauer, J. M., Bergstrøm, R., & Foss-Madsen, R. (2021). Are you sure, you want a cookie? The effects of choice architecture on users' decisions about sharing private online data. *Computers in Human Behavior, 120*, 106729.
11. Bellekens, X., Hamilton, A., Seeam, P., Nieradzinska, K., Franssen, Q., & Seeam, A. (2016). Pervasive eHealth services a security and privacy risk awareness survey. In *2016 International Conference on Cyber Situational Awareness, Data Analytics and Assessment (CyberSA)* (pp. 1–4). IEEE.
12. Bellotti, V., & Edwards, K. (2001). Intelligibility and accountability: Human considerations in context-aware systems. *Human–Computer Interaction, 16*(2–4), 193–212.
13. Bourka, A. (2018). Exploring the "design" in privacy by design.
14. Brandimarte, L., Acquisti, A., & Loewenstein, G. (2013). Misplaced confidences: Privacy and the control paradox. *Social Psychological and Personality Science, 4*(3), 340–347.
15. Caine, K., & Hanania, R. (2013). Patients want granular privacy control over health information in electronic medical records. *Journal of the American Medical Informatics Association, 20*(1), 7–15.
16. Cate, F. H. (2010). The limits of notice and choice. *IEEE Security & Privacy, 8*(2), 59–62.
17. Choi, H., Park, J., & Jung, Y. (2018). The role of privacy fatigue in online privacy behavior. *Computers in Human Behavior, 81*, 42–51.
18. Council, N. C. (2018). *Deceived by design—how tech companies use dark patterns to discourage us from exercising our rights to privacy*. Norwegian Consumer Council.
19. Cranor, L. F. (2012). Necessary but not sufficient: Standardized mechanisms for privacy notice and choice. *Journal on Telecommunications & High Technology Law, 10*, 273.
20. Danezis, G., Domingo-Ferrer, J., Hansen, M., Hoepman, J.-H., Metayer, D. L., Tirtea, R., & Schiffner, S. (2015). Privacy and data protection by design-from policy to engineering. arXiv preprint arXiv:1501.03726.

21. De Hert, P., Papakonstantinou, V., Malgieri, G., Beslay, L., & Sanchez, I. (2018). The right to data portability in the GDPR: Towards user-centric interoperability of digital services. *Computer Law & Security Review, 34*(2), 193–203.
22. Degeling, M., Utz, C., Lentzsch, C., Hosseini, H., Schaub, F., & Holz, T. (2018). We value your privacy... now take some cookies: Measuring the GDPR's impact on Web privacy. arXiv preprint arXiv:1808.05096.
23. Desai, D. (2021). Role of privacy concern and control to build trust in personalized social networking sites. In A. Sheth, A. Sinhal, A. Shrivastava, & A. K. Pandey (Eds.), *Intelligent systems* (pp. 91–100). Springer.
24. Dinev, T., & Hart, P. (2006). An extended privacy calculus model for e-commerce transactions. *Information Systems Research, 17*(1), 61–80.
25. Doing sociolegal research in design mode. Retrieved August 30, 2022, from https://www.routledge.com/Doing-Sociolegal-Research-in-Design-Mode/Perry-Kessaris/p/book/9780367177652
26. EDPB. (2020). Guidelines 4/2019 on article 25 data protection by design and by default version 2.0, 5 adopted on 20 October 2020.
27. Eijk, R. (2021). Cookies and tracking technologies: Risks, challenges, and future outlook (presentation slides). Technical report #id 3773624, Social Science Research Network.
28. Englehardt, S., & Narayanan, A. (2016). Online tracking: A 1-million-site measurement and analysis. In *Proceedings of the 2016 ACM SIGSAC Conference on Computer and Communications Security*, CCS '16 (pp. 1388–1401). Association for Computing Machinery.
29. European Data Protection Board. (2018). Guidelines on Transparency under Regulation 2016/679 RN.36. Technical report #WP260REV.01., WP29.
30. Feng, Y., Yao, Y., & Sadeh, N. (2021). A design space for privacy choices: Towards meaningful privacy control in the Internet of Things. In *Proceedings of the 2021 CHI Conference on Human Factors in Computing Systems*, CHI '21. Association for Computing Machinery.
31. Fischer-Hübner, S., Angulo, J., & Pulls, T. (2014). How can cloud users be supported in deciding on, tracking and controlling how their data are used? In M. Hansen, J.-H. Hoepman, R. Leenes, & D. Whitehouse (Eds.), *Privacy and identity management for emerging services and technologies* (pp. 77–92). Springer.
32. Gerber, N., Reinheimer, B., & Volkamer, M. (2018). Home sweet home? investigating users' awareness of smart home privacy threats. In *Proceedings of an Interactive Workshop on the Human Aspects of Smarthome Security and Privacy (WSSP).*
33. Gerber, N., Reinheimer, B., & Volkamer, M. (2019). Investigating people's privacy risk perception. *Proceedings on Privacy Enhancing Technologies, 2019*(3), 267–288.
34. Gerl, A., Meier, B., & Becher, S. (2020). Let users control their data—privacy policy-based user interface design. In T. Ahram, R. Taiar, S. Colson, & A. Choplin (Eds.), *Human interaction and emerging technologies* (pp. 790–795). Springer.
35. Goodwin, N. C. (1987). Functionality and usability. *Communications of the ACM, 30*(3), 229–233.
36. Graßl, P., Schraffenberger, H., Zuiderveen Borgesius, F., & Buijzen, M. (2021). Dark and bright patterns in cookie consent requests. *Journal of Digital Social Research, 3*(1), 1–38.
37. Gray, C. M., Kou, Y., Battles, B., Hoggatt, J., & Toombs, A. L. (2018). The dark (patterns) side of UX design. In *Proceedings of the 2018 CHI Conference on Human Factors in Computing Systems*, CHI '18 (pp. 1–14). Association for Computing Machinery.
38. Gray, C. M., Santos, C., Bielova, N., Toth, M., & Clifford, D. (2021). Dark patterns and the legal requirements of consent banners: An interaction criticism perspective. In *Proceedings of the 2021 CHI Conference on Human Factors in Computing Systems*, CHI '21. Association for Computing Machinery.
39. Habib, H., Li, M., Young, E., & Cranor, L. (2022). "Okay, whatever": An evaluation of cookie consent interfaces. In *Proceedings of the 2022 CHI Conference on Human Factors in Computing Systems*, CHI '22. Association for Computing Machinery.
40. Hartung (2018). *Datenschutz-Grundverordnung, Bundesdatenschutzgesetz: DS-GVO/BDSG. Kommentar* (14–16th Ed.).

41. Hauff, S., Veit, D., & Tuunainen, V. (2015). Towards a taxonomy of perceived consequences of privacy-invasive practices. In *ECIS 2015 completed research papers* (p. 16).
42. Heckmann, D., & Paschke, A. (2018). Art. 12 Rn. 20. DS-GVO.
43. International, Human Factors. (2001). HFI helps staples.com boost repeat customers by 67%.
44. ISO. ISO/TC 159/SC 4 2006. ISO/DIS 9241-110.
45. Jakobi, T., Alizadeh, F., Marburger, M., & Stevens, G. (2021). A consumer perspective on privacy risk awareness of connected car data use. In *Proceedings of Mensch Und Computer 2021*, MuC '21 (pp. 294–302). Association for Computing Machinery.
46. Jakobi, T., Patil, S., Randall, D., Stevens, G., & Wulf, V. (2019). It is about what they could do with the data: A user perspective on privacy in smart metering. *ACM Transactions on Computer-Human Interaction, 26*(1), 1–44.
47. Jakobi, T., Stevens, G., Castelli, N., Ogonowski, C., Schaub, F., Vindice, N., Randall, D., Tolmie, P., & Wulf, V. (2018). Evolving needs in IoT control and accountability: A longitudinal study on smart home intelligibility. *Proceedings of the ACM on Interactive, Mobile, Wearable and Ubiquitous Technologies, 2*(4), 1–28.
48. Jakobi, T., Stevens, G., Seufert, A.-M., Becker, M., & von Grafenstein, M. (2020). Web tracking under the new data protection law: Design potentials at the intersection of jurisprudence and HCI. *i-com, 19*(1), 31–45.
49. Jakobi, T., Stevens, G., von Grafenstein, M., Pins, D., & Boden, A. (2020). User-friendly formulation of data processing purposes of voice assistants: A user perspective on the principle of purpose limitation. In *Proceedings of Mensch Und Computer 2020*, MuC '20 (pp. 361–372). Association for Computing Machinery.
50. Jakobi, T., von Grafenstein, M., & Schildhauer, T. (2021). *The machine age of customer insight, chapter data privacy: A driver for competitive advantage*. Emerald Publishing Limited.
51. Jakobi, T., von Grafenstein, M., Smieskol, P., & Stevens, G. (2022). A taxonomy of user-perceived privacy risks to foster accountability of data-based services. *Journal of Responsible Technology, 10*, 100029.
52. Judgment of the Court of Justice in Case C-673/17 Planet49. Technical report. Technical Report #No 125/2019. European Court of Justice.
53. Karwatzki, S., Trenz, M., Tuunainen, V. K., & Veit, D. (2017). Adverse consequences of access to individuals' information: An analysis of perceptions and the scope of organisational influence. *European Journal of Information Systems, 26*(6), 688–715.
54. Karwatzki, S., Trenz, M., & Veit, D. (2018). Yes, firms have my data but what does it matter? Measuring privacy risks.
55. Kelley, P. G., Bresee, J., Cranor, L. F., & Reeder, R. W. (2009). A "nutrition label" for privacy. In *Proceedings of the 5th Symposium on Usable Privacy and Security*, SOUPS '09. Association for Computing Machinery.
56. Kelley, P. G., Cesca, L., Bresee, J., & Cranor, L. F. (2010). Standardizing privacy notices: An online study of the nutrition label approach. In *Proceedings of the SIGCHI Conference on Human Factors in Computing Systems*, CHI '10 (pp. 1573–1582). Association for Computing Machinery.
57. Kröger, J. L., Lindemann, J., & Herrmann, D. (2020). How do app vendors respond to subject access requests? a longitudinal privacy study on IOS and android apps. In *Proceedings of the 15th International Conference on Availability, Reliability and Security*, ARES '20. Association for Computing Machinery.
58. Kröger, J. L., Lutz, O. H.-M., & Ullrich, S. (2021). The myth of individual control: Mapping the limitations of privacy self-management. Available at SSRN.
59. Kulyk, O., Hilt, A., Gerber, N., & Volkamer, M. (2018). "This website uses cookies": Users' perceptions and reactions to the cookie disclaimer. In *European Workshop on Usable Security (EuroUSEC)* (Vol. 4).
60. Leenes, R., & Kosta, E. (2015). Taming the cookie monster with Dutch law—a tale of regulatory failure. *Computer Law & Security Review, 31*(3), 317–335.

61. Liu, B., Andersen, M. S., Schaub, F., Almuhimedi, H., Zhang, S. A., Sadeh, N., Agarwal, Y., & Acquisti, A. (2016). Follow my recommendations: A personalized privacy assistant for mobile app permissions. In *Twelfth Symposium on Usable Privacy and Security (SOUPS 2016)* (pp. 27–41).
62. Mathur, A., Acar, G., Friedman, M. J., Lucherini, E., Mayer, J., Chetty, M., & Narayanan, A. (2019). Dark patterns at scale: Findings from a crawl of 11k shopping websites. *Proc. ACM Hum.-Comput. Interact.*, 3(CSCW).
63. Matte, C., Bielova, N., & Santos, C. (2020). Do cookie banners respect my choice? Measuring legal compliance of banners from IAB Europe's transparency and consent framework. In *2020 IEEE Symposium on Security and Privacy (SP)* (pp. 791–809). IEEE.
64. McDonald, A., & Cranor, L. (2008). The cost of reading privacy policies. *ISJLP, 4*, 543.
65. Milne, G. R., Culnan, M. J., & Greene, H. (2006). A longitudinal assessment of online privacy notice readability. *Journal of Public Policy & Marketing, 25*(2), 238–249.
66. Okoyomon, E., Samarin, N., Wijesekera, P., Elazari Bar On, A., Vallina-Rodriguez, N., Reyes, I., Feal, Á., Egelman, S., et al. (2019). On the ridiculousness of notice and consent: Contradictions in app privacy policies. In *Workshop on Technology and Consumer Protection (ConPro 2019), in Conjunction with the 39th IEEE Symposium on Security and Privacy*.
67. Olausson, M. (2018). User control of personal data: A study of personal data management in a GDPR-compliant graphical user interface. Bachelor's thesis, Linnaeus University, Faculty of Technology, Department of Computer Science and Media Technology (CM).
68. Omeiza, D., Web, H., Jirotka, M., & Kunze, L. (2021). Towards accountability: Providing intelligible explanations in autonomous driving. In *2021 IEEE Intelligent Vehicles Symposium (IV)* (pp. 231–237). IEEE.
69. Parliament, E., & Council (2016). Regulation (EU) 2016/679 on the protection of natural persons with regard to the processing of personal data and on the free movement of such data, and repealing directive 95/46/EC (general data protection regulation).
70. Peffers, K., Tuunanen, T., Rothenberger, M. A., & Chatterjee, S. (2007). A design science research methodology for information systems research. *Journal of Management Information Systems, 24*(3), 45–77.
71. Pins, D., Jakobi, T., Boden, A., Alizadeh, F., & Wulf, V. (2021). Alexa, we need to talk: A data literacy approach on voice assistants. In *Designing Interactive Systems Conference 2021*, DIS '21 (pp. 495–507). Association for Computing Machinery.
72. Pins, D., Jakobi, T., Stevens, G., Alizadeh, F., & Krüger, J. (2022). Finding, getting and understanding: The user journey for the GDPR's right to access. *Behaviour & Information Technology, 41*(10), 2174–2200.
73. Ploug, T., & Holm, S. (2013). Informed consent and routinisation. *Journal of Medical Ethics, 39*(4), 214–218.
74. Pötzsch, S. (2009). Privacy awareness: A means to solve the privacy paradox? In V. Matyáš, S. Fischer-Hübner, D. Cvrček, & P. Švenda (Eds.), *The future of identity in the information society* (pp. 226–236). Springer.
75. Raschke, P., Küpper, A., Drozd, O., & Kirrane, S. (2017). Designing a GDPR-compliant and usable privacy dashboard. In *IFIP International Summer School on Privacy and Identity Management* (pp. 221–236). Springer.
76. Reidenberg, J. (2014). Disagreeable privacy policies: Mismatches between meaning and users' understanding. Technical report #ID 2418297. Social Science Research Network.
77. Rosni, K., Shukla, M., Banahatti, V., & Lodha, S. Consent recommender system: A case study on LinkedIn settings. In *Central Europe Workshop Proceedings*.
78. Sadeh, N., Acquisti, A., Breaux, T. D., Cranor, L. F., McDonald, A. M., Reidenberg, J. R., Smith, N. A., Liu, F., Russell, N. C., Schaub, F., et al. (2013). The usable privacy policy project. In *Technical report, Technical Report, CMU-ISR-13-119*. Carnegie Mellon University.
79. Sanchez-Rola, I., Dell'Amico, M., Kotzias, P., Balzarotti, D., Bilge, L., Vervier, P.-A., & Santos, I. (2019). Can I opt out yet? GDPR and the global illusion of cookie control. In *Proceedings of the 2019 ACM Asia Conference on Computer and Communications Security*, Asia CCS '19 (pp. 340–351). Association for Computing Machinery.

80. Santos, C., Rossi, A., Sanchez Chamorro, L., Bongard-Blanchy, K., & Abu-Salma, R. (2021). Cookie banners, what's the purpose? Analyzing cookie banner text through a legal lens. In *Proceedings of the 20th Workshop on Workshop on Privacy in the Electronic Society*, WPES '21 (pp. 187–194). Association for Computing Machinery.
81. Schaub, F., Balebako, R., Durity, A. L., & Cranor, L. F. (2015). A design space for effective privacy notices. In *Eleventh Symposium on Usable Privacy and Security (SOUPS 2015)* (pp. 1–17).
82. Schaub, F., Marella, A., Kalvani, P., Ur, B., Pan, C., Forney, E., & Cranor, L. F. (2016). Watching them watching me: Browser extensions' impact on user privacy awareness and concern. In *NDSS Workshop on Usable Security* (Vol. 10).
83. Soe, T. H., Nordberg, O. E., Guribye, F., & Slavkovik, M. (2020). Circumvention by design—dark patterns in cookie consent for online news outlets. In *Proceedings of the 11th Nordic Conference on Human-Computer Interaction: Shaping Experiences, Shaping Society*, NordiCHI '20. Association for Computing Machinery.
84. Steinfeld, N. (2016). 'I agree to the terms and conditions': (How) do users read privacy policies online? An eye-tracking experiment. *Computers in Human Behavior, 55*, 992–1000.
85. The European Parliament and the Council Proposal for a Regulation of the European Parliament and of the Council Concerning the Respect for Private Life and the Protection of Personal Data in Electronic Communications and Repealing Directive 2002/58/EC. (Regulation on Privacy and Electronic Communications).
86. Tolsdorf, J., Fischer, M., & Lo Iacono, L. (2021). A case study on the implementation of the right of access in privacy dashboards. In N. Gruschka, L.F.C. Antunes, K. Rannenberg, & P. Drogkaris (Eds.), *Privacy technologies and policy* (pp. 23–46). Springer.
87. Urban, T., Tatang, D., Degeling, M., Holz, T., & Pohlmann, N. (2019). A study on subject data access in online advertising after the GDPR. In *Data privacy management, cryptocurrencies and blockchain technology* (pp. 61–79). Springer.
88. Utz, C., Degeling, M., Fahl, S., Schaub, F., & Holz, T. (2019). (un)informed consent: Studying GDPR consent notices in the field. In *Proceedings of the 2019 ACM SIGSAC Conference on Computer and Communications Security*, CCS '19 (pp. 973–990). Association for Computing Machinery.
89. Voigt, P., & Von dem Bussche, A. (2018). *EU-Datenschutz-Grundverordnung (DSGVO): Praktikerhandbuch*. Springer-Verlag.
90. von Grafenstein, M. (2019). Co-regulation and the competitive advantage in the GDPR: Data protection certification mechanisms, codes of conduct and the "state of the art" of data protection-by-design. In G. González-Fuster, R. van Brakel, & P. De Hert (Eds.), *Research handbook on privacy and data protection law: Values, norms and global politics*. Edward Elgar Publishing.
91. von Grafenstein, M., Jakobi, T., & Stevens, G. (2022). Effective data protection by design through interdisciplinary research methods: The example of effective purpose specification by applying user-centred UX-design methods. *Computer Law & Security Review, 46*, 105722.
92. Warner, R., & Sloan, R. H. (2014). Beyond notice and choice: Privacy, norms, and consent. *Journal of High Technology Law, 14*, 370.
93. Wong, J., & Henderson, T. (2018). How portable is portable? Exercising the GDPR's right to data portability. In *Proceedings of the 2018 ACM International Joint Conference and 2018 International Symposium on Pervasive and Ubiquitous Computing and Wearable Computers*, UbiComp '18 (pp. 911–920). Association for Computing Machinery.
94. Woodruff, A., Pihur, V., Consolvo, S., Brandimarte, L., & Acquisti, A. (2014). Would a privacy fundamentalist sell their DNA for 1000... if nothing bad happened as a result? The Westin categories, behavioral intentions, and consequences. In *10th Symposium on Usable Privacy and Security (SOUPS 2014)* (pp. 1–18).
95. Zimmermann, V., Gerber, P., Marky, K., Böck, L., & Kirchbuchner, F. (2019). Assessing users' privacy and security concerns of smart home technologies. *i-com, 18*(3), 197–216.
96. Żywiołek, J., & Nedeliaková, E. (2020). Personal data protection as an element of competitive advantage. *System Safety: Human-Technical Facility-Environment, 2*(1), 55–61

Expert Opinions as a Method of Validating Ideas: Applied to Making GDPR Usable

Johanna Johansen and Simone Fischer-Hübner

1 Introduction

We present a novel method for validating ideas. In science, one can say that ideas are validated through the "test of time" [4, 18], where the ideas the author publishes in a research article are, in time, either adopted or forgotten by the community. Here we develop and show how to use a method that, metaphorically, "speeds up time" so that the ideas are immediately validated with a representative selection of the community, i.e., with experts from relevant fields. Our method is an adaptation of existing methods that are currently used to validate the data that studies are carried on. We adapt these methods, as explained in Sect. 2, to validate, instead of data, the research ideas and concepts proposed in [9, 10].

We consider the method detailed here a contribution to the research community in general, but maybe even more valuable are the results that each of its applications would bring in terms of validation of particular ideas. For our case here, we find it very important to validate the ideas behind the research program started in [9]. An interesting outcome of our study is that expert opinions are a very good method for bringing out open problems.

Most of this chapter will be spent on applying the expert opinions method in a study for validating the following five (types of) ideas or concepts.

J. Johansen (✉)
Department of Computer Science, Norwegian University of Science and Technology, Gjøvik, Norway
e-mail: johanna.johansen@ntnu.no

S. Fischer-Hübner
Department of Mathematics and Computer Science, Karlstad University, Karlstad, Sweden
e-mail: simone.fischer-huebner@kau.se

N. Gerber et al. (eds.), *Human Factors in Privacy Research*,
https://doi.org/10.1007/978-3-031-28643-8_7

First, we validate a rather general idea, of the type often found in position papers, which usually propose or motivate a direction of research. In [9], a case is made for the need to produce measurable evaluations of the usability with which privacy goals of data protection are reached. Having a scale showing *how well* a product respects the privacy of its users, and *how easy* it is for the user to understand the level of privacy protection that a product offers, works toward fulfilling the goal expressed in Recital (100) of the European General Data Protection Regulation (GDPR), i.e., that of "allowing data subjects to quickly assess the *level* of data protection of relevant products and services."

Second, we validate the *definition of Usable Privacy*,[1] which extends and adapts the definition of usability from the ISO 9241-11:2018 [6] to privacy.

> *Usable privacy* refers to the extent to which a product or a service protects the privacy of the users in an efficient, effective, and satisfactory way by taking into consideration the particular characteristics of the users, goals, tasks, resources, and the technical, physical, social, cultural, and organizational environments in which the product/service is used.

Third, we evaluate a list of 30 Usable Privacy Goals (UP Goals)[1] extracted from the GDPR text. One such goal is, e.g., found in the Article 12:

> ... any information ... and communication ... relating to processing [to be provided] to the data subject in a *concise*, *transparent*, *intelligible* and *easily accessible form*, using *clear and plain language*, ...

How concise, transparent, or intelligible the form of presentation is can be determined by measurements of efficiency, effectivity, and satisfaction, in a respective context of use. The emphasized words are those that can be interpreted differently based on the context they are used in and can result in objective and perceived measurements when evaluated using usability methods.

Validating this list involves looking at properties such as adequacy, completeness, or coverage, e.g., whether the list covers well the GDPR document from which it was extracted. Such lists often appear, for example, in surveys.

Fourth, we validate how appropriate is the set of 24 Usable Privacy Criteria[1] (UP Criteria), which are meant in [9, 10] to produce measurable evaluations of usability of privacy that can be translated into scales to be used in certifications. For example, the above goal is associated a criterion that contains several specific sub-criteria worded so to produce measurements, such as the one below that requires to measure efficiency:

> How much time/effort/financial and material resources does the data subject need to invest in order to access the information related to the processing of his/her personal data?

Finally, we validate a model called the Usable Privacy Cube (UP Cube) model,[1] which is proposed in [9] with the purpose of guiding the process of evaluation of usability in privacy certifications. The UP Cube model depicted in Fig. 1 has three

[1] Short videos used during our interviews are good introductions to these concepts: the definition of Usable Privacy and UP Goals in https://vimeo.com/569510999, the UP Criteria in https://vimeo.com/556133682, whereas the UP Cube model is in https://vimeo.com/571358474.

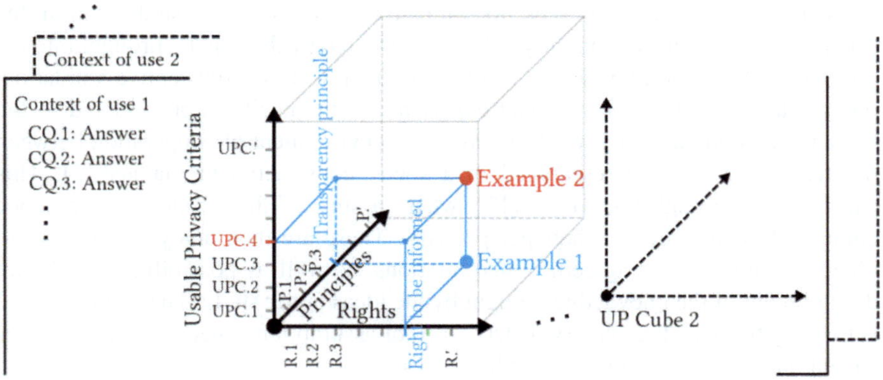

Fig. 1 Usable Privacy Cube model from [9]

axes of variability, with the two at the base containing the existing criteria of the European certification body EuroPriSe, reorganized into:

(i) Rights of the data subjects
(ii) Data protection principles

These two axes also capture the two usual perspectives on privacy, i.e.:

(i) The perspective of the users of whom private information is being collected (and the ones that the regulations usually seek to protect)
(ii) The perspective of the industry/controllers developing products or services that collect and process private information (the ones who must conform with regulations such as GDPR and show compliance by going through certifications such as the EuroPriSe)

The third vertical axis is composed of Usable Privacy Criteria intended for measuring the usability level of privacy in a specific context of use. The UP Cube model comes with additional concepts beneficial for certification processes, such as allowing/asking for ordering and prioritization of the criteria on each axis, or the possibility to identify intersections between the axes.

2 Method

To validate the type of concepts described above, we employ a critical qualitative research [3], where we take an interrogative stance toward the experts' meanings and experiences expressed in the opinions we collect through interviews. Special for our method is that the participants are not brought to discuss the data, but to discuss the ideas and concepts under study. Their meanings then represent the data that one analyzes to obtain a validation result.

With this approach we seek to validate the ideas under study within the scientific and practice community [1], as represented by the experts brought into the discussion. With the intention to reach both an ethical and substantive validation, one recruits experts who have had experience with specific topics related to the ideas under study and then works to create an environment of cooperation between the researcher and the researched in a social constructivism manner [1]. This can be done through interviews [12], where a slice of the research and practice community could present their perspectives. These are then analyzed in order to identify conflicting or agreeing interpretations, as well as possibilities for future development of the knowledge brought by the ideas under study. The technical goal is to bring forth a "disciplinary matrix" [13] of assumptions, theories, and practices shared on the topics around the studied ideas.

For our specific study to validate the above five ideas from [9, 10], we are particularly interested in how usability is understood by the broader communities that the participants' expertise are representing. To achieve this, we involve three different theoretical perspectives in a "theory triangulation" manner [14]. We have thus grouped our participants around one of three specific kinds of expertise that we consider important for validating the above concept, namely into:

(A) "Usability group," used to study/validate the Usable Privacy Definition and the Usable Privacy Criteria
(B) "Certifications group," used for the Usable Privacy Cube model
(C) "Law group" for the Usable Privacy Goals

2.1 Collecting Interview Data

Conducting interviews for collecting our data (i.e., experts' opinions) is the best suited method for our case where we need to explore understandings, perceptions, and mental constructions (for our specific study, these will refer to topics related to usability in data protection). Moreover, this would generate rich and detailed responses when the chosen participants have a personal stake in the study topics; for our study, most of our participants work with privacy certifications and standards, and thus they most probably need to address on a regular basis aspects covered in the interviews. Since [9] use methods and terminology from the field of Ergonomics of human–system interaction to evaluate usability of data protection, we also invited experts from this field, especially those that have been working at the interaction between usability and privacy, e.g., from the field known as usable privacy and security.

The interviews were semi-structured, having a list of questions to guide the conversation, while the participants were encouraged to talk freely on the main topics of the interview. The topics and questions were adapted to the different expertise the participants had, concerning different aspects of the study ideas.

All three interview types had two main parts:

(I) A first part is used to learn about the participants' current understanding of (and their relation with) the ideas under study, without biasing them by presenting the views expressed by these ideas.
(II) The second, and larger, part of the interview starts with presenting the concepts under study, in our case through a short video. Then the participants are asked to express their opinion directly in relation to what was presented.

Before the interview, we informed the participants only about the general research topic, as we did not want to influence them with our opinions. We also wanted spontaneous and not preconceived responses, so to reflect ingrained knowledge of the respective fields and areas of practice the participants represent.

We started each interview with a topic common for all three groups, where the participants presented their understanding and experience with usability in data protection (see [11, Appendix B]). The intentions for this common first part were: (i) to reveal the current understanding of usability in the respective domains, (ii) whether there are differences or overlaps between these, and (iii) to collect unswayed opinions from which we could analyze how their perspectives are (or could be) related to the definition of "usable privacy" that we were validating. Afterwards we had specific topics for each of the groups:

(A) With the "certifications group," we discussed topics related to evaluating and measuring usability, as well as the Usable Privacy Cube model, because these participants have knowledge about processes and methods currently used in evaluation and certification of privacy. For the exact topics and questions, see [11, Appendix E.1 and E.2].
(B) The "usability group" addressed topics related to the definition of usable privacy and Usable Privacy Criteria because this group is acquainted with the ISO 9241-11:2018 standard on usability that was used as a basis for the definition of usable privacy in [9]. Moreover, this group knows well methods and processes of evaluating usability of digital products in general, as well as the process of formulating goals into evaluation criteria. For the exact topics and questions, see [11, Appendix C.1 and C.2].
(C) The participants in the "law group," being well-acquainted with the GDPR text, were asked to check the completeness of the Usable Privacy Goals list from [10], and whether the goals were correctly chosen to represent usability aspects. For the exact topics and questions, see [11, Appendix D.1].

2.2 Participants

The participants were sampled using convenience and snowball methods. Most of the participants have a composite background, a mixture of computer science, law, and human factors. Common for all is that they are working on aspects related to privacy and European data protection applied to IT services/products, thus all having knowledge of information technology.

The "certifications group" consists of six people working with standards, certifications, and data protection organizations. This is confirmed by their answers to our demographic questions: 4 out of 6 have this as their main field of expertise, with the remaining two working for DPAs. Moreover, all these participants have law/data protection as part of their expertise (one as primary and 5 as secondary). The years of experience range from 6 to 32, and the gender is equally represented. The work experience ranges from leadership and research for DPAs, consulting, audit, or technical assessment for certification bodies and other governmental organizations, or board membership and other functions for standardization committees. We consider these backgrounds to represent well our target group.

The "usability group" contains seven people working with usability (sometimes also called HCI/IxD/UX), confirmed by their answers: 6 out of 7 have this as their main field of expertise. Their secondary expertise was somewhat more diverse, including: law/data protection, privacy and security, cybersecurity, contract design, design thinking, and information systems development from an organizational perspective. The years of experience range from 3 to 28, among 4 females and 3 males. Three of the participants have experience with work in industry as: freelance consulting on privacy as a competitive advantage, CEO and head designer for legal design consultancy, and member of task group of usable security and privacy. Even though all participants have academic positions ranging from PhD student to Professor, we consider these backgrounds to represent well our target group.

The "law group" consists of four people, three having law/data protection as their main field of expertise. As the second field of expertise, one chose again law/data protection, another chose certifications/ISO standards/regulations, and the other two chose usability/HCI/IxD/UX. The fourth participant chose usability/HCI/IxD/UX as primary field of expertise and law/data protection as secondary expertise. The years of experience range from 5 to 14, with 3 females and one male. The balance here is skewed toward academic roles (three out of four) ranging from PhD student to Professor, with one participant working for a privacy consultancy firm. For this group, it was more difficult to find people who had knowledge of usability, besides privacy and data protection (for a discussion of interdisciplinary HCI and law research, please refer to the chapter "What HCI Can Do for (Data Protection) Law—Beyond Design").

2.3 Thematic Analysis

We use thematic analysis (TA) for analyzing the opinions from the interviews (representing our data), following [3]. We identify the themes in a "top-down" fashion, where we use data to explore the concepts of interest, related to the ideas being validated. Since the analysis is guided by existing theoretical concepts, as well as by our standpoints, disciplinary knowledge, and epistemology, we adopt a theoretical variant of TA. However, we also employ experiential and constructionist variants of TA. For example, a critical and constructionist analysis is used to identify the concepts and ideas that underpin the assumptions and meanings in our data (e.g., we look at how the field of expertise of the participants influences the way they define and understand usability of privacy). We also use TA to develop a detailed descriptive account of usable privacy and related concepts such as processes and criteria for evaluating usable privacy. At the same time, in an experiential TA fashion, we are interested in the participants' standpoints toward, and how they experience and make sense of, the presented privacy aspects as related to evaluating and measuring usability.

We adopted a researcher-derived approach while performing our coding. When analyzing the opinions, we focused on identifying answers that could be used to (dis)prove the validity of the five ideas we are studying. The themes have been created based on how meaningful the specific comments of the participants are, how many of the participants have mentioned the specific aspect, as well as on how strongly an opinion was articulated and argued for.

3 The Need to Evaluate and Measure Usability of Privacy

This section shows how to validate the first of the fine study ideas from the introduction. The following sections are each dedicated to one of the remaining concepts, respectively. The results of this study are summed up in Sect. 8.

3.1 Evaluating Usability of Privacy

The first idea under validation—*"evaluating and measuring on scales the usability of privacy"*—was formulated in interview questions [11, Appendix E.1] that were addressed specifically to the certifications group, as they are best acquainted with the existing certifications, their needs, and practices. One of the interview questions aimed to elicit whether they find it important to evaluate usability aspects when certifying for compliance with data protection. The answers all fall into a theme that we called: *"we need evaluations of usability of privacy"*:

> we need evaluations of usability, All the GDPR certification programs or schemas need to also look at usability (CertP1).

Moreover, all participants identified several areas where the evaluation of usability is of special importance, or that evaluation should be done "at least" in these instances that they exemplified.

One outstanding example (i.e., mentioned by three out of four participants that specified cases where usability is important) is that *"usability is important for exercising data subjects' rights."* Usable transparency and usable intervenability are presented by one of the participants as preconditions for the users to exercise their rights. At this point, we can conclude that a sub-theme representative for the "certifications group" is that:

> evaluating usability is important for data subjects to exercise their rights and for data controllers to comply with the transparency principle.

3.2 Measuring Usability of Privacy

The other side of our first idea under study—*"measuring on scales the usability of privacy"*—is important for making evaluations of usability of privacy more objective and easier to follow by both the companies wanting to be certified and by the certification organizations and lay persons. During the interview, the respondents were asked whether they see as useful to concretely measure and evaluate how well the usability of privacy is dealt with by companies wishing to be GDPR compliant. We also explained to each participant that by measurements it is meant some form of scale or score of the type used to indicate energy consumption for home appliances. The theme representative for the answers at this question is: *"Measuring is definitely useful but where do we start?"*.

> Yeah, I think measurement is a good thing. It is something everybody or those who are in the community agree on. (CertP1)

That the community is favorable toward scale-based measurements, such as traffic lights, is also exemplified through research work such as [2, 17] or by the work done on privacy icons [5, 8].

Even though the respondents were in favor of measuring privacy, they all brought up several challenges. These are indicative to where the community is at the moment in terms of measuring (usability of) privacy, and what are possible solutions that the community sees.

One of the discussed challenges for measuring (usability aspects of) privacy is the fact that in privacy we do not deal with "stabilized knowledge" (citing one of the respondents). One example from a respondent is of actors such as the Stiftung Warentest[2] who compare products/services based on aspects such as usefulness, functionality, or environmental impact, and that are using a scoring system based on percentages. However, usability of privacy is not as easy to measure as, for example, the "consistency for the shampoo" (CertP1). One conclusion from several of the participants is that we are still in a rather initial phase regarding measuring usability of privacy, where one still asks basic questions such as:

> how do you measure it and what do you measure (CertP2)

[2] Stiftung Warentest is a German consumer organization and foundation involved in investigating and comparing goods and services.

Other aspects that were brought up by the participants and, similar to the above one, are relevant to some of the five ideas under study are the context of use and the target group:

> what do you measure in what kind of context and who is the target group (CertP3);

This is related to the concept of Usable Privacy Cube model [9] and the Usable Privacy Criteria [10] that account for the specific context of use, as well as the users with their goals and specific environments.

4 Usable Privacy Definition Adapts Well ISO 9241-11:2018

Since the definition of usable privacy from [9] adapts the ISO standard 9241-11:2018 to privacy, we validate this definition here primarily with experts from the HCI/IxD/UX community, as these are supposedly more acquainted with this ISO standard. During the interviews, we presented the definition and explained how it is relevant for GDPR, after which the respondents had to answer whether this definition captures their own (or their community's) current understanding of usable privacy. The multiple-choice answers (i.e., "completely," "partially," "not at all") were followed by an explanation of their choice [11, Appendix C.1]. In addition to asking directly the usability experts to validate our definition, the participants in all three groups have been asked to explain their understanding of usability in the context of data protection and to also anchor it in the reality of their practice [11, Appendix B]. In order to gather unswayed perspectives, these questions were asked in the beginning of the interview, before presenting our definition; we refer to these as the "unswayed perspectives on usable privacy."

All participants agreed that adapting the definition of the ISO 9241-11:2018 to privacy captures (the choice "completely" being used by the majority, while the remaining chose the alternative "partially"), the current understanding of usable privacy in their field, e.g.:

> I would say that it is a complete coverage of the different concepts that one could expect within the usable privacy domain because I think indeed there is quite a resemblance to the definition that comes from the ISO standard. (UsabilityP2)

Moreover, besides agreeing with the definition itself, one of the participants also appreciated our exemplification of how the definition applies to GDPR.

> This was a more marvelous thing to see how well you related to the GDPR and to the ISO standard. (UsabilityP7)

We thus formulate the following theme where all answers fit: *"We trust the usability definition from the ISO standard 9241-11:2018"*.

For the respondents that checked the "partially" choice, we can group their answers under the theme *"Instances of the usable privacy definition"*, as these are more specific cases or occurrences of the aspects that are represented at a higher level by the definition. For example, the following comment can be mapped to the part of the definition "...taking into consideration the particular characteristics of the users ...".

> People would be able to do so [understand the privacy policies if they are written in a non legal way], but in practice they don't [read] because it just doesn't work with their lives and it doesn't match the current goal of just signing up for the service and using it. (UsabilityP5)

The literature on usable privacy and security covers well this topic, e.g., [7, 16] speak of a privacy gap between what the user says that would do when asked or tested in the laboratory and what it actually does when in a real situation.

5 A Comprehensive List of Usable Privacy Goals

In the same sense as in the previous section, what is called Usable Privacy Goals in [9, 10] can be considered "instances" of the usable privacy definition. Here we validate the UP Goals with the "law group," since this is well-acquainted with the GDPR text from which the UP Goals were extracted.

The participants were given a list with all 28 UP Goals (see [11, Appendix D.1]) and were asked to choose the ones that they thought relate to usability. We then discussed their choices and opinion about this list, whether they thought it was exhaustive, and whether they could provide additional goals.

After counting the numbers of goals checked by the participants, the mean is 21,75 choices out of 28, giving a 77,67% coverage. Thus, the participants generally agree with our UP Goals, where particularly LawP1 checked all the goals, whereas LawP3 and LawP4 expressed directly their satisfaction with how well the list covers usability aspects.

> ...your list was very complete. I cannot think of something that is not on this list. ... I think this list here is very broad and very comprehensive regarding usability. I cannot think of anything else. (LawP3)

Therefore, we can derive the following theme (see also [11] for more details): *"I am happy with the list of Usable Privacy Goals"*.

6 Ways to Meet the Usable Privacy Criteria

Having established the list of usability goals that GDPR stipulates, the practice in the Interaction Design field is to operationalize these by turning them into usability criteria formulated as questions [15]. Criteria can be seen as specific objectives to be reached by those that aim to reach the set of goals that the criteria relate to. In our case, the Usable Privacy Criteria enable one to assess the privacy-related features that a product or system provides in terms of how much these improve the control that the data subjects have over their data. Examples of commonly used usability criteria (i.e., not specific to privacy) are:

(i) Time used to complete a task (efficiency), such as reading a privacy statement
(ii) The number of errors made when carrying out a given task (effectiveness), such as when choosing desired privacy settings

Usability criteria can provide quantitative indicators of the extent to which, for example, the data subjects understand the implications for their privacy from using a certain technology.

The UP Criteria are validated in this study with the "usability group," as they are most acquainted with the process of formulating criteria to meet goals such as efficiency, effectiveness, and satisfaction.

The participants were given examples of the UP Criteria and were asked to comment on them (see [11, Appendix C.2]). The UP Criteria have been assessed as good by most participants, using quick and simple statements, such as:

> I definitely see the reasoning behind it and it makes sense for me. (UsabilityP2)

However, the participants were keen on the discussion to quickly turn toward another related topic that seems to be preoccupying the community at the moment, that of establishing standards, recommendations, and creating guidelines or design patterns, to help with meeting such criteria.

That such more concrete guidance is needed is confirmed by the participants in the "certifications group" as well. Their assent is especially valuable as they are the ones that are actually performing the evaluation in practice. The focus of the participants was thus more on the particularities of the evaluation, addressing questions such as who would perform the evaluation, what kind of expertise the evaluators would need to have, or which specific HCI methods should they use.

Since the UP Criteria functioned more as a trigger for discussing other more particular aspects of the privacy evaluation processes, a theme that would characterize best the type of feedback that we received from the participants is *"Ways to meet the Usable Privacy Criteria"*.

7 Usable Privacy Cube Model as an Abstraction of Known and Implied Principles of Privacy Evaluations

Finally, we validate whether the Usable Privacy Cube (or UP Cube) model reflects the existing privacy and data protection evaluation processes, and to what extent (i.e., totally, partially, or not at all). Specifically, we discuss with the participants the following features of the UP Cube model:

(i) Represents the perspectives of both data subjects and controllers/processors
(ii) Grouping, prioritization, and organization of the criteria

(iii) Interactions between the different criteria
(iv) Context of use (or context of processing, as a term often used in GDPR)

To our question "Does the UP Cube model represent, at a high level, the existing data protection and privacy evaluation processes?," two out of the five participants chose "Completely," while three chose "Partially."

The answer of CertP1 is exemplary:

> What I know best is EuroPriSe and the previous data protection seals from ULD [Landeszentrum für Datenschutz Schleswig-Holstein[3]], so it's quite very much related, but I think not completely. So I would say partially, although on the abstract level will be the same as the Standard Data Protection model that also uses the different axes for something like that. So the general principle I think is quite well known ...

The Standard Data Protection model[4] has the notion of allocating the legal requirements of the German Federal Data Protection Act—BDGS (Data minimization, Availability, Integrity, Confidentiality, Unlinkability, Transparency, Intervenability) to the protection goals, in a tabular manner. A cube (like the UP Cube model) can be understood as a three-dimensional tabulation mechanism, i.e., represents three tables, each based on the combination of two of the axes. Therefore, the model mentioned by the respondent can replace the EuroPriSe in the base square of the UP Cube, and it is already fitting, to some extent, within the two axes of organization in rights and principles.

As in the case of the Usable Privacy Definition, here too we had questions preceding the presentation of the model, asking if the participants know whether the certifications or standards that they are acquainted with have a high-level model to guide the process of evaluation. The conclusion from these answers is that it does not exist a published or well-established model to guide the process of the evaluation, but there are some main guiding pillars. These are following the GDPR text, or in the case of the standards for evaluating management systems, the risk management or the Privacy Impact Analysis (PIA) is the focal point. The theme that we extract from all the answers is that *"UP Cube is an abstract representation of known, but implied or covert practices."*

8 Summarizing the Results of the Validation Study

Figure 2 collects the main themes that we identified regarding the validation of the five ideas/concepts under study (full details of this study are in [11]). We present these themes hierarchically starting at the top with the theme regarding the first and most general study concept.

---

[3] https://www.datenschutzzentrum.de/guetesiegel/.

[4] https://www.datenschutzzentrum.de/uploads/sdm/SDM-Methodology_V1.0.pdf.

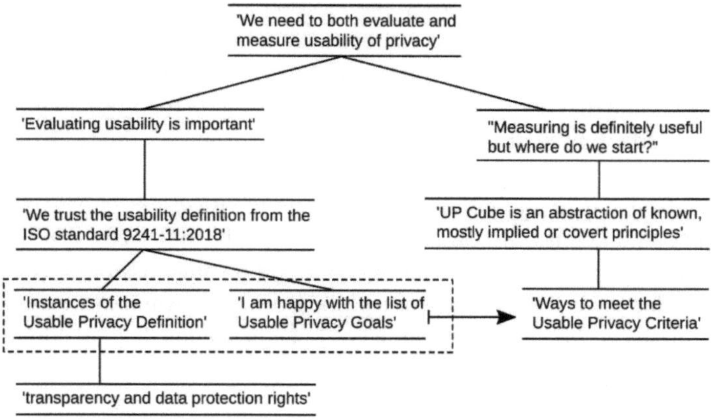

Fig. 2 Overview of hierarchical and lateral themes and their relations

At the second level, we place two lateral themes about the importance of "evaluating" and "measuring" usability of privacy. For evaluations, one needs clear definitions, and for our purposes, the Usable Privacy Definition is well-accepted by the study participants as an adaptation of the usability definition from the ISO standard 9241-11:2018. One important outcome of this study is that the participants were preoccupied more with finding instances of the Usable Privacy Definition, for example, related to *"transparency and data protection rights"* (the theme appearing at the bottom level of Fig. 2). This confirms the importance of the work on identifying Usable Privacy Goals done in [10]. Another outstanding finding (schematized on the right side of the tree in Fig. 2) is that the experts were often wondering about "where to start" with the measuring. In the end, it is generally agreed that starting points can be the Usable Privacy Criteria, integrated in the UP Cube model that is considered a good abstraction of known, but implied principles of existing privacy evaluations.

We interviewed experts from three relevant fields of practice and research: data protection law, privacy/data protection certifications and standardization, and usability (spanning fields such as Human–Computer Interaction, Usable Privacy and Security, or User Experience). The experts were asked to share their knowledge, understanding, and opinions on the studied concepts. The study plan used one group of experts to address one specific topic; the expertise of the group was thus thought to match the topic. Therefore, the analysis of each of the five concepts is done within the frame of one group. Nevertheless, the topic of usability in privacy, being more general, was addressed by all participants and was therefore analyzed across the groups. Moreover, we sometimes found answers in one group to be relevant for a different study concept than the one in focus. We thus often use such additional opinions to strengthen the findings from a group.

A second design aspect of the interviews was to have two main parts: *(i)* in the first part, the interviewees present their opinions without being influenced by the ideas under validation, whereas *(ii)* in the second part, we present to them the study concepts after which we ask them to directly comment on what was presented. The answers from the first part were used to corroborate the responses from the second part, and we found that the participants were consistent in their opinions, the change being only in adapting their answers to what was relevant for the study concepts.

A characteristic of all the participants was to look for more concrete, particular, and practical aspects to address in the future and to suggest possible solutions. For example, in conjunction with validating the UP Criteria, they were pointing out what is yet to be done to meet these criteria, and even proposing possible solutions. Some of these overlap with the further work proposed in [9]. Among the answers from the experts, one can find a substantial list of open problems that the community can address.

9 Conclusion

We have shown how to validate five types of ideas (or we can call these also concepts) that have been proposed and described in depth in [9, 10]. This provides researchers with a method to use when needing to validate their research results that are similar to the ones we have studied here, namely:

(i) Models (in our case, we validated the components of the UP Cube model).
(ii) Definitions (we have validated the Usable Privacy definition, which was an extension/adaptation of a standard definition of usability) (see also chapter "Data Collection Is Not Mostly Harmless: An Introduction to Privacy Theories and Basics" for privacy definitions).
(iii) Prescriptive lists (our list of Usable Privacy Goals was extracted from a well-known legal text, the GDPR).
(iv) A set of criteria (the Usable Privacy Criteria have been built using HCI methods out of the above list of goals).
(v) General research ideas (e.g., described in position papers) that are not always as focused or as clearly stated as the above four types; in our case, this idea was stating "the need for evaluations and measuring of usability of privacy".

To do these in a systematic and controlled manner, we devised a method for validating such research ideas. The method that we have presented and then applied to the above five types of ideas uses (i.e., combines) several methods for interacting with human respondent. In particular, we take a method that is normally used to analyze data and adapt it to analyze opinions from a special type of human, namely experts in a domain relevant to the idea under validation. Thus, for us the experts' opinions form the data, which we then analyze with respect to the ideas under study. But care must be taken along the way, from the selection of the experts to be interviewed, to the design of the interviews and then running the interviews (e.g., taking care to guide the discussions so not to lose focus on the studied ideas), and in the end when coding and extracting the information from the content produced by the interviews. A much welcomed side effect of this method is that experts quite often focus on the open problems, i.e., on expanding and going beyond the ides of the study. This can potentially be a gold mine of avenues of research; therefore, the application of this method of validating ideas with expert opinions can sometimes prove more valuable in identifying the next problem the researchers can focus on.

References

1. Angen, M. J. (2000). Evaluating interpretive inquiry: Reviewing the validity debate and opening the dialogue. *Qualitative Health Research, 10*(3), 378–395 (2000).
2. Bal, G. (2014). Designing privacy indicators for smartphone app markets: A new perspective on the nature of privacy risks of apps. In *Americas Conference on Information Systems (AMCIS)* (Vol. 6).
3. Clarke, V., & Braun, V. (2013). *Successful qualitative research: A practical guide for beginners*. Sage.
4. Dawkins, R. (1976). *The selfish gene*. Oxford University Press.
5. Efroni, Z., Metzger, J., Mischau, L., & Schirmbeck, M. (2019). Privacy icons: A risk-based approach to visualisation of data processing. *European Data Protection Law Review, 5*(3), 352–366.
6. Ergonomics of Human-System Interaction—Part 11: Usability: Definitions and concepts (2018). Standard ISO 9241-11:2018.
7. Gerber, N., Gerber, P., & Volkamer, M. (2018). Explaining the privacy paradox: A systematic review of literature investigating privacy attitude and behavior. *Computers & Security, 77*, 226–261, 2018.
8. Holtz, L.-E., Nocun, K., & Hansen, M. (2011). Towards displaying privacy information with icons. In *Privacy and identity management for life* (pp. 338–348). Springer.
9. Johansen, J., & Fischer-Hübner, S. (2020). Making GDPR usable: A model to support usability evaluations of privacy. *IFIP Advances in Information and Communication Technology, 576*, 275–291
10. Johansen, J., & Fischer-Hübner, S. (2019). Making GDPR usable: A model to support usability evaluations of privacy. Technical report, arXiv.
11. Johansen, J., & Fischer-Hübner, S. (2022). Expert opinions on making GDPR usable. Technical report, arXiv.
12. Kvale, S. (1994). *Interviews: An introduction to qualitative research interviewing*. Sage Publications.
13. Mishler, E. G. (1990). Validation in inquiry-guided research: The role of exemplars in narrative studies. *Harvard Educational Review, 60*(4), 415–443.
14. Patton, M. Q. (1999). Enhancing the quality and credibility of qualitative analysis. *Health Services Research, 34*(5 Pt 2), 1189–1208.
15. Preece, J., Rogers, Y., & Sharp, H. (2015). *Interaction design: Beyond human-computer interaction*. Wiley.
16. Spiekermann, S., Grossklags, J., & Berendt, B. (2001). E-privacy in 2nd generation e-commerce: Privacy preferences versus actual behavior. In *Proceedings of the 3rd ACM Conference on Electronic Commerce* (pp. 38–47), 2001.
17. Tesfay, W. B., Hofmann, P., Nakamura, T., Kiyomoto, S., & Serna, J. (2018). PrivacyGuide: Towards an implementation of the EU GDPR on Internet privacy policy evaluation. In *Proceedings of the Fourth ACM International Workshop on Security and Privacy Analytics*, IWSPA '18 (pp. 15–21). Association for Computing Machinery.
18. Weinberg, S. (2015). *To explain the world: The discovery of modern science*. Penguin Books.

Part III
Application Areas

Privacy Nudges and Informed Consent? Challenges for Privacy Nudge Design

Verena Zimmermann

1 Introduction to Nudging

Nudges, a term coined by Thaler and Sunstein [49], describe small decision interface tweaks supposed to support decision-making without restricting the choice set and by activating automatic cognitive processes. Much-cited examples include the image of a fly in urinals to avoid spilling or the formulation of opt-in defaults to increase the number of organ donors [49].

As several definitions of a nudge have been suggested [33] and to distinguish the nudge from related concepts such as information provision or feedback, the definition box provides an overview on common features of a nudge. For a detailed discussion and derivation of these aspects, the reader is referred to Zimmermann and Renaud [56]. First of all, a nudge is supposed to be applied for the good of the nudgee as opposed to, e.g., the good of the nudge designer or service provider [49]. Furthermore, a nudge should not restrict the choice set, i.e., no choice should be removed or prohibited. Here, it is important to distinguish between choices and options (also see [33] for a discussion of that aspect). For example, removing large plates at a buffet and only leaving small plates to reduce calorie intake instead would limit the number of options (large and small plates versus only small plates). However, the choice to eat as much as one likes would not be restricted if people were still allowed to refill their plate at the buffet without additional charge. This brings us to the next feature: Nudges should not make one choice significantly more costly than the others, be that in terms of money, time, effort, or social sanctions [25]. This feature distinguishes the nudge from the concept of financial incentivization. Next, nudges as an intervention should be implemented

V. Zimmermann (✉)
ETH Zürich, Zürich, Switzerland
e-mail: verena.zimmermann@gess.ethz.ch

N. Gerber et al. (eds.), *Human Factors in Privacy Research*,
https://doi.org/10.1007/978-3-031-28643-8_8

with care and purpose to reach an intended and predicted outcome [23, 49]. Thus, nudges should predictably influence decisions as compared to arbitrary deployed nudges producing unintended outcomes or side effects. Finally, nudges make use of automatic, cognitive processes to encourage a certain choice [12, 22, 25, 49]. Thus, with regard to dual information theories [27, 39, 45] that generally distinguish between System 1 (fast, automatic, and implicit information processing) and System 2 (slow, rational, and explicit information processing), nudges primarily target System 1 information processing. Automatic cognitive processes comprise biases, heuristics, norms, and learned associations. An example is the human tendency to comply with social norms.

Social norm nudges may thus show that one choice is socially more acceptable or that the majority of users tend to make the same choice. The chapter "The Hows and Whys of Dark Patterns: Categorizations and Privacy" provides further explanations on System 1/System 2 information processing and provides a table with more examples of heuristics and biases.

Definition: Nudges

- Are intended for the good of the nudgee.
- Retain the original choice set.
- Do not make one choice significantly more costly than the others.
- Predictably influence toward a predicted outcome.
- Target automatic cognitive processes.

see [12, 22, 25, 25, 33, 49, 56]

2 An Overview on Privacy Nudges

Digital privacy decisions are very complex for users as it is very difficult to determine what kind of data is actually collected, processed, and what future consequences and vulnerabilities may potentially arise from the decision [1]. Furthermore, privacy decisions include making nuanced trade-offs with other factors such as convenience, usability, or functionality. Besides, privacy is seldom the user's primary task [1]. Given that users are confronted with a plethora of decisions every day and that their cognitive resources to evaluate all options are limited [43], nudges appear to be a promising approach to facilitate privacy decisions for the user. Indeed, nudges have successfully been deployed beyond the physical context to support users in making a "wise" choice with regard to digital privacy decisions.

Privacy-related nudge examples include Choe et al.'s [15] use of framing nudges to encourage privacy-friendly app choices. The authors visualized the app's privacy rating and framed it either in a positive or a negative way [14]. The visualizations generally were effective in influencing the users' decisions. The framing played a role for apps with a low privacy rating, e.g., the trustworthiness for apps with a low privacy rating was lower when the privacy rating was framed positively. Apart from users, also the app developers' perspective was analyzed with regard to privacy by Balebako et al. [7]. Based on interviews and a survey with developers, the authors conclude that nudges might be a promising way to help developers overcome privacy-related hurdles such as difficulties with reading privacy policies.

Other privacy nudges analyzed by Balebako et al. [8] or Almuhimedi et al. [4] aimed to discourage unintended location disclosure. Balebako et al. [8] studied an application called Locaccino that supports users in controlling when they make their location visible to others. Almuhimedi et al. [4] provided smartphone users with an app permission manager that also included privacy nudges. For example, one privacy nudge made users aware of how many times the location has been shared with which app to encourage users to make changes to the settings. The study results showed that the implemented privacy nudges can increase the utility of the permission manager.

Masaki et al. [34] used social nudges to reduce potentially risky choices in terms of privacy, such as image disclosure, in social network services. Similar to Choe et al. [15], Masaki et al. [34] also studied framing effects in this context. The social nudges were formulated as, e.g., "90% of users would not share..." as compared to "10% of users would share...." They found that people were less likely to make potentially risky choices when presented with negative framing. However, the authors also found that the nudges can be helpful in scenarios in which people have polarized opinions but that the nudges were not effective in scenarios in which people already support privacy-concerned choices. This finding indicates challenges in designing nudges across application scenarios.

Wang et al. [52, 53] also trialed privacy nudges to discourage disclosures on social networks that users might regret later. The analyzed nudges included visual reminders of the audience of the post, a time delay before posting, and feedback about how other users might perceive the post. While time delay and the visual reminder of the audience overall have been found to be a promising way to prevent unintended disclosure, especially the time delay nudge has not only been rated as beneficial but also annoying and intrusive. A potential explanation might be the higher "cost" in terms of time related to that nudge. This example also shows the challenge to design nudges that do not make one option significantly more costly than the others. For example, while a time delay of 10 s as implemented in the study by Wang et al. [52] might be rated as a burden, a time delay of five seconds might have been found more acceptable. The authors also found that the perceived benefit of the privacy nudges depended on how a person used social networks. For example, it was perceived as beneficial by individuals posting personal thoughts but less so by people who actively aimed to share information, e.g., for commercial purposes. This finding hints at different user preferences posing a challenge to design a nudge that is unanimously perceived as good by the users as intended by the nudge definition. For a description of further applications of security- and privacy-related nudges, the reader is referred to Acqusiti et al. [1].

3 Ethical Considerations

Despite the various well-intended and often successful examples of privacy nudges described above, the application of nudges is associated with several challenges. Nudging is often labeled as a soft paternalistic approach [1]. That is because nudges encourage a certain choice but do not restrict the original choice set to retain freedom of choice. In contrast, bans or laws would actively limit the choice set or require a certain choice.

However, a general criticism concerns the potential manipulation of users by nudges targeting automatic and perhaps unaware cognitive processes [23]. One concern is that nudges might lead users to make choices they might not have made without the nudge [55]. For example, a default nudge in a software wizard might be difficult to detect and go unnoticed by the users leading them to automatically installing unnecessary and unwanted software features.

A related concern is that the intended freedom of choice and human autonomy are actually endangered if users are not fully aware of their choices and the reasons for them [29, 35]. Furthermore, the role and the power of the choice architect, i.e., the person who designs and implements the decision interface including the nudge, is questioned [36]. Who is to say what the "wise" choice for the user is? In the context of security and privacy decisions, the selection of the wise choice might well change over time with technological advancements (e.g., server capacity), depend on the sensitivity of the data (e.g., banking data vs. a forum), or the target user group (e.g., lay users vs. experts).

As with many other technologies or mechanisms, the power of the choice architect or that of the nudge itself can be misused to nudge users away from what is good for them and toward what is good for the service provider or choice architect. Examples of nudges not applied for the good of the user—the so-called sludges [48] or dark patterns [37]—include attempts to sell products not needed by the user or to make the user provide personal information not necessary for a service. For the interested reader, the chapter "The Hows and Whys of Dark Patterns: Categorizations and Privacy" deals with dark patterns as a strategy to make users select a privacy choice that is beneficial for the service provider but not necessarily for users.

A prominent argument for nudging, however, is that nudges are inevitable [1, 10, 47]. Every design decision, purposefully made or arbitrary, can influence the user decision. Examples include the positioning of options, the use of colors and visualizations, or the formulation of instructions. The supporters of nudging thus argue that nudges should better be purposefully and ethically designed for the good of the user rather than influencing in unintended and perhaps negative ways. Another argument for the active use of nudges is that these can be helpful in supporting users to navigate the huge amount of complex decisions they are confronted with on a daily basis [9].

Yet, even the supporters of nudging argue for the use of transparent nudges [49] to counteract unethical deployments and to address the concerns associated with manipulation through the nudge's potentially hidden influence. Hansen and Jespersen [23] propose a taxonomy of transparent vs. non transparent and Type 1 vs. Type 2 nudges. While Type 1 nudges primarily target automatic cognitive processes, Type 2 nudges engage reflective thinking via activating automatic cognitive processes. As an example for transparent Type 2 nudges, Hansen and Jespersen [23] list green footprints leading to dustbins that aim to encourage people to use the bins rather than throw rubbish into the environment. The green footprints are easily visible for people, and their intention becomes clear when reflecting on the green color (i.e., green may be associated with something good or nature protection) and their path leading to the dustbins. In terms of ethical considerations, Hansen and Jespersen argue for the use of transparent Type 2 nudges.

Yet, the discussion calls for guidance that supports the choice architects such as service providers in designing ethically favorable and transparent nudges. Therefore, the following sections review and present guidelines for the design of ethical privacy nudges as detailed in Renaud and Zimmermann [41]. They are based on ethical guidelines for psychological research such as described by the British Psychological Society [50] or the American Psychological Association [5].

Respect for Persons Nudges should be designed in a way that they acknowledge all people regardless of individual differences such as age, gender, or religion. They should not treat certain groups of people unfairly. Ethical checklist questions addressing this principle described in [41] include whether the user is aware of the nudge or that an experiment is undertaken in case of nudge research, respectively. If the user is somehow deceived or not informed beforehand, this should be well justified. In addition, users should then be debriefed.

Beneficence Nudges should be beneficial. Furthermore, users should be protected from harm or risks. Researchers or practitioners implementing nudges should thus check whether the benefit of the intended nudge has already been analyzed and if not, evaluate their benefit. Further consideration should be given to who benefits from the nudge, e.g., individuals or society at large. Users should further have the option to contact the choice architects if the nudge is not perceived as beneficial.

Justice Nudges should be just in that all people should be eligible to benefit without having to overcome undue burdens. Ethical checklist questions for this criterion thus ask whether all users can indeed benefit equally and which measurements have been undertaken in this regard. When conducting research on or using the nudge, potential concerns should be analyzed. These may, for example, be concerned with accessibility or unintended side effects of the nudge for certain groups.

Scientific Integrity The design and evaluation of the nudge should be informed by ethical and scientific standards. Based on this ethical criterion and the nudge definition's aspect to predictably influence, the design of the nudge should be based on previous research, e.g., previous empirical results or theoretic models. The designed nudge should match the implementation context such as the type of the targeted decision (e.g., a simple A/B decision vs. a complex decision).

Social Responsibility The design of nudges involves a social responsibility that should be considered, e.g., in terms of expected as well as unexpected consequences of the nudge. In terms of the ethical checklist detailed in [41], this means that researchers and practitioners should give thought to the nudge's consequences beyond the intended immediate influence on the decision. For example, also the long-term consequences should be monitored, and measures to avoid or decrease potential negative effects should be implemented. There should also be an option to deal with potential negative effects such as removing or replacing the nudge.

4 Challenges of Designing Privacy Nudges

Besides the challenges discussed above, privacy nudges require additional considerations. Identifying the "wise" choice that the user should be nudged to is challenging per se as this can vary between different groups of users, with technological advancements or new scientific insights. For example, what has been considered a good password ten years ago, might not apply any more as technologies for guessing passwords greatly advanced. The choice architect thus bears a great responsibility. The case of privacy nudges, however, is especially challenging in this regard.

For example, in terms of security decisions, such as the choice of an encrypted versus unencrypted public Wi-Fi, it is often clear which option is the more secure and thus the "wiser" choice for the user from a security perspective. Likewise, it is often easy to distinguish the more privacy-preserving option from the less privacy-preserving one. Examples are provided by the privacy nudge studies described above, such as location disclosure versus non-disclosure [4, 8] or the choice of a privacy-friendly as compared to a privacy-invasive smartphone application [15].

However, with regard to privacy, the choice is less clear when considering legal requirements. Current EU regulations such as EU-GDPR [18] suggest data minimization as a principle (EU-GDPR Article 5), i.e., the collection of data that are adequate, relevant, and necessary for the intended purpose. However, GDPR neither prohibits the collection of personal data nor prescribes the automatic selection of the more privacy-preserving option. Instead, the decision to consent to the data processing rests with the user (EU-GDPR Article 6). For the user to be able to make an informed decision, the processor needs to provide the relevant information in a transparent, concise, intelligible, and accessible way (EU-GDPR Article 12). Consenting should be as easy as withdrawing (EU-GDPR Article 7).

What does that mean for the design of privacy nudges? In line with the current legislation, already [23] Sunstein and Thaler agreed: Nudges should be applied "for good" [23]—as considered by the users themselves. Yet, from that aspect, a challenge that has also been discussed by others including Acquisti et al. [1], Albrecht [2], and Hagman et al. [21] arises: How can the "for good" aspect of the nudge be measured? One distinction of nudges is into nudges that are intended for the good of the individual user, i.e., pro-self, or for societal goals, i.e., pro-social [21]. The informed consent suggestion of EU-GDPR suggests that when it comes to privacy, the individual good is concerned. However, this might not necessarily be the most privacy-preserving option. Of course, users can choose—and might often be willing to do so—to withdraw or to select the more privacy-preserving option. However, users might also decide to consent to more excessive data processing considering convenience, functionality, social aspects, or other factors. For example, users might knowingly prefer a more privacy-invasive messenger to a privacy-friendly one if the privacy-invasive one is easier to use, provides more features, or is used by most friends and relatives. Yet, what is the criterion for measuring the success of the nudge then? The happiness of the user with the decision (or minimum regret, respectively [1])? The majority of users agreeing to the choice nudged to or the alignment of individual stated preferences with the decision as suggested by Acquisti et al. [1]? The short-term or the long-term preferences? These questions mirror the discussion in the section on ethical considerations about the power and responsibility of the choice architects to design and evaluate the nudge in line with the users' intentions.

An additional challenge with the informed consent approach lies with the term "informed." First, even though required by GDPR, can we assume that users always read and understand the provided information to make an informed decision? Previous research indicates that this is unlikely: Privacy information is often lengthy, complicated and thus seldom read [38]. Second, nudges might not be the ideal mechanism to address or change that. As defined in the introduction, nudges primarily target automated cognitive processes such as heuristics and biases rather than targeting rational information processing. Thus, as also criticized by the opponents of nudging, a "nudged" decision is not necessarily an informed one depending on the nudge design. The next section therefore discusses several approaches to designing privacy nudges in line with ethical considerations and the GDPR approach for informed consent.

5 Discussion of Approaches

Apart from privacy-preserving nudges, this chapter also discusses options for and challenges associated with designing privacy nudges that align with the suggestion for informed consent.

5.1 Design of Privacy-Preserving Nudges

So far, many privacy nudges described in the literature have been designed as preventative nudges that aim to encourage the more privacy-preserving choice, such as preventing unintended disclosure in social networks. And there seems to be a good reason for that. First, granting access to personal information or disclosing personal data cannot always be reversed. For example, when disclosing privacy-invasive information in posts within social networks, it can be stored or shared by others even before the user has the option to delete the information. Likewise, when the user agrees to sharing personal data with service providers who might again share the information with third parties, it might be difficult to impossible to revoke that later. Also, research showed that users sometimes regret their choice to disclose later [54].

Second, service providers that have an interest in the user's personal information for financial or marketing reasons might deploy strategies to encourage users to choose the more privacy-invasive option, the so-called sludges [48] or dark patterns [20, 37]. They have, for example, been studied in detail in the context of cookie banners that nudge users to accept all even if they are not necessary for the functionality of the service [19, 20, 30, 37, 44]. Thus, to protect the users from unintentionally disclosing information or to counteract existing dark patterns, it might make sense to nudge users toward the privacy-preserving option. Along with the mentality to rather be safe than sorry, it might be "wiser" for the users to first select the privacy-preserving option that can often easily be changed later rather than the privacy-invasive option that is not always easily reversible. Furthermore, privacy-preserving nudges can be helpful in identifying the privacy-preserving option in the first place in cases in which this is not easily visible for the user. For example, highlighting the privacy-preserving option can provide support for users searching for that option within the often lengthy and complicated privacy information.

However, as outlined in the section above, the privacy-preserving option might not always be the option perceived as most favorable by the user. As shown in the study by Wang et al. [52], the privacy-preserving nudges were not unanimously perceived as beneficial by all users, but less so by users who actively aimed to share information for financial reasons. Thus, when considering additional factors such as commercial interests, convenience, or functionality, users might willingly tend toward the more privacy-invasive option.

Therefore, when considering the GDPR requirement for informed consent, several implications for the design of privacy-preserving nudges arise:

- Privacy-preserving nudges should be transparent and easily visible for the user so that they are not nudged toward the privacy-preserving option unawares. The design of a nudge toward the privacy-preserving option bears the same ethical considerations as the design of a nudge toward other options. Here, the reader is referred to Hansen's and Jespersen's proposal of transparent Type 2 nudges [23] as described in the Ethical Considerations section. For example, labeling the privacy-preserving option as such or rating the privacy invasiveness of different options might be easily visible and understandable approaches allowing for an informed decision. In contrast, a default selection of the privacy-preserving option with the other options not easily visible or hidden behind a button might lead users to accept the default selection without being aware what they agreed to.

- The selection of the more privacy-invasive option should be as easy as the selection of the more privacy-preserving option. Following the above example, hiding the privacy-invasive options behind buttons, or forwarding users to separate pages, would pose an additional effort for the user.
- Ideally, measures should be in place to detect a potential mismatch between the implemented nudge and the users' wishes. For example, testing the nudge in a study before its actual implementation in practice might reveal deviations between the researcher's and the users' intentions. In real-life settings, users might have the option to express thoughts or concerns concerning the nudge design via provided contact details or survey instruments. If a mismatch or unintended side effects are detected, the nudge design can be adapted accordingly.

5.2 Design of Nudges that Target Reflective Thinking

Another option to address the requirement for informed privacy decisions might be to design nudges that do not directly target either the more or the less privacy-preserving option, but the interaction or engagement with the decision as such. The question is: Can we design nudges that encourage users to read privacy policies? Or can we design nudges that make users reflect on their choice? As described in the definition section, the nudges per se do not primarily target reflection and rational information processing. Thus, measures that directly prompt reflection on the decision might exceed the definition of the nudge. Examples might be an intervention that asks users to reflect on their choice before they can proceed and to rate all options in terms of their perceived privacy invasiveness on a scale ranging from 1 to 10, or to ask users to write down a reason for their choice. This does not mean that these interventions are not feasible, but only that they might not be classified as a nudge. For a discussion on ideas for combining nudges with other approaches, see Sect. 5.4.

However, nudges might still be used as a tool to encourage users to choose options that include reflective elements. Furthermore, certain types of nudges, i.e., transparent Type 2 nudges [23], might have the potential to activate reflective information processes via automatic cognitive processes. Even though further research on these questions is definitely needed, examples from related research areas provide ideas on what nudges that target reflective thinking could look like. For example, Caraban et al. [13] conducted a literature research on nudging in the HCI domain and categorized the nudges according to their mechanism such as facilitation, confrontation, or reinforcement.

The following list details ideas on nudges that may foster engagement with privacy information, reflection on the decision, or throttle quick unthinking choices.

Engaging with privacy information:

- In general, the same nudge mechanisms deployed to encourage, e.g., the privacy-preserving choice might be applicable to encourage users to read a short text, to look at a graphical description of the privacy policy, or to click on a button labeled "more information." These may include visual highlighting (e.g., bold text or green color), positioning (e.g., upmost or central position), social comparisons (e.g., an indication that reading the information is socially desirable), or the default selection of the choice (e.g., button "more information" is pre-selected). However, it remains unclear whether nudging users to, e.g., click on a button labeled "more information" actually leads to users engaging with the text behind the button or rather to frustration that the privacy decision is delayed. Thus, nudges in this regard should be designed carefully with a focus on the effort for the user and evaluated in future work.

Reflecting on the decision:

- As outlined above, designing nudges that target reflection on the decision is a challenging task as the original definition of the nudge includes targeting automatic cognitive processes instead of reflective cognitive processes.
- In the context of information disclosure, some studies successfully tested nudges that made users aware of and potentially reflect on the consequences of their choices. For example, Wang et al. [52, 53] confronted users with visual reminders of the audience of their social media post to prevent them from disclosures they might regret later. Harbach et al. [24] made users aware of the potential consequences of the app permissions granted. For example, if an app had been granted access to the user's photos, the user was shown a random photo stored on their phone along with the message that the specific app had access to this photo.
- A common password nudge is a so-called password meter [17, 51, 56]. It often takes the form of a bar that dynamically provides visual and textual feedback on the strength of the currently selected password. It is supposed to nudge users to increase password strength and close potential gaps between user's security perception and technical security requirements. This type of nudge can be classified as a transparent Type 2 nudge as it is not only easily visible for the user but also triggers reflective processes. For example, users might ask themselves why their password score is low and try to enhance their score so that the bar fills and changes its color from red to green. Thereby, users might reflect on the changes they made to their password. Similar feedback meters have already been applied to other authentication mechanisms such as pattern unlock [46] and might also be helpful for supporting informed privacy decisions. For example, users might receive feedback on the "privacy score" of their selected option that might trigger them to rethink their choice and to try different options to see how the score changes. A similar approach has already been tested in the context of privacy risks related to app permissions by Kang et al. [28].

Throttling mindless choices:

- In the contexts of phishing [6] and information disclosure in social networks, nudges [52, 53] have been trialed that aim to prevent quick, unthinking choices, e.g., by implementing a timer. After users have made a selection, a timer delays the realization of the choice for some seconds providing users with the option to cancel the process or change their selection. While Wang et al. [52, 53] generally evaluated the delay nudge as a promising approach, it was not unanimously liked by all users, but also rated as annoying. Further research might be necessary to find a good balance for the timer, i.e., the time should be long enough to rethink and change the selection while not being perceived as significant burden. Also, options to skip the timer as implemented by Wang et al. might be a suitable compromise.

5.3 Ask the Users

Several researchers have argued that nudges are not "one-size-fits-all" solutions [11, 13, 26], but that their effectiveness depends on the characteristics of the individual user, their aims, and the context the nudge is deployed in. As such, it cannot be assumed that all users in all contexts favor the same privacy decision or might benefit similarly from the same choice. As an example, the study by Wang et al. [52] revealed that users who had a financial interest in disclosing information rated the privacy-preserving nudges differently than people who had no financial interests. Therefore—and in line with the requirement for informed consent—some researchers suggest personalization of nudges. The following list provides some examples:

- Acquisti et al. [1] suggest designing nudges for disclosures that users are likely to regret later (e.g., when made under the influence of alcohol) or that align behavior with stated preferences. As an example, they describe that many users are concerned about disclosing their political or religious affiliation with potential employers. These specific cases might thus be contexts in which privacy-preserving nudges are warranted as compared to deploying privacy-preserving nudges across all types of data disclosure.
- Another option for personalizing nudges is provided by personalized privacy assistants (PPAs) that first ask users for their preferences and needs before supporting them in implementing these preferences across decisions or services. Examples are provided by Liu et al. [32] who implemented and tested a PPA for mobile app permissions that also included daily privacy nudges. Das et al. [16] summarize current research on PPAs for the Internet of Things with a focus on the infrastructure that is needed to detect nearby sensors and devices and to inform users about their data-handling practices (see also the chapter "Increasing Users' Privacy Awareness in the Internet of Things: Design Space and Sample Scenarios" for a discussion of this topic). Salem et al. [42] designed a nudge-based recommender system for social media use. It balances recommendations for privacy protection with individual preferences and sharing needs. The system objectively evaluates risks and compares these with the users' personal willingness to share personal information, i.e., their subjective privacy

threshold. The users' behavior following the system's recommendations is then again used to update the subjective threshold.

5.4 Choose a Combination of Approaches

Finally, nudges are not the only or the exclusive way forward. Even though they have been shown to be effective measures across many physical and digital decision contexts, including security and privacy decisions, other measures might be equally or even more suitable for certain cases. This includes interactive approaches that support users in reaching their aims, such as the use of gamification or persuasive technologies. Furthermore, when focusing on the "informed" in informed consent, measures that primarily target rational information processing, such as information provision, feedback mechanisms, or reflection might be beneficial. Here, it is important to mention that sometimes the border between nudges and other forms of interventions is not crystal clear. Certain types of nudges such as password meters also provide users with feedback. Others, such as privacy ratings of app permissions, also transport privacy information. Likewise, nudges are often included in larger gamified environments as motivational elements as illustrated in the examples below. However, when we understand nudges and related interventions as a toolbox to support users in making privacy-related decisions, this is not a problem but can be an advantage. The combined power of approaches may lead to positive outcomes that cannot be achieved by the exclusive use of one strategy. Depending on the deployment context and the aim of the researcher, it is just important to be aware of the limits of certain strategies and of potential side effects triggered by the combination of approaches. There might be combinations of strategies that contradict each other or that reduce the impact of the other strategy. Thus, careful consideration is necessary not only when designing a nudge but also when combining nudges with other approaches.

For example, nudges are known to make use of automatic and perhaps unaware cognitive processes. This raises the question of whether the power of nudges is reduced when combining them with information that targets rational and aware information processing. Research in this regard has shown that the combination of nudges and information provision, also labeled as hybrid nudge [56], can have beneficial rather than adverse effects as outlined in the examples below. Also Sunstein agrees that nudges can be educative and that nudges and education do not contradict but can complement each other [47]. By targeting both System 1 and System 2 information processing, the combination of approaches may be a suitable option for nudging toward informed consent to privacy decisions.

The following list illustrates some examples of combining nudges with other mechanisms but is of course not exhaustive. Other combinations have already been trialed or are well possible and should be further investigated:

- Kroese, Marchiory and de Ridder [31] combined nudges and information provision outside the privacy and security context: To encourage healthy food choices, they repositioned food in a store. They found that healthy food choices increased, regardless of whether the intervention was not disclosed to customers or transparently combined with an information sign that explained the intervention. Thus, even though customers were aware of the nudge, this did not diminish its effectiveness. Furthermore, many customers agreed with the intervention as it aligned with their own intention for healthy food choices. This implies that bringing nudges to the users' awareness and combining it with information may also have the advantage of facilitating the detection of mismatched nudges.
- Zimmermann and Renaud [56] tested the impact of no intervention, a nudge, information provision, and a combination of a nudge and information, i.e., a hybrid nudge, in the context of four different security- and privacy-related decisions. This included password selection, the choice to encrypt one's smartphone, the choice of a public WiFi, and the selection of a cloud service provider. Across all decisions and nudges deployed, the study revealed that the hybrid nudge was always at least as or even more effective in encouraging secure choices as compared to single nudging or information provision.
- In a study by Petrykina, Schwartz-Chassidim and Toch [40], nudges were included into a gamification environment called security bot that rewards secure online behavior. Their results revealed a reduction of downloaded malware without reducing productivity.
- Alemany et al. [3] included personalized privacy nudges into an online social network called PESEDIA that also had the purpose to educate users about privacy and to enhance awareness for privacy risks.

6 Summary

Overall, the key points with regard to designing privacy nudges can be summarized as follows:

- Privacy nudges aim to support users in making complex privacy decisions by purposefully altering the decision interface to encourage the "wise" choice. They are intended for the good of the user and work by targeting automatic cognitive processes. Nudges do not limit the choice set nor do they make one option significantly more costly.
- Numerous examples from the literature show that privacy nudges can successfully influence users' privacy decisions, e.g., by increasing awareness for data sharing practices or visualizing privacy ratings.
- The design of privacy nudges requires ethical considerations given that nudges target automatic cognitive processes and thus might not always be visible or comprehensible for the user. Ethical guidelines therefore call for transparent

nudges designs that are noticeable for the user so that they can resist their influence in case it does not align with their intentions.

- Another challenge is the selection of the "wise" or "good" choice, respectively. Given rapid technological advancements and legal guidance suggesting informed consent rather than the automatic selection of the most privacy-preserving option, it is difficult to determine which option is actually intended for the good of the user.
- This chapter discusses four approaches to address this challenge:
  - *Design of privacy-preserving nudges:* Often privacy nudges are designed to encourage the privacy-preserving option as the one protecting users from potentially unintended data disclosure. These nudges should be designed transparently so that users can easily identify the most privacy-preserving option but can also easily select another option.
  - *Design of nudges that target reflective thinking:* Certain types of nudges can activate reflective System 2 information processing via targeting automatic System 1 information processing. These nudges might be used to nudge users toward engaging with the privacy decision rather than toward a final decision.
  - *Ask the users:* User intentions can vary depending on individual preferences and needs. One option would thus be to first ask the users for their privacy preferences before implementing nudges that align with the users' aims.
  - *Choose a combination of approaches:* Nudges can be successfully combined with other approaches such as information provision or feedback. These combinations have the potential to encourage a certain choice while informing users on the reasons for or implications of that choice.

References

1. Acquisti, A., Adjerid, I., Balebako, R., Brandimarte, L., Cranor, L. F., Komanduri, S., Leon, P. G., Sadeh, N., Schaub, F., Sleeper, M., Wang, Y., & Wilson, S. (2017). Nudges for privacy and security: Understanding and assisting users' choices online. *ACM Computing Surveys (CSUR), 50*(3), 1–41.
2. Albrecht, L. (2017). How behavioral economics is being used against you. MarketWatch https://www.marketwatch.com/story/nobel-prize-winning-economist-richard-thalers-nudge-theory-has-a-dark-side-too-2017-10-17
3. Alemany, J., del Val, E., & García-Fornes, A. (2020). Assisting users on the privacy decision-making process in an OSN for educational purposes. In *International Conference on Practical Applications of Agents and Multi-Agent Systems* (pp. 379–383). Springer.
4. Almuhimedi, H., Schaub, F., Sadeh, N., Adjerid, I., Acquisti, A., Gluck, J., Cranor, L. F., & Agarwal, Y. (2015). Your location has been shared 5,398 times! A field study on mobile app privacy nudging. In *Proceedings of the ACM Conference on Human Factors in Computing Systems (CHI)* (pp. 787–796). ACM.
5. American Psychological Association. (2016). Ethical principles of psychologists and code of conduct. http://www.apa.org/ethics/code/

6. Antonucci, A. E., Levy, Y., Dringus, L. P., & Snyder, M. (2022). Experimental study to assess the impact of timers on user susceptibility to phishing attacks. *Journal of Cybersecurity Education, Research and Practice, 2021*(2), 6.
7. Balebako, R., & Cranor, L. (2014). Improving app privacy: Nudging app developers to protect user privacy. *IEEE Security & Privacy, 12*(4), 55–58.
8. Balebako, R., Leon, P. G., Almuhimedi, H., Kelley, P. G., Mugan, J., Acquisti, A., Cranor, L. F., & Sadeh, N. (2011). Nudging users towards privacy on mobile devices. In *Proceedings of the CHI Workshop on Persuasion, Nudge, Influence and Coercion* (pp. 1–4). ACM.
9. Blumenthal-Barby, J. S., & Naik, A. D. (2015). In defense of nudge–autonomy compatibility. *The American Journal of Bioethics, 15*(10), 45–47.
10. Brooks, T. (2013). Should we nudge informed consent? *The American Journal of Bioethics, 13*(6), 22–23.
11. Brown, P. (2012). A nudge in the right direction? Towards a sociological engagement with libertarian paternalism. *Social Policy and Society, 11*(3), 305–317.
12. Calo, R. (2014). Code, nudge or notice? *Iowa Law Review, 99*, 773.
13. Caraban, A., Karapanos, E., Gonçalves, D., & Campos, P. (2019). 23 ways to nudge: A review of technology-mediated nudging in human-computer interaction. In *Proceedings of the 2019 CHI Conference on Human Factors in Computing Systems*, CHI '19 (pp. 1–15). Association for Computing Machinery.
14. Castano, E., Yzerbyt, V., Paladino, M.-P., & Sacchi, S. (2002). I belong, therefore, I exist: Ingroup identification, ingroup entitativity, and ingroup bias. *Personality and Social Psychology Bulletin, 28*(2), 135–143.
15. Choe, E. K., Jung, J., Lee, B., & Fisher, K. (2013). Nudging people away from privacy-invasive mobile apps through visual framing. In *Proceedings of the IFIP Conference on Human-Computer Interaction* (pp. 74–91). Springer.
16. Das, A., Degeling, M., Smullen, D., & Sadeh, N. (2018). Personalized privacy assistants for the Internet of Things: Providing users with notice and choice. *IEEE Pervasive Computing, 17*(3), 35–46.
17. Dupuis, M., & Khan, F. (2018). Effects of peer feedback on password strength. In *Proceedings of the APWG Symposium on Electronic Crime Research (eCrime)* (pp. 1–9). IEEE.
18. EU GDPR Compliant (2018). Cookies consent under the GDPR. February 14 https://eugdprcompliant.com/cookies-consent-gdpr/
19. Graßl, P., Schraffenberger, H., Zuiderveen Borgesius, F., & Buijzen, M. (2021). Dark and bright patterns in cookie consent requests. *Journal of Digital Social Research, 3*(1), 1–38.
20. Gray, C. M., Santos, C., Bielova, N., Toth, M., & Clifford, D. (2021). Dark patterns and the legal requirements of consent banners: An interaction criticism perspective. In *Proceedings of the 2021 CHI Conference on Human Factors in Computing Systems*, CHI '21. Association for Computing Machinery.
21. Hagman, W., Andersson, D., Västfjäll, D., & Tinghög, G. (2015). Public views on policies involving nudges. *Review of Philosophy and Psychology, 6*(3), 439–453.
22. Hansen, P. G. (2016). The definition of nudge and libertarian paternalism: Does the hand fit the glove? *European Journal of Risk Regulation, 7*, 155–174.
23. Hansen, P. G., & Jespersen, A. M. (2013). Nudge and the manipulation of choice: A framework for the responsible use of the nudge approach to behaviour change in public policy. *European Journal of Risk Regulation, 4*(1), 3–28.
24. Harbach, M., Hettig, M., Weber, S., & Smith, M. (2014). Using personal examples to improve risk communication for security & privacy decisions. In *Proceedings of the SIGCHI Conference on Human Factors in Computing Systems*, CHI '14 (pp. 2647–2656). Association for Computing Machinery.
25. Hausman, D. M., & Welch, B. (2010). Debate: To nudge or not to nudge. *Journal of Political Philosophy, 18*(1), 123–136.
26. Johnson, E. J., Shu, S. B., Dellaert, B. G., Fox, C., Goldstein, D. G., Häubl, G., Larrick, R. P., Payne, J. W., Peters, E., Schkade, D., Wansink, B., & Weber, E. U. (2012). Beyond nudges: Tools of a choice architecture. *Marketing Letters, 23*(2), 487–504.

27. Kahneman, D. (2011). *Thinking, fast and slow*. Farrar, Straus and Giroux.
28. Kang, J., Kim, H., Cheong, Y. G., & Huh, J. H. (2015). Visualizing privacy risks of mobile applications through a privacy meter. In *International Conference on Information Security Practice and Experience* (pp. 548–558). Springer.
29. Kelly, D., & Morar, N. (2016). Nudging and the ecological and social roots of human agency. *The American Journal of Bioethics, 16*(11), 15–17.
30. Krisam, C., Dietmann, H., Volkamer, M., & Kulyk, O. (2021). Dark patterns in the wild: Review of cookie disclaimer designs on top 500 German websites. In *European Symposium on Usable Security 2021* (pp. 1–8). Association for Computing Machinery.
31. Kroese, F. M., Marchiori, D. R., & de Ridder, D. T. (2015). Nudging healthy food choices: A field experiment at the train station. *Journal of Public Health, 38*(2), e133–e137.
32. Liu, B., Andersen, M. S., Schaub, F., Almuhimedi, H., Zhang, S. A., Sadeh, N., Agarwal, Y., & Acquisti, A. (2016). Follow my recommendations: A personalized privacy assistant for mobile app permissions. In *Twelfth Symposium on Usable Privacy and Security (SOUPS 2016)* (pp. 27–41).
33. Marchiori, D. R., Adriaanse, M. A., & De Ridder, D. T. (2017). Unresolved questions in nudging research: Putting the psychology back in nudging. *Social and Personality Psychology Compass, 11*(1), e12297.
34. Masaki, H., Shibata, K., Hoshino, S., Ishihama, T., Saito, N., & Yatani, K. (2020). Exploring nudge designs to help adolescent SNS users avoid privacy and safety threats. In *Proceedings of the 2020 CHI Conference on Human Factors in Computing Systems*, CHI '20 (pp. 1–11). Association for Computing Machinery.
35. Mitchell, G. (2004). Libertarian paternalism is an oxymoron. *Northwestern University Law Review, 99*, 1245–1277.
36. Murray, P. R. (2017). Who will nudge the nudgers. *Regulation, 40*, 55.
37. Nouwens, M., Liccardi, I., Veale, M., Karger, D., & Kagal, L. (2020). Dark patterns after the GDPR: Scraping consent pop-ups and demonstrating their influence. In *Proceedings of the 2020 CHI Conference on Human Factors in Computing Systems*, CHI '20 (pp. 1–13). Association for Computing Machinery.
38. Obar, J. A., & Oeldorf-Hirsch, A. (2016). The biggest lie on the Internet: Ignoring the privacy policies and terms of service policies of social networking services. In *Proceedings of the Research Conference on Communication, Information and Internet Policy (TPRC 44)*.
39. Osman, M. (2004). An evaluation of dual-process theories of reasoning. *Psychonomic Bulletin & Review, 11*(6), 988–1010.
40. Petrykina, Y., Schwartz-Chassidim, H., & Toch, E. (2021). Nudging users towards online safety using gamified environments. *Computers & Security, 108*, 102270.
41. Renaud, K., & Zimmermann, V. (2018). Ethical guidelines for nudging in information security & privacy. *International Journal of Human-Computer Studies, 120*, 22–35.
42. Salem, R. B., Aïmeur, E., & Hage, H. (2020). A nudge-based recommender system towards responsible online socializing. In *OHARS@ RecSys* (pp. 23–39).
43. Simon, H. A. (1957). *Models of man; social and rational*. Wiley
44. Soe, T. H., Nordberg, O. E., Guribye, F., & Slavkovik, M. (2020). Circumvention by design—dark patterns in cookie consent for online news outlets. In *Proceedings of the 11th Nordic Conference on Human-Computer Interaction: Shaping Experiences, Shaping Society* (pp. 1–12). Association for Computing Machinery.
45. Stanovich, K. E., & West, R. F. (2000). Individual differences in reasoning: Implications for the rationality debate? *Behavioral and Brain Sciences, 23*(5), 645–665.
46. Sun, C., Wang, Y., & Zheng, J. (2014). Dissecting pattern unlock: The effect of pattern strength meter on pattern selection. *Journal of Information Security and Applications, 19*(4–5), 308–320.
47. Sunstein, C. R. (2015). Nudges do not undermine human agency. *Journal of Consumer Policy, 38*(3), 207–210.
48. Thaler, R. H. (2018). Nudge, not sludge. *Science, 361*(6401), 431–431.

49. Thaler, R. H., Sunstein, C. R., & Leonard, T. C. (2008). Nudge: Improving decisions about health, wealth, and happiness. *Constitutional Political Economy, 19*(4), 356–360.
50. The British Psychological Society (2014). Code of human research ethics. https://cms.bps.org.uk/sites/default/files/2022-06/BPS%20Code%20of%20Human%20Research%20Ethics%20%281%29.pdf
51. Ur, B., Alfieri, F., Aung, M., Bauer, L., Christin, N., Colnago, J., Cranor, L. F., Dixon, H., Emami Naeini, P., Habib, H., Johnson, N., & Melicher, W. (2017). Design and evaluation of a data-driven password meter. In *Proceedings of the ACM Conference on Human Factors in Computing Systems (CHI)* (pp. 3775–3786). ACM.
52. Wang, Y., Leon, P. G., Acquisti, A., Cranor, L. F., Forget, A., & Sadeh, N. (2014). A field trial of privacy nudges for Facebook. In *Proceedings of the ACM Conference on Human Factors in Computing Systems (CHI)* (pp. 2367–2376). ACM.
53. Wang, Y., Leon, P. G., Scott, K., Chen, X., Acquisti, A., & Cranor, L. F. (2013). Privacy nudges for social media: An exploratory Facebook study. In *Proceedings of the 22nd International Conference on World Wide Web*, WWW '13 Companion (pp. 763–770). Association for Computing Machinery.
54. Wang, Y., Norcie, G., Komanduri, S., Acquisti, A., Leon, P. G., & Cranor, L. F. (2011). "I regretted the minute I pressed share": A qualitative study of regrets on Facebook. In *Proceedings of the Seventh Symposium on Usable Privacy and Security*, SOUPS '11. Association for Computing Machinery.
55. Wilkinson, T. M. (2013). Nudging and manipulation. *Political Studies, 61*(2), 341–355.
56. Zimmermann, V., & Renaud, K. (2021). The nudge puzzle: Matching nudge interventions to cybersecurity decisions. *ACM Transactions on Computer-Human Interaction (TOCHI), 28*(1), 1–45.

The Hows and Whys of Dark Patterns: Categorizations and Privacy

Agnieszka Kitkowska

1 Introduction

Interaction with any technologies that possess a user interface (UI) is usually influenced by how such an interface is designed. In principle, a designer should produce an interface that guides a user and helps complete desired tasks and goals. Different guidelines exist that designers should follow to make the interaction experience smooth, seamless, and easy. For instance, there are some usability and user experience (UX) recommendations originating from usability heuristics [48], as well as standards such as ISO 9241 "Ergonomics of human-system interaction," and others detailing the rules for the design of interactive systems [32] (see also the chapter "Achieving Usable Security and Privacy Through Human-Centered Design"). Through education, designers learn about UI structure. They are usually informed that even standard elements, such as navigation menus, footers, and similar, should be presented in a specific way, so the user would not struggle during interactions. Considering this, one could say that no design is entirely neutral and that most UI designers "nudge" users to interact with designs in specific and, most likely, predictable ways.

The construct of nudging and its use in the context of privacy is discussed in detail in the chapter "Privacy Nudges and Informed Consent? Challenges for Privacy Nudge Design". The applicability of nudging is, at times, perceived as controversial, and nudges might be regarded as designs that negatively affect an individual's autonomy. For more on this, we refer the reader to the above-mentioned chapter and recommend reading the discussion by Hansen and Jespersen [30]. **In this chapter**,

A. Kitkowska (✉)
Karlstad University, Karlstad, Sweden

Jönköping University, Jönköping, Sweden
e-mail: agnieszka.kitkowska@ju.se

N. Gerber et al. (eds.), *Human Factors in Privacy Research*,
https://doi.org/10.1007/978-3-031-28643-8_9

we focus on what can be described as nudge's sibling, which, contradictory to nudges, exploits human nature and deceives users—the phenomenon called dark patterns. In principle, dark patterns are deceptive designs that trick users into specific choices that they did not initially desire. What makes them related to nudges is that the mechanisms used in dark patterns are similar to the mechanisms underlying the design of nudges. Dark patterns exploit human psychology, particularly how people make judgments and decisions and the different heuristics and biases that these decisions are predisposed by.

2 Dark Patterns

The UI design relates to different attributes through which people can interact physically (e.g., pressing a button may result in haptic feedback), perceptually (e.g., elements displayed on the screen, sounds), and conceptually (e.g., people try to work out what is the device's purpose, and find information about it in the device) with technologies [8]. The most influential elements that affect users are the visual elements of UI (e.g., layout, color, size of the font, buttons) and content (e.g., image, text). These UI components influence users' behavior, and, in particular, they may steer users' choices. It might be impossible to design an entirely neutral choice architecture. However, the design's effects are not always intended or directed toward "any well-defined or consistent end" [30, p.9]—meaning that UI designers or choice architects do not always have all the possible users' goals in mind while designing. Still, it is a common practice in the digital market that companies implement designs that intend to direct users into specific and predictable choices, i.e., nudge them. However, such designs are often unlikely to consider users' interests as an ultimate goal. Instead, they may concentrate on maximizing the company's benefits, primarily financial (e.g., collecting more information about user behavior to sell them additional products or target them with personalized advertising).

Such designs are referred to in the literature as dark patterns. Dark patterns term was first time used by Harry Brignull, who most recently amended the terminology to deceptive design. He defined it as follows: "Deceptive design patterns (also known as 'dark patterns') are tricks used in websites and apps that make you do things that you didn't mean to, like buying or signing up for something" [10]. Therefore, dark patterns (or deceptive designs) are designed to confuse users purposefully or guide them toward specific choices. Users exposed to dark patterns can no longer follow their own desires or preferences or might become subject to manipulation [43].

When considering the topic of manipulation, it is essential to note that not all dark patterns are manipulative. Susser et al. [59] discussed it and defined manipulation as "an attempt to change the way someone would behave absent the manipulator's interventions," concluding that manipulation might be leaning toward persuasion—therefore being more acceptable. However, it may also be deceptive or coercive, becoming more dangerous and harmful to an individual. All four constructs, persuasion, manipulation, deception, and coercion, have been applied in the research on dark patterns and will be referred to in the subsequent parts of the chapter.

2.1 Why Do Dark Patterns Work?

Not without reason, the term dark patterns (deceptive designs) was introduced by Brignull—a UX practitioner and cognitive researcher. The premises that dark patterns rely on are similar to the fundamentals of nudging, stemming from cognitive science and psychology. It is crucial to understand underlying psychological processes and how people make physical and digital decisions to comprehend what dark patterns are and why and when they could be successful (for a discussion on psychological behavior models, please refer to the chapter "From the Privacy Calculus to Crossing the Rubicon: An Introduction to Theoretical Models of User Privacy Behavior").

Traditionally, economic theories are the most common approach to explaining how people make decisions. These assume that people tend to be "rational" and make decisions aiming to maximize their benefit—construct often referred to as *homo economicus*. According to economic-based theories, such utility maximization, or benefit utilization, is always the most preferred outcome [50]. Historically, traditional economic theories have been subject to change due to their predictive inaccuracy. For instance, Kahneman and Tversky proposed their prospect theory to explain decision-making processes under risk and uncertainty [39]. The prospect theory considers value function differently from other economic-based theories, implying that value function is concave for gains but convex for losses, and the slope of the loss is steeper than the slope for gains [63].

Additionally, prospect theory suggests that the probability scale is nonlinear—people tend to underweight high probability and overweight low probability. To take this further, Tversky and Kahneman proposed cumulative prospect theory, enabling modeling decisions that recognize different phenomena of choice, listing framing effects, nonlinear preferences, source dependence, risk-seeking, and loss aversion [63]. Nevertheless, the theory failed to explain decision-making processes 100% accurately, and other developments were proposed, such as the contrast-weighting theory, stochastic difference model, or regret theory [50].

Still, the concept of *homo economicus*—rational, in an economic sense, decision-maker—did not prevail, and many anomalies around the decision-making process were identified in the research. One theoretical approach tries to explain why the anomalies exist through the class of theories referred to as dual-process theories. Dual-process theories imply that there are two kinds of thinking involved in decision-making, in the literature often referred to as System 1 and System 2, or Type 1 and Type 2 as suggested by Evans and Stanovich [24] (to ensure that these are not the only two processes that occur during the decision-making). Fast, intuitive, and automatic processes characterize Type 1 thinking, while slow, computational, and analytical processes characterize Type 2 [24, 35]. The intuitiveness of Type 1 thinking implies that it is autonomous. Therefore, it does not need to engage working memory (memory used to plan and carry behavior, including short-term memory and some other processes that help to make use of short-term memory [18]). The situation is inverted in Type 2 thinking, which is reflective and engages working memory. Additionally, Type 2 relies on cognitive decoupling (i.e., when testing hypotheses, people can prevent confusion of the actual world representations with some imaginary situations) and mental simulation [58].

A common and not entirely correct view is that decisions based on Type 1 thinking must be "bad"—i.e., less optimal, "irrational." On the contrary, the same view implies that Type 2 thinking must lead to a "better" and more optimal decisional outcomes. Such presumptions might be a reminiscence of the economic approach to decision-making, where people are "rational" and desire to maximize benefits. The research suggests that these common beliefs are incorrect, indicating that the goodness or badness of decisional outcomes is interchangeable between the two modes of thinking [24]. More importantly, there is an implication that the optimization of Type 1 thinking depends on the environment, particularly its hostility [24]. The importance of hostile vs. benign environments is crucial for understanding issues related to the design of UI. For Type 1 processing, the hostile environment characterizes by no cues that might be useful or known. Even more critical for the purposes of the present chapter, the hostile environment contains simple cues that are injected into the environment by other agents to trigger Type 1 processing of information (e.g., deliberately developing space to ensure that the company's profit is maximized) [24].

In a hostile environment, decisions are made based on attribute substitution, relying on heuristics and biases. Kahneman and Fredrick explain that such reliance exists when "an individual assesses a specific target attribute of a judgment object by substituting another propriety of that object—the heuristic attribute—which comes more readily in mind" [37]. For example, when people are asked, separately, about the level of happiness in their lives and the number of dates they had last month, the correlation between the answers to these two questions is minimal. However, research shows that when these questions are being asked together, people tend to automatically associate the level of their happiness with dating, and the correlation increases. As implied by Kahneman and Fredrick, such correlation results from heuristic-based or biased thinking [37].

Heuristics and Biases

According to Gigerenzer, heuristics can be defined as strategies that people use to make decisions faster, more frugally, and at times, more accurately if they were to use different decision-making methods [26]. Often this is achieved by ignoring certain information and relying on what could be colloquially named "rules of thumb." Cognitive biases stem from heuristics and Type 1 reasoning and can be defined as systematic patterns of deviation from norm or rationality in judgment and decision-making. Individuals perceive an input based on their "subjective" rationality, and such perception dictates how they behave.

Research from different disciplines, e.g., psychology, economics, and neuroscience, demonstrates that intuitive Type 1 thinking, particularly heuristics and biases, affects decision-making [20, 36, 53]. Hence, the assumptions of how heuristics and biases are triggered have been applied in nudging. As a result, nudging has been applied in real-life contexts and was successful; for instance, companies utilize status quo bias (a preference for no change) to ensure that employees sign up for pension plans [34, 60, 61]. Other examples are when governments use a default effect (a pre-set course of action) for organ donation or shops display first the healthy products in the cafeteria to increase their consumption [34, 60, 61]. The applicability of Type 1 thinking also expands to the digital environment, particularly to UI and choice architecture design, often in the form of dark patterns.

While making decisions based on mental shortcuts might be beneficial in some contexts, as argued by Gigerenzer and Gaissmaier [26], e.g., in sport, medicine, and law, the effect that such thinking has on users of the digital environment is predominantly adverse. These negative effects are particularly valid when people make privacy-related decisions because the way of information processing (whether it is Type 1 or Type 2) is crucial for users that often automatically overshare their personal, and sometimes sensitive, information [12].

There are many mental shortcuts and biases that dark patterns can exploit. For the present chapter, Table 1 illustrates some of the biases most prominent in privacy-related decision-making in the digital context.

2.2 *Privacy Decision-Making*

As hinted above, the psychological underpinnings, including biases and heuristics, have been shown to govern also privacy-related decision-making. For instance, many privacy researchers applied normative, neoclassical economic theories to explain how people decide about their privacy in the digital context. Predominantly, such research focused on information disclosure, considering the transactional dimensions of behavior—supposedly, people disclose information that has some value to gain the desired benefit. Sometimes, information protection has been given a monetary value, showing that the value of information online might be worth less than the offline information [14]. Some studies investigated privacy calculus, following neoclassical economics where people calculate the trade-off between the risks and benefits of information disclosures [21, 29]. Often, privacy calculus models were used to improve understanding of privacy concerns, as these were presumed to be central to explaining privacy-related decisions [21, 22]. Building on economic approaches to decision-making and psychological theories such as the theory of planned behavior or theory of reasoned action, researchers proposed the APCO (Antecedents—Privacy Concerns—Outcomes) framework that attempts to explain

Table 1 Definitions of heuristics and biases likely to be exploited in the design of privacy dark patterns. Note: This list is not exhaustive, and other psychological effects might be used to design privacy dark patterns

Affect heuristic. When people make judgments, they use representations of objects or events, which they tag in their minds to different levels of affect; they add the so-called goodness or badness experienced as a feeling or demarcate a positive or negative quality of stimulus [55]. For example, affective images (eliciting positive emotions) were shown to influence risk-taking: risk is perceived higher when affective images are presented to users [40].
Anchoring. Under conditions of uncertainty (e.g., when the information about a given decision is incomplete), decisions might be skewed toward the point that people used to calculate estimates [62]. Presenting people with arousing imagery was shown to affect their information disclosures: people were more likely to disclose personal information because pictures "anchored" what is appropriate to disclose [15].
Choice overload. The many options to choose from may bring negative consequences, e.g., decreased motivation or commitment to choice, a decrease of satisfaction with a choice, and negative emotions (for instance, regret, frustration) [51]. For instance, too many data-sharing options resulted in adverse emotional reactions regarding the decision-making process, making people feel overwhelmed [42].
Contrast effect. People making decisions tend to evaluate an event/person by comparing them to another event/person instead of relying on objective criteria [65]. In the UI design, designers may manipulate color to hide privacy-related information; for instance, when a privacy policy link is presented in a color low-contrasting against the background color, and, therefore, it is less likely to be noticed by users.
Default effect. A preference for the default option over changing it or accepting the status quo [54]. For example, people are unlikely to deselect a pre-selected checkbox for accepting a website's privacy policy.
Framing. The decision frame might be designed to control the decision problem's presentation, thus influencing the final decision [47]. For example, a small font can be used to make disadvantageous information about privacy less visible to the user and more likely to be overlooked.
Functional fixedness. People tend to fixate on a specific use of an object or one of its parts [33]. For instance, designs commonly associated with privacy, such as shield icons, could be applied on the website, making users believe that the service is privacy-protective. However, the service might lack privacy.
Hyperbolic discounting. Individuals behave inconsistently over time and value smaller and present rewards more than future and larger gains [61]. For instance, immediate gains from disclosures in e-commerce settings (e.g., recommendations of products) led participants with high privacy concerns to disclose their personal information [1].
Instant gratification. People tend to sacrifice future gain for immediate pleasure/satisfaction. In this trade-off, the reward is quick, and the cost is delayed [7]. For instance, people might disregard changing privacy settings when they sign up for a digital well-being application, as they may be looking for immediate support.
Loss aversion. People tend to value an object more when asked to give it up than when they acquire the same object [38]. Past research shows that people are willing to accept more money for their information disclosures than they would be keen to pay to regain control over the same data [1].

(continued)

Table 1 (continued)

Optimism bias (unrealistic optimism). Individuals tend to think that their chances of experiencing negative events (e.g., getting a divorce) are lower than the average and that they are unvulnerable [64]. For instance, people might think that they have "nothing to hide" and that their personal information would not be profitable, unlike personal information from celebrities. This may result in over-disclosures.
Social norms. Unwritten rules and standards that are understood by members of a group that guide and/or constrain behavior [47]. For instance, showing information about other people granting permission for an app decreased people's privacy concerns and increased their disclosure behaviors [66].
Status quo. People often prefer an option that causes no change over an option that would affect the current state of the world [4]. For instance, users of privacy-protective tools might believe that the tools' settings are sufficient to protect their privacy and do not change anything within such settings [1]. However, it is possible that some settings in these tools are not matching their privacy preferences.

relationships between the different factors that may affect privacy behaviors [56]. Here, the antecedents of privacy affect privacy concerns, which in turn lead to specific behavioral reactions (e.g., information disclosure). Privacy calculus, privacy information display, and trust are also considered influencers of either behavior or privacy concerns. Still, both the initial framework's authors and empirical evidence showed that economic approaches are insufficient to explain privacy decisions [2, 3]. As suggested in the redefined APCO framework, privacy decisions can also be affected by other factors, such as the level of cognitive effort required to make a decision that includes affect, motivation, time constraints, and similar [23]. For a more detailed discussion of behavioral frameworks, please refer to the chapter "From the Privacy Calculus to Crossing the Rubicon: An Introduction to Theoretical Models of User Privacy Behavior" . Furthermore, the biases, heuristics, peripheral cues, and misattribution effects influence privacy behaviors. The effects of some of these additional factors affecting privacy decisions are apparent in the research on privacy nudges, as discussed in the chapter "Privacy Nudges and Informed Consent? Challenges for Privacy Nudge Design".

2.3 *Categorization of Dark Patterns*

Researchers attempted to categorize dark patterns in various ways to make understanding and counteract the phenomena easier. For instance, as early as 2010, Sobiesk and Conti proposed categorizing malicious interfaces—often utilized to increase revenue [16]. Their paper presents a taxonomy consisting of eleven categories containing different subcategories: (1) Coercion—threat and mandate users' compliance (required form fields; user-threatening messages); (2) Confusion—questions or information that users cannot comprehend; (3) Distraction—driving user attention away by exploitation of pre-attentive processes (different media formats: video/animation/blinking, etc.; color); (4) Exploiting errors—taking advantage of errors to facilitate designers' goals (typing errors); (5) Forced work—increasing users' workload (delay of work effort; difficult de-installation); (6) Interruption—interfering the task flow (force viewing; "hot" interface elements); (7) Manipulating navigation—guiding users toward the designers' goals (dead-end trails; important information hidden deep in navigation); (8) Obfuscation—hiding desired information (low-contrast colors; mask warning messages); (9) Restricted functionality—only some controls needed by the user are present (omit controls; hide desired UI elements); (10) Shock—presentation of disturbing content (controversial content); (11) Trick—mislead or deceive users (silent/invisible behavior; lie; spoof content).

Gray et al. [27], based on the dark patterns identified by Brignull, provided an overview of user experience dark patterns. They categorized patterns into five groups: (1) Nagging (redirection to expected functionality); (2) Obstruction (adding difficulties to the interaction process); (3) Sneaking (hide information that might be useful, delay it so it would be disregarded at the decision time); (4) Interface interference (UI design manipulation that emphasizes some aspects over others); (5) Forced action (user must perform some action in order to access some functionality). Similarly, Cara [13] focused their review on dark patterns in the context of user experience.

Luguri and Strahilevitz [43] summarized the existing taxonomies of dark patterns, describing different categories and variants within those categories. Finally, Mathur et al. [44] defined the taxonomy of dark patterns based on empirical research of real e-commerce websites, identifying seven categories: (1) Sneaking (e.g., hiding information that could have affected users' choice); (2) Urgency (e.g., patterns that place deadlines on the decision); (3) Misdirection (e.g., use of design proprieties to steer users toward a specific choice); (4) Social proof (e.g., the specific choice is driven by the behavior of others); (5) Scarcity (there is limited availability of something, and therefore, its value increases); (6) Obstruction (some choices might be more challenging to make than others); (7) Forced action (additional action is required to complete a task).

3 Privacy Dark Patterns

Similarly, there were attempts to classify dark patterns in the context of privacy. However, the research in this context is more scarce. Bosch et al. [12] reviewed dark privacy strategies and dark patterns in their research, describing also contrasting privacy patterns (designs encouraging privacy-protective interactions). Their article suggests eight dark strategies that could be used in the design. (1) Maximize—design that enhances collection, storage, and processing of as much data as possible, instead of focusing on data that are required to ensure the system's functionality. (2) Publish—personal information might be visible to the public; no mechanism protects access to such data. (3) Centralize—personal information is stored/processed in a central entity, enabling linkability that might create a clearer picture of an individual(s). (4) Preserve—data are kept in the original state; it does not undergo processing that could affect interrelationships between data. (5) Obscure—users cannot assess what happens to their data, e.g., via complex terminology used in privacy policies. (6) Deny—users cannot control their data, e.g., the service provider might deny account deletion. (7) Violate—a service provider might have a privacy policy. However, it is not upheld by the provider. (8) Fake—a service provider claims robust data protection techniques and practices; however, none is actually implemented in a given service.

Beyond the academic research, the Norwegian Consumer Agency (Forbruker-Radet) published a report focusing on dark patterns that deceive consumers [25]. In particular, patterns that prevent people from exercising privacy rights. The report based on the research on the three tech giants—Facebook, Google, and Windows 10 (Microsoft product)—presents an analysis of how these companies implement malicious designs and nudge users toward privacy-intrusive actions. The report's aim was not to categorize dark patterns but to present a real-life analysis of how some tech companies violate privacy by utilizing deceiving designs in their pop-up windows. As a result, six such design techniques have been identified among the analyzed software: (1) Privacy-intrusive default settings; (2) Unequal ease (the number of clicks) for privacy-friendly options; (3) Visual design (color and symbols) that leads toward intrusive privacy option; (4) Language that leads toward intrusive privacy option; (5) Privacy unfriendly option presented without "warnings"; (6) Users cannot postpone the decision while accessing the service in the meantime.

Another non-academic categorization comes from CNIL (the French National Commission on Informatics and Liberty) in a report focusing on deceiving design and its effects on the privacy of individuals based on the GDPR principles. Although the report is not exclusively dedicated to dark patterns, it proposes a typology of deceptive design practices. There are four suggested categories containing different techniques that design is based on—enjoy, seduce, lure, complicate, and ban. (1) The first category contains designs that push the individual to accept sharing more than what is strictly necessary. (2) The second is designs that influence consent. (3) The third are designs that create friction in data protection actions. (4) And lastly, designs that divert individuals.

Another attempt to classify dark patterns comes from the legal field. Jarovsky [33] tried to create a taxonomy of dark patterns and suggested a new definition for dark patterns. Focusing on the legal aspects of design and the need for improvement of the privacy protection of data subjects (users), Jarovsky argues that privacy dark pattern "consists of user interface design choices that manipulate the data subject's decision-making process in a way detrimental to his or her privacy and beneficial to the service provider" [33, p.8]. Moreover, she proposes a dark pattern taxonomy that addresses the legal challenges of such deceptive designs. There are four categories in the taxonomy. (1) The first one is pressure, meaning the designs that pressure users to share more data to continue using a service/product. (2) The second one is hinder, meaning designs that delay, hide, or make it cumbersome for users to take action and protect their privacy. (3) The third is mislead, meaning designs that use language or different UI elements to mislead users during privacy-protective interactions. (4) The last, fourth category is misrepresent—designs that misrepresent facts to drive users toward sharing more (or more in-depth) personal data than needed.

3.1 Examples of Privacy Dark Patterns

A considerable share of the above categorizations of dark patterns, particularly categorizations of privacy dark patterns, overlaps to some extent, thus making it hard to systematize dark patterns. In this section, we attempt to group privacy dark patterns based on the similarity of mechanisms that these patterns employ. Moreover, we point out the dark patterns defined in the literature as different, yet it seems that it is mainly the nomenclature that differentiates them. Their descriptions imply the same phenomena (we refer to them as subgroups). In Table 2, we present the overall grouping, and the detailed descriptions are presented in the subsequent sections. The list aims to draw connections between previously identified dark patterns and relate them to psychological biases they might be exploiting. Note that the list is not exhaustive.

Invisible to the Human Eye

This group of privacy patterns contains dark patterns that are impossible to spot by a user's eyes. Their mechanisms are hidden, working at the "back-end" of a given technology:

1. *Address book leeching.* This pattern uses contact lists that are uploaded to the service from a user's device (predominantly a mobile phone). However, users are unaware that their contacts are being stored and processed by the service provider. Importing contacts may expose information to different third parties and place users' privacy at risk [6].

   Related dark patterns or other interchangeable phenomena:

   - Shadow user profiles—information about users not registered for a given service, for instance, on social networks, might be collected [12]. A service provider might manage and process such information without users' knowledge.

   Associated categories/design types: maximize, preserve, centralize [12].
   Associated psychological effects: N/A due to lack of perceptual interaction.
2. *Camouflaged advertising.* Also known as disguised ads [10]. The dark patterns that disguise adverts as other elements of UI [52]. Users might interact with such hidden advertisements, and as a result, they may be exposed to unwanted ads

Table 2 Groups of privacy dark patterns based on the similarities between the premises they built on

Group	Privacy dark pattern	Related pattern(s)
Invisible to human eye	Address book leeching	Shadow user profile
	Camouflaged advertisements	
UI design tricks	Attention diversion	Bad visibility
		False hierarchy
	Chameleon strategy	Bait and change
	Wrong signal	Twist
Constrained actionability	Comparison obfuscation	
	Forced action	False continuity
		Forced registration
		Impenetrable wall
		Pressure to receive marketing
	Immortal accounts	Difficult deletion
Emotion-related	Confirmshaming	Blaming the individual
	Toying with emotions	Framing
Affecting comprehension	Hidden legalese stipulations	
	False necessity	
	Just between you and us	Improving experience
	Trick questions	Ambiguity
		Double negative
Time-related	Last-minute consent	Repetitive incentive
	Safety blackmail	Cannot postpone decision
Affecting privacy options	Bad defaults	Default sharing
		Pressure to share
		Privacy-invasive defaults
	Privacy Zuckering	Difficult settings
		Hidden settings
		Making it fastidious to adjust confidentiality settings
		Obfuscation
		Obfuscating settings
		Unequal ease

and motivated to distribute their data by signing up for new services or buying products.

Associated categories/design types: diverting the individual/lure [52]; interface interference [27].

Associated psychological effects: Because ads are not visible, this dark pattern does not directly exploit any psychological effects. However, it relies on an automatic processing mode, assuming that users depend on quick decisions and will click on the ad.

UI Design Tricks

The dark patterns in this group rely only on the UI design mechanisms that aim to distract and mislead the user, sometimes using commonly recognizable UI elements and misusing them. For instance, an icon with a specific meaning is used for purposes other than what the original meaning might imply. Figure 1 presents an example of a dark pattern (attention diversion) from this group:

1. *Attention diversion.* This dark pattern exploits visual design proprieties to draw users' attention to something other than privacy-related parts of UI [52]. For instance, when signing up for an e-commerce application, within the privacy settings, changes to the settings could be presented in smaller and less-contrasting font, while the button for special discounts could be more prominent. The user will likely focus on the new desire to obtain a product at a discounted rate than on managing their privacy.

   Related dark patterns or other interchangeable phenomena:

   - Bad visibility—low contrasting, light colors, and small fonts make privacy-protective options less visible [33].
   - False hierarchy—some of the options appear more prominent than others [6].

   Associated categories/design types: influence consent/enjoy [52]; mislead [33]; interface interference [27].

   Associated psychological effects: anchoring, framing.
2. *Chameleon strategy.* This dark pattern occurs when a third-party service uses the style and visual appearance of the website browsed to make it look like a natural continuation [52]. For instance, a hotel booking suddenly becomes part of booking a train ticket. If the user follows through with such booking, their personal information might be automatically transferred to the rail service provider without informing users about the privacy implications of such transfer or asking for explicit consent.

   Associated categories/design types: diverting the individual/lure [52]; interface interference [27].

Fig. 1 Example of attention diversion privacy dark pattern, where users' attention is likely to be driven toward the discounts advertisement

Associated psychological effects: Because the pattern is not visible, it does not directly exploit any psychological effects. However, it relies on the Type 1 decision-making mode, which assumes that users will automatically select additional services.

3. *Wrong signal.* This dark pattern misuses commonly recognizable patterns, symbols, and similar, to create confusion related to the choice that a user makes [52]. For example, a service might be using a padlock icon in the UI design, yet the service lacks privacy and security protections.

   Related dark patterns or other interchangeable phenomena:

   - Bait and change—users' choice produces unexpected consequences. For instance, "giving acceptance value to a button with a cross, which in users' minds is synonymous with 'close and move on'" [52].
   - Twist—colors and symbols used in a way that misguides users [33].

   Associated categories/design types: mislead [33]; influence consent/lure, diverting the individual/lure [52]; interface interference [27].

   Associated psychological effects: anchoring, framing, affect heuristic, social norms (indirectly).

Constrained Actionability

This group of dark patterns includes designs that affect users' actions. Users are prevented from taking specific actions either by not having the possibility to act or by making actions challenging to carry on. Figure 2 presents an example of dark pattern (immortal account/difficult to delete) from this group:

1. *Comparison obfuscation.* When this pattern is applied, users struggle with comparing the different service providers or specific settings or rules within a service [52]. For example, when the changes in the service privacy policy content are implemented in a way that forbids users to compare these changes with the original content of the policy.

   Associated categories/design types: influence consent/complicate [52]; forced action and timing [25]; obstruction [27].

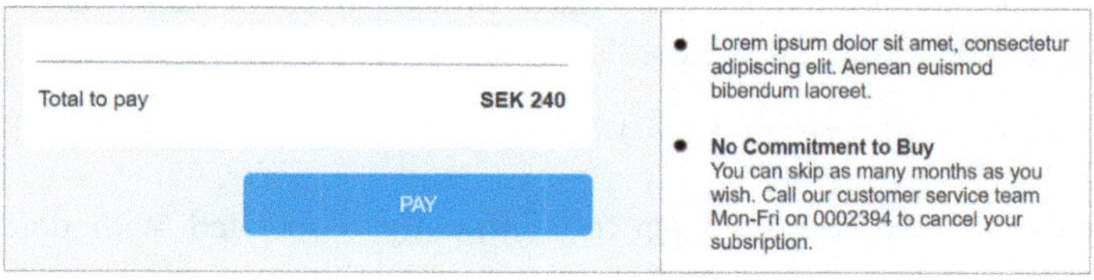

Fig. 2 Example of immortal account/difficult to delete privacy dark pattern (from [6]). A user has to call customer service within a specific time frame to cancel the subscription

Associated psychological effects: Exploiting the information asymmetry between the service providers and users, this pattern may trigger anchoring or optimism biases.

2. *Forced action.* This pattern forces users to make choices on the spot [6]. For instance, it may nudge users to agree to all the T&Cs when purchasing a product or service. As a result, they may blindly accept all the terms and be unaware of potential risks to privacy.

   Related dark patterns or other interchangeable phenomena:

   - False continuity—user is asked to provide personal information, such as the email address, to read an article, yet they are not warned that this might be a subscription to a newsletter [52].
   - Forced registration—forces users to register in order to use a service/product [12]. As a result, a company might gain access to personal information about a user, tracking their behavior.
   - Impenetrable wall—access to a service is blocked by a cookie wall or account creation, while it is not needed for service to function (also known as take-it-or-leave-it) [52].
   - Pressure to receive marketing—users must check the box "receive marketing offers per email" to complete a purchase or sign up for a service [33].

   Associated categories/design types: maximize [12]; creating friction on data protection actions/ban, pushing the individual to accept sharing more than what is strictly necessary/lure [52]; pressure [33]; sneaking [27]; forced action [27, 43].

   Associated psychological effects: instant gratification, framing, status quo. These dark patterns rely on automatic, Type 1 information processing, in which decisions can be constrained by external factors, such as situational context or time pressure.

3. *Immortal accounts.* This dark pattern appears when users have already created an account with a given service provider and want to delete their account and any associated data [12]. However, the service provider makes the deletion process cumbersome by not providing a straightforward deletion option. Instead, the user is confronted with a long process, which, in the end, if the user manages to complete deletion, may still trick the user and retain some personal information with the service.

   Related dark patterns or other interchangeable phenomena:

   - Difficult deletion—making it hard or inconvenient (e.g., call customer service—therefore, switch media) to delete an account [33].

   Associated categories/design types: deny, obscure [12]; hinder [33]; obstruction [27].

   Associated psychological effects: Difficulty associated with deletion, the prolonged process, etc., contribute to preventing the use of Type 2 thinking. Instead, the heuristic-based mode is activated, in which people tend to use mental shortcuts and come to conclusions quickly.

Emotion-Related

In this group, privacy dark patterns exploit human nature's emotional aspects. They target emotions, often connecting them with the different social aspects of life. Figure 3 presents an example of dark pattern (confirmshaming) from this group:

1. *Confirmshaming.* This pattern makes the user feel guilty about not opting into something or opting out of something. It uses the power of language to steer users into making a specific and undesired choice [6]. For instance, when users do not want to get tracked, companies might use language such as "No, I do not want to save money and receive discount codes" to shame users' choices.

   Related dark patterns or other interchangeable phenomena:

   - Blaming the individual—makes users feel guilty about their choices [52].
   - Toying with emotions—language, style, color, and other UI design elements can be used to evoke a particular emotional state. These design elements are explicitly applied to persuade the user into specific actions [6].
   - Framing—privacy-invasive features might be described in a positive way, preventing users from reflecting on these features' negative effects [33].

   Associated categories/design types: creating friction on data protection actions/enjoy [52]; interface interference [27]; misdirection [44]; mislead [33]; interface interference [27].

   Associated psychological effects: affect heuristic, optimism bias, contrast effect, default effect, framing, anchoring.

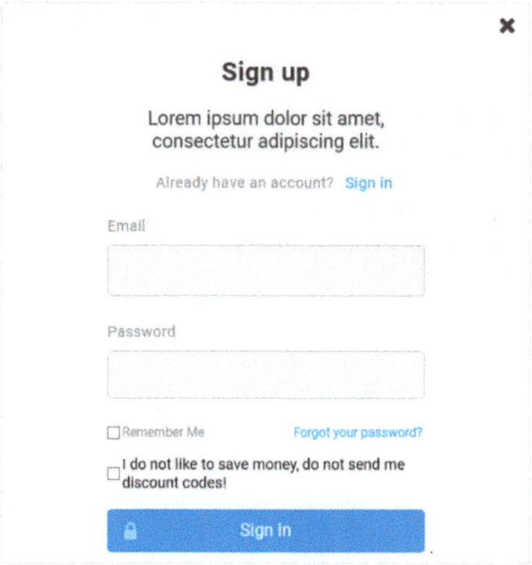

Fig. 3 Example of confirmshaming privacy dark pattern where service provider wants to further process users' information by shaming users for not wanting to receive discounts

Affecting Comprehension

This group of dark patterns affects information understanding. They either use language that is difficult to comprehend, inject false information that is impossible to verify by a user, or similar. Figure 4 presents an example of dark pattern (trick questions) belonging to this group:

1. *Hidden legalese stipulations.* The pattern is often used in legally required documents, such as privacy policies and terms and conditions. Often, such texts are written in legal jargon, difficult to understand by an average user, who intentionally skips reading the long texts [12]. Simultaneously, these legally binding texts may include a stipulation that targets users' privacy. For example, information that policy may change without further notice.

   Associated categories/design types: obscure [12].

   Associated psychological effects: The likelihood of users missing the opportunity to comprehend all the details increases the probability of Type 1 information processing.
2. *False necessity*—Falsely informing users that certain types of data are legally necessary or required for the system to function [33]. For instance, a social network mobile application may ask for access to contacts' email addresses to ensure the application's full functionality, while such information is not truly needed for the application to function.
3. *Just between you and us.* This dark pattern makes false promises. For instance, a service provider might request additional information, promising that such information will remain "invisible" and users will have full control, and it will allow better service [52]. Similar to this dark pattern are:

   Related dark patterns or other interchangeable phenomena:

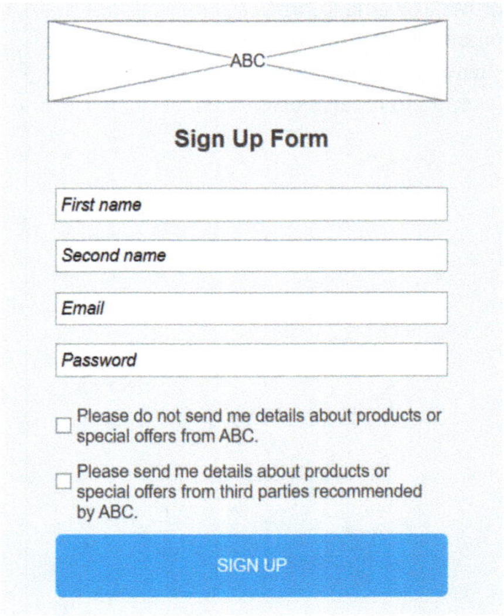

Fig. 4 Example of trick questions privacy dark pattern where the service provider applies sentences that purposefully confuse users (from [6])

- Improving the user experience—encouraging users to share more personal information to improve services and their experiences [52].

Associated categories/design types: pushing the individual to accept sharing more than what is strictly necessary/seduce [52]; deny [12].

Associated psychological effects: instant gratification, optimism bias, framing.

4. *Trick questions.* This pattern is usually formed as a question that appears to be one thing while meaning something else. This dark pattern may rely on confusing wording, double negatives, or other similar tricks that could confuse users [6].
Related dark patterns or other interchangeable phenomena:

- Ambiguity—confusing language, e.g., "do not share my data with third parties" and options to choose from "yes" and "no" [33, p.31].
- Double negative—using double negative in sentences makes it harder to grasp the meaning[33].

Associated categories/design types: mislead [33]; influence consent/lure [52]; interface interference [27, 44]; misdirection [44].

Associated psychological effects: default effect, framing, anchoring.

Time-Related

In this group are the dark patterns that, to a different extent, rely on temporal aspects of decision-making. They often exploit that decisions are made in a hurry, on the spot, which prevents users from engaging in a more analytical decision-making process. Figure 5 presents an example of privacy dark pattern (last-minute consent) from this group:

1. *Last-minute consent.* This dark pattern is time- and context-dependent. It seeks consent for the data collection at a specific moment when users are in a hurry or close to finishing a given task [52]. For instance, a service provider might add a new opt-in for information transfer to a third party at the end of the purchasing process. Users pursuing a goal of completing a transaction might provide their consent since they have already invested a long time and effort into purchasing.
Related dark patterns or other interchangeable phenomena:

- Repetitive incentive—different incentives on data sharing can be inserted, repetitively, to interfere with users' tasks [52].
- Users cannot postpone decision—users are urged to act without the possibility of postponing their decision. For example, pop-ups force users to select privacy preferences when they install software, which might prevent them from reflecting on their choices [25].

Associated categories/design types: influence consent/enjoy, creating friction on data protection actions/complicate [52]; forced action [25, 27]; nagging, obstruction [27]; urgency [43].

Associated psychological effects: loss aversion, status quo.

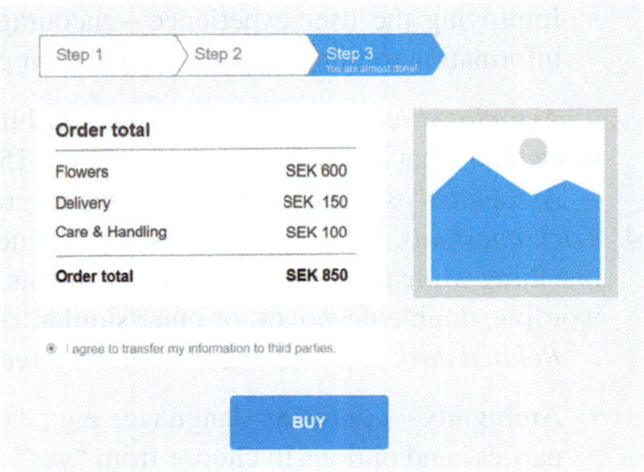

Fig. 5 Example of a last-minute consent privacy dark pattern where the service provider asks for additional consent when users are at the end of the task after dedicating much effort to achieving it (adapted from [6])

2. *Safety blackmail.* This dark pattern occurs during the login process as a request for additional information [52]. At such times, users' actions are time-driven and under pressure, and users want to complete the task and move on, accepting anything. For instance, a user might be tricked into providing their phone number, thinking it will be used for two-factor authentication, while it is only used for telemarketing.

   Associated categories/design types: pushing the individual to accept sharing more than what is strictly necessary/enjoy [52]; nagging [27]; misrepresent [33].

   Associated psychological effects: functional fixedness, loss aversion, restraint bias, instant gratification.

Affecting Privacy Options

This group of dark patterns consists of designs that provide users with privacy options. However, these options are often presented in a way that purposefully confuses users, makes it hard to choose, make it challenging to change options, or similar. Figure 6 presents an example of privacy dark pattern (bad default) from this group:

1. *Bad defaults.* This dark pattern exists when the options (particularly related to a user account) are predefined (e.g., selected checkboxes) in a way that encourages over-sharing personal information [12]. As a result, users might share information they have not intended to at the start of an interaction.

   Related dark patterns or other interchangeable phenomena:

   - Default sharing—pre-checked options for information sharing [52].
   - Pressure to share—users are obliged to share specific personal data with other users in order to use a service; they are given no alternative option [33].
   - Privacy-invasive defaults—applications that share certain data by default (e.g., a social network app that, per default, shares users' videos publicly) [33].

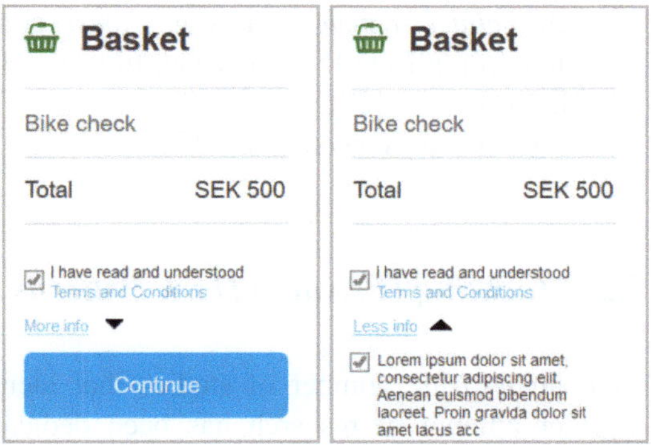

Fig. 6 Example of a bad default privacy dark pattern [6]. A default option (e.g., about data processing) is hidden, and users must perform additional action (click on "More info") to discover it

Associated categories/design types: obscure [12]; pushing the individual to accept sharing more than what is strictly necessary/enjoy [52]; pressure, hinder [33]; covert, asymmetric, misdirection [44].

Associated psychological effects: default effect; status quo; loss aversion; instant gratification.

2. *Privacy Zuckering.* This dark pattern exists when the service provider allows changing privacy-related settings [12]. Still, these settings are purposefully designed to be unnecessarily complex and difficult to understand by the user. For example, using a layered design to present users with an application's privacy settings—users have to open many sub-menus and, following different links, open new pages to reach the setting they desire to change.

   Related dark patterns or other interchangeable phenomena:

- Difficult settings—privacy settings are complex, multilayered; contain links (including third-party links), sub-menus, which all make it less likely for user to read/use [33].
- Hidden settings—privacy settings are placed on the least-expected and not intuitive part of a UI [33].
- Making it fastidious about adjusting confidentiality settings—consent is quick, but the process of data-protective actions is long and complex. For example, the continue button is given to accept all opt-ins, while the alternative choice requires many different interactions with "find out more" and similar options [52].
- Obfuscation—injecting privacy-unrelated settings to the section of UI where privacy settings are presented [33].
- Obfuscating settings—to reach the desired privacy settings, users have to go through a long and cumbersome process, reducing the chance that the user will change the settings and increasing the chance that they will give up before achieving their task [52].
- Unequal ease—the number of clicks to reach the privacy-protective options is higher, requiring more effort from the user [25].

Associated categories/design types: creating friction on data protection actions/complicate [52]; mislead, hinder [33]; obscure [12]; forced action [27, 43]; obstruction [27].

Associated psychological effects: choice overload; status quo; framing.

3.2 Tackling (Privacy) Dark Patterns

Compared to the number of studies that identify and categorize dark patterns, a lesser amount of research has been dedicated to answering the question of how to prevent dark patterns. One such attempt is given by Mathur et al. [45], categorizing existing research on dark patterns into two types of choice architecture: (1) modifying the decision space (dark patterns attributes: asymmetric, restrictive, disparate treatment, and covert) and (2) manipulating the information flow (dark patterns attributes: deceptive, and information hiding). Based on the new categorization and providing arguments about the normative principles of why dark patterns should be of concern (collective and individual welfare, regulatory objectives, autonomy), their article expands on what could be done in the field of human–computer interaction to tackle dark patterns. In particular, to assess the "darkness" of deceptive designs, they propose to analyze dark patterns through the above-mentioned normative lenses, specifying how such analysis should look. For instance, focusing on who might be affected by a specific dark pattern—general public or a particular group of users—should be considered when designing studies assessing dark patterns.

Nonetheless, the approach proposed by Mathur et al. [45] applies mainly to the research on dark patterns and aims to help identify the phenomena. A different aspect of dark patterns research that did not receive much attention, even though it could help to gain a better understanding of the phenomena and to develop solutions preventing it, relates to the harms that deceptive designs cause to users [6, 28, 45]. Kitkowska et al. [41] attempted to investigate this in the expert interviews-based study. Their research asked which dark patterns are more or less harmful and how to prevent the detrimental effects of these malicious designs. The experts identified two broad categories of dark patterns to assess their harmfulness. The first one, called *first generation*, contains "traditional," easier to identify, often leading to clear economic loss dark patterns. The second one, called *second generation*, consists of designs that are complex, hard to identify, and related to extensive data collection, leading to economic and less tangible loss. The latter has been classified as more harmful and requiring greater attention from regulators. In their research, privacy matters emerged because of the increasing insights about individuals gathered by companies and the potential use of such insights to target vulnerable users (here, everyone can be vulnerable as it might be a temporal state) and personalize dark patterns. Although their research did not focus on privacy dark patterns, it is essential to note that they attempted to identify ways to tackle dark patterns. Using the behavioral change framework (COM-B) [46], they identified policy categories (regulation, legislation, guidelines, service provision) and intervention functions (education, coercion, modeling, training) that could be applied to prevent companies from applying deceptive designs and bring balance to the digital market.

Another way to tackle dark patterns is through legislation and regulation, as recommended in research [6, 9, 41]. In various geographical regions, lawmakers introduce regulations that could help prevent dark patterns. The General Data Protection Regulation (GDPR) could be adapted, as argued by Jarovsky [33], to identify dark patterns, focusing on the principle of fairness in data protection as well as the lawful basis for obtaining consent. Some of the regulations directly targeting dark patterns already exist. For instance, European Directive 2005/29/EC on unfair commercial practices (UCPD) [17] adds dark patterns to the category of manipulative practices. It states: "If dark patterns are applied in the context of business-to-consumer commercial relationships, then the Directive can be used to challenge the fairness of such practices, in addition to other instruments in the EU legal framework, such as the GDPR." In the USA, California Consumer Privacy Act (CCPA) [5, 49]) defines a dark pattern as a design that "subvert user autonomy, decisionmaking, or choice" and considers acquiring consent obtained through the use of a dark pattern invalid. Still, these new regulations have their limitations. For instance, the UCPD lists only some designs that could be dark patterns, and the sanctions are still left to be defined by the EU members independently and might be too little.

Another issue regarding legal regulations is the problematic enforcement, for instance, how the digital market should be surveyed or how to identify and correctly recognize dark pattern designs. Some researchers propose automated tools that could help identify dark patterns. For instance, techniques used by Mathur et al. [44] to scrap web pages for dark patterns. However, the authors recognize the shortcomings of their methods, e.g., that images are not considered, yet they might be used as a part of dark pattern design. Curley et al. [19] proposed a framework for automated dark pattern detection, concluding that some deceptive designs are easier to detect. In contrast, others are impossible to identify through automated means.

Similarly, Soe et al. [57] attempted to use machine learning to identify dark patterns, and, achieving relatively good accuracy, they concluded that more research has to be done to improve it. Particularly, it might be challenging to create a good-enough data set that could yield better-promising results. Also, recognizing the difficulties around the variety of dark patterns and the ways they work, researchers suggest decomposing dark patterns into elements that could be automated (or quickly processed) and focusing on one specific domain. Both of these recommendations have the potential to improve automated detection tools.

Still, policymaking, regulation, and enforcement might be insufficient to entirely prevent the use of dark patterns in the digital market. One other way that surfaced in findings from [6] is a need for guidance and education for service providers. Such an approach, however practical, might also be inadequate since the profits that companies gain through the utilization of dark patterns might be too significant. Nevertheless, suppose appropriate regulations are in place and high fines for lack of compliance (e.g., similar to the fines that GDPR places on incompliant companies).

In that case, the guidelines and education can reduce the number of dark patterns applied in the digital market.

Additionally, some researchers proposed that users' education might be another way of reducing the adverse effects of dark patterns. As much as such an approach might be helpful to some extent, the early empirical findings suggest otherwise. For instance, Bongard-Blanchy et al. [9] examined how awareness of heuristics and biases applied in deceptive designs affects users. The study showed that a number of participants, conscious of psychological tricks used to deceive them, still fail and are influenced by dark patterns, yet they admit that they are aware of deceptive designs.

3.3 Dark Patterns and Implications on Businesses

Most of the research discusses dark patterns and their potential implications on users. However, only scarce investigations seem to tackle the effects that utilization of dark patterns might have on businesses that employ them. Moreover, none of the research papers considered in the present chapter examines privacy dark patterns' effects on companies.

In their meta-analyses, Hummel et al. [31] tried to compare the effects of different nudging practices in digital and physical contexts. They found that digital nudging does not differ significantly from nudging in other contexts. However, the analysis showed that approximately one-third of the effect sizes reported in the existing nudging studies were insignificant. Notably, the default nudges (corresponding to the bad default dark patterns) seem to be the most effective. Brown and Jones [11], in their longitudinal study, investigated the effectiveness of dark patterns and other changes in the UI design. Based on data collected from e-commerce websites that implemented A/B testing (comparative testing of different versions of a product to identify which one consumers prefer) of various designs between 2014 and 2017, they showed that only specific design changes carry the potential to increase revenue per visitor. Unfortunately, some categories of dark patterns—scarcity, social proof, urgency, abandonment (persuading the user not to leave the site, which indicates abandonment behavior), and product recommendation—seem to increase revenue. Although the revenue increase identified in this research was relatively small and ranged from +0.4% to +2.9%, from the business point of view, the implementation of deceptive designs might still be worth it, as any increase in the revenue adds to the business growth.

Considering the very little research conducted in the context of dark patterns' effects on businesses, it seems necessary to identify which dark patterns cause the most damage and regulate their applicability in the digital environment, similar to how the GDPR regulates data protection. Perhaps, this benefit-based approach could guide regulators to ban specific designs entirely or place high financial fines for using such deceptive designs. On the other hand, if the results of such proposed research would show minimal business benefits, such findings could be used to convince companies to stop the manipulative practices as they do not significantly impact their revenue. Instead, such research could show that users, becoming more aware of dark patterns, might be less loyal to a given business, perceive it as less trustworthy, and resign from the business's services. Hence, this chapter proposes that more research on the effects of dark patterns on businesses and consumers could help regulators and policymakers.

4 Concluding Remarks

This chapter aimed to explain the mechanisms through which dark patterns work based on existing research. It presented a list of psychological biases and heuristics that privacy dark patterns exploit. Moreover, it provided a non-exhaustive list of privacy dark patterns, grouping them into patterns that relate to each other. Although numerous attempts at dark pattern categorizations exist, many researchers seem to describe very similar phenomena yet name them differently (e.g., obfuscation vs. obfuscating settings, immortal accounts vs. difficult deletion). In this chapter, such similarities were pointed at to improve the understanding of how specific designs exploit psychological vulnerabilities. Nevertheless, such a plethora of categories and different dark patterns might make the use of knowledge about privacy dark patterns challenging to digest and utilize, e.g., in automated dark patterns recognition tools.

Further, the existing research on dark patterns, particularly in the context of privacy, is somewhat limited. It is not entirely clear what harms deceptive designs cause and how (if possible) such harms could be classified. Such classification could prove helpful in assessing the severity of harm and the potential need to develop measures against specific privacy dark patterns, which effects might be more detrimental than others. To summarize, the scarce and not entirely systematic research on privacy dark patterns call for future work, mainly empirical, to feed policymaking, help companies dodge deceptive designs, and produce automated tools helping in recognition of digital manipulation.

References

1. Acquisti, A., Adjerid, I., Balebako, R., Brandimarte, L., Cranor, L. F., Komanduri, S., Leon, P. G., Sadeh, N., Schaub, F., Sleeper, M., et al. (2017) Nudges for privacy and security: Understanding and assisting users' choices online. *ACM Computing Surveys, 50*(3), 1–41.
2. Acquisti, A., & Grossklags, J. (2005). Privacy and rationality in individual decision making. *IEEE Security & Privacy, 3*(1), 26–33.
3. Acquisti, A., Taylor, C., & Wagman, L. (2016). The economics of privacy. *Journal of Economic Literature, 54*(2), 442–492.
4. Anderson, C. J. (2003). The psychology of doing nothing: Forms of decision avoidance result from reason and emotion. *Psychological Bulletin, 129*, 139–167.
5. Attorney General Rob Bonta - Press Release. (2021). Attorney General Becerra Announces Approval of Additional Regulations That Empower Data Privacy Under the Califor-

nia Consumer Privacy Act, https://oag.ca.gov/news/press-releases/attorney-general-becerra-announces-approval-additional-regulations-empower-data
6. Barriers to a well-functioning digital market. Effects of visual design and information disclosures on consumer detriment. https://www.konsumentverket.se/globalassets/publikationer/produkter-och-tjanster/ovriga-omraden/underlagsrapport-2021-1-barriers-digital-market-konsumentverket.pdf Tech report, 2021
7. Baumeister, R. F., & Bushman, B. J. (2013). *Social psychology and human nature*. Cengage Learning.
8. Benyon, D. (2010). *Designing interactive systems: A comprehensive guide to HCI and interaction design*. Pearson.
9. Bongard-Blanchy, K., Rossi, A., Rivas, S., Doublet, S., Koenig, V., & Lenzini, G. (2021). *"I am Definitely Manipulated, even When I am Aware of it. It's Ridiculous!"- Dark patterns from the end-user perspective* (Vol. 1). Association for Computing Machinery.
10. Brignull, H. Deceptive design. https://www.deceptive.design
11. Browne, W., & Jones, M. S. (2017). *What works in e-commerce-a meta-analysis of 6700 online experiments* (p. 21). Qubit Digital Ltd.
12. Bösch, C., Erb, B., Kargl, F., Kopp, H., & Pfattheicher, S. (2016). Tales from the dark side: Privacy dark strategies and privacy dark patterns. *Proceedings on Privacy Enhancing Technologies, 2016*, 237–254.
13. Cara, C. (2019). Dark patterns in the media: A systematic review. *Network Intelligence Studies, VII*, 105–113.
14. Carrascal, J. P., Riederer, C., Erramilli, V., Cherubini, M., & De Oliveira, R. (2013). Your browsing behavior for a Big Mac: Economics of personal information online. In *Proceedings of the 22nd International Conference on World Wide Web* (pp. 189–200).
15. Chang, D., Krupka, E. L., Adar, E., & Acquisti, A. (2016). Engineering information disclosure: Norm shaping designs. In *Proceedings of the 2016 CHI Conference on Human Factors in Computing Systems*, CHI '16 (pp. 587–597). Association for Computing Machinery.
16. Conti, G., & Sobiesk, E. (2010). Malicious interface design: Exploiting the user. In *Proceedings of the 19th International Conference on World Wide Web, WWW '10* (pp. 271–280).
17. Council of European Union. (2005). Directive 2005/29/EC of the European parliament and of the council. https://eur-lex.europa.eu/legal-content/EN/TXT/PDF/?uri=CELEX:32005L0029&from=EN
18. Cowan, N. (2006) Chapter 20 what are the differences between long-term, short-term, and working memory. In *Progress in brain research* (pp. 323–338). Elsevier.
19. Curley, A., O'Sullivan, D., Gordon, D., Tierney, B., & Stavrakakis, I. (2021). The Design of a framework for the detection of web-based dark patterns. In *ICDS 2021: The 15th International Conference on Digital Society*
20. de Martino, B., Kumaran, D., Seymour, B., & Dolan, R. J. (2006). Frames, biases, and rational decision making in the human brain. *Science, 313*, 684–687.
21. Dinev, T., & Hart, P. (2004). Internet privacy concerns and their antecedents-measurement validity and a regression model. *Behaviour & Information Technology, 23*(6), 413–422.
22. Dinev, T., & Hart, P. (2006). An extended privacy calculus model for e-commerce transactions. *Information Systems Research, 17*(1), 61–80.
23. Dinev, T., McConnell, A. R., & Smith, H. J. (2015). Research commentary—informing privacy research through information systems, psychology, and behavioral economics: thinking outside the "APCO" box. *Information Systems Research, 26*(4), 639–655.
24. Evans, J. S. B. T., & Stanovich, K. E. (2013). Dual-process theories of higher cognition: Advancing the debate. *Perspectives on Psychological Science, 8*, 223–241.
25. ForbrukerRadet. (2018). Deceived by design. https://fil.forbrukerradet.no/wp-content/uploads/2018/06/2018-06-27-deceived-by-design-final.pdf
26. Gigerenzer, G., & Gaissmaier, W. (2011). Heuristic decision making. *Annual Review of Psychology, 62*, 451–482.
27. Gray, C. M., Kou, Y., Battles, B., Hoggatt, J., & Toombs, A. L. (2018). The dark (patterns) side of UX design. In *Proceedings of the 2018 CHI Conference on Human Factors in Computing Systems* (pp. 1–14).

28. Gunawan, J., Choffnes, D., Hartzog, W., & Wilsom, C. (2021). Towards an understanding of dark pattern privacy harms. In *Position Paper at the CHI 2021 Workshop: What Can CHI Do About Dark Patterns*.
29. Hann, I-H., Hui, K.-L., Lee, S.-Y. T., & Png, I. P. (2007). Overcoming online information privacy concerns: An information-processing theory approach. *Journal of Management Information Systems, 24*(2), 13–42.
30. Hansen, P. G., & Jespersen, A. M. (2013). Nudge and the manipulation of choice. *European Journal of Risk Regulation, 4*, 3–28.
31. Hummel, D., & Maedche, A. (2019). How effective is nudging? A quantitative review on the effect sizes and limits of empirical nudging studies. *Journal of Behavioral and Experimental Economics, 80*, 47–58.
32. ISO9241-11. (1998). *Ergonomics of human-system interaction*. Standard, International Organization for Standardization.
33. Jarovsky, L. (2022). Dark patterns in personal data collection: Definition, taxonomy and lawfulness. In *Taxonomy and lawfulness*.
34. Johnson, E. J., & Goldstein, D. (2003). Do defaults save lives? *Science, 302*, 1338–1339.
35. Kahneman, D. (2003). A perspective on judgment and choice. *American Psychologist, 3*, 7–18.
36. Kahneman, D. (2011). *Thinking, fast and slow*. Macmillan.
37. Kahneman, D., & Frederick, S. (2014). Representativeness revisited: Attribute substitution in intuitive judgment. In T. Gilovich, D. Griffin, & D. Kahneman (Eds.), *Heuristics and biases: The psychology of intuitive judgment* (pp. 49–81). Cambridge University Press.
38. Kahneman, D., Knetsch, J. L., & Thaler, R. H. (1991). Anomalies: The endowment effect, loss aversion, and status quo bias. *Journal of Economic Perspectives, 5*, 193–206.
39. Kahneman, D., & Tversky, A. (1979). Prospect theory: An analysis of decision under risk. *Econometrica, 47*, 263–292.
40. Kehr, F., Kowatsch, T., Wentzel, D., & Fleisch, E. (2015). Blissfully ignorant: The effects of general privacy concerns, general institutional trust, and affect in the privacy calculus. *Information Systems Journal, 25*(6), 607–635.
41. Kitkowska, A., Högberg, J., & Wästlund, E. (2022). Barriers to a well-functioning digital market: Exploring dark patterns and how to overcome them. In *55th Hawaii International Conference on System Sciences*.
42. Korff, S., & Böhme, R. (2014). Too much choice: End-user privacy decisions in the context of choice proliferation. In *SOUPS '14: Proceedings of the Tenth Symposium on Usable Privacy and Security* (pp. 69–87).
43. Luguri, J., & Strahilevitz, L. J. (2021). Shining a light on dark patterns. *Journal of Legal Analysis, 13*(1), 43–109.
44. Mathur, A., Acar, G., Friedman, M. J., Lucherini, E., Mayer, J., Chetty, M., & Narayanan, A. (2019). Dark patterns at scale: Findings from a crawl of 11k shopping websites. In *Proceedings of the ACM on Human-Computer Interaction* (Vol. 3).
45. Mathur, A., Mayer, J., & Kshirsagar, M. (2021). What makes a dark pattern... dark? Design attributes, normative considerations, and measurement methods. In *ACM reference format*.
46. Michie, S., van Stralen, M. M., & West, R. (2011). The behavior change wheel: A new method for characterising and designing behavior change interventions. *Implementation Science, 6*(42), 1–12.
47. Mirsch, T., Lehrer, C., & Jung, R. (2017). Digital nudging: Altering user behavior in digital environments. In *Proceedings of 13th International Conference on Wirtschaftsinformatik* (pp. 634–648).
48. Nielsen, J. (1994). Heuristic evaluation. In *Usability inspection methods* (pp. 25–62).
49. Office of the Attorney General, California Department of Justice. California Consumer Privacy Act of 2018. https://leginfo.legislature.ca.gov/faces/codes_displayText.xhtml?division=3.&part=4.&lawCode=CIV&title=1.81.5
50. Oppenheimer, D. M., & Kelso, E. (2015). Information processing as a paradigm for decision making. *Annual Review of Psychology, 66*(1), 277–294.

51. Scheibehenne, B., Greifeneder, R., & Todd, P. M. (2010). Can there ever be too many options? A meta-analytic review of choice overload. *Journal of Consumer Research, 37*, 409–425.
52. Shaping choices in the digital world IP reports innovation and foresight n°06 from dark patterns to data protection: The influence of UX/UI design on user empowerment (2019). www.cnil.fr
53. Sharot, T., Riccardi, A. M., Raio, C. M., & Phelps, E. A. (2007). Neural mechanisms mediating optimism bias. *Nature, 450*, 102–105.
54. Simmons, J. P., Nelson, L. D., & Simonsohn, U. (2011). False-positive psychology: Undisclosed flexibility in data collection and analysis allows presenting anything as significant. *Psychological Science, 22*, 1359–1366.
55. Slovic, F. (2002). *Heuristics and biases; the psychology of intuitive judgement*. Cambridge University Press.
56. Smith, H. J., Dinev, T., & Xu, H. (2011). Information privacy research: An interdisciplinary review. *MIS Quarterly, 35*, 989–1015.
57. Soe, T. H., Santos, C. T., & Slavkovik, M. (2022). Automated detection of dark patterns in cookie banners: How to do it poorly and why it is hard to do it any other way. Preprint arXiv:2204.11836.
58. Stanovich, K. E., & Toplak, M. E. (2012). Defining features versus incidental correlates of Type 1 and Type 2 processing. *Mind and Society, 11*, 3–13.
59. Susser, D., Roessler, B., & Nissenbaum, H. (2019). Online manipulation: Hidden influences in a digital world. *Georgetown Law Technology Review, 4*, 1.
60. Thaler, R., & Sunstein, C. (2008). *Nudge. Improving decisions about health, wealth, and happiness*.Penguin.
61. Thaler, R. H., & Benartzi, S. (2004). Save more tomorrow^{TM}: Using behavioral economics to increase employee saving. *Journal of Political Economy, 112*, S164–S187.
62. Tversky, A., & Kahneman, D. (1974). Judgment under uncertainty: Heuristics and biases. *Science, 185*, 3–20.
63. Tversky, A., & Kahneman, D. (1992). Advances in prospect theory: Cumulative representation of uncertainty. *Journal of Risk and Uncertainty, 5*, 297–323.
64. Weinstein, N. D. (1980). Unrealistic optimism about future life events. *Journal of Personality and Social Psychology, 39*, 806–820.
65. Yang, X. J., Wickens, C. D., & Hölttä-Otto, K. (2016). How users adjust trust in automation: Contrast effect and hindsight bias. In *Proceedings of the HFES 60th Annual Meeting* (pp. 196–200). Human Factors and Ergonomics Society.
66. Zhang, B., & Xu, H. (2016). Privacy nudges for mobile applications: Effects on the creepiness emotion and privacy attitudes. In *Proceedings of the 19th ACM Conference on Computer-Supported Cooperative Work & Social Computing* (pp. 1676–1690).

"They see me scrollin"—Lessons Learned from Investigating Shoulder Surfing Behavior and Attack Mitigation Strategies

Alia Saad, Jonathan Liebers, Stefan Schneegass, and Uwe Gruenefeld

1 Introduction

People interact with an evergrowing number of mobile computing devices in everyday life. Nowadays, these devices have become ubiquitous and are commonly used in various places such as buses, trains, airports, coffee shops, and restaurants [3, 14]. As a result of the continuous growth, privacy and security challenges of these devices are becoming increasingly pressing. For example, smartphones hold sensitive information about users, including business records, financial interactions, personal details, and many more that should be kept hidden from others. Nevertheless, finding privacy-preserving solutions is not restricted to smartphones only. These solutions need to consider a variety of personal devices (e.g., smartwatches and tablets) as well as public or shared devices (e.g., ATMs and ticket machines).

All these devices are subject to various types of attacks. For instance, thermal attacks, where intruders use thermal cameras to analyze the heat traces of the entered authentication [1] or attacks that analyze the smudges on the screen for password reconstruction and gaining illegitimate access [49, 52]. However, smudge attacks are mainly focused on the authentication period, and thermal attacks require technical support and proper planning for a person to take a photo, feed it to a recognizer, and gain unauthorized access. On the other hand, observation attacks, commonly known as *shoulder surfing attacks*, are directly performed by humans and usually do not require additional hardware to be successfully completed. Despite a large body of work on these observation attacks, shoulder surfing remains a significant unresolved problem that requires more attention.

A. Saad (✉) · J. Liebers · S. Schneegass · U. Gruenefeld
University of Duisburg-Essen, Essen, Germany
e-mail: alia.saad@uni-due.de; jonathan.liebers@uni-due.de; stefan.schneegass@uni-due.de; uwe.gruenefeld@uni-due.de

N. Gerber et al. (eds.), *Human Factors in Privacy Research*,
https://doi.org/10.1007/978-3-031-28643-8_10

Fig. 1 Sketched example of a spontaneous shoulder surfing attack taking place during daily commute

Observation attacks are not limited to a specific device, location, or acquaintance level. Shoulder surfer can gaze at a person interacting with their personal phone or at someone's PIN, while they authenticate themselves after getting the phone out of the pocket. They do not need an extra device and can quickly memorize entered PINs or passwords. They could be standing in a train [46], or sitting next to the victim in an office [2] (see Fig. 1). The incident could occur between two closely tied people or with total strangers. Previous works confirm that observation attacks are widespread and highly likely to occur [14].

With this pervasiveness, nearly everyone is both *attacker* and *victim*. Albeit, recent studies showed that shoulder surfing incidents often take place opportunistically, and without malicious intent. To this end, we consider a person looking at the *user's* interaction as an *observer*, as we are not sure of their motives. Many researchers focused on understanding the occurrence of the observation attacks. However, regardless of the intentions of the observers, researchers also worked on various approaches to mitigate the risk of being observed, either by detection of the observer, or by providing novel solutions to prevent the looker from perceiving the content displayed.

Chapter Overview In the next section, we define the term shoulder surfing, describe different dimensions relevant for shoulder surfing attacks, and present key findings from previous research. Thereafter, we look at proposed strategies to mitigate shoulder surfing attacks. Here, we start by looking at threat models and algorithmic detection of shoulder surfers. Finally, we outline challenges and future research directions for shoulder surfing research.

2 Investigating the Phenomenon

In this section, we first define shoulder surfing to set the scope for this chapter. After that, we describe different methods with which researchers have investigated the phenomenon and discuss their advantages and disadvantages. Finally, we highlight the key findings from studies investigating shoulder surfing behavior.

2.1 *Defining Shoulder Surfing (Attacks)*

Observation attacks, commonly known as *shoulder surfing attacks*, are directly performed by humans and usually do not require additional technology to be successful. Farzand et al. [16] define shoulder surfing as observing someone's device screen without their consent. There are technology-based approaches to investigate observation attacks using machine vision, commonly referred to as recording attacks or video-based observation attacks (e.g., [30, 61]). Nonetheless, this chapter primarily focuses on shoulder surfing attacks performed by humans.

To be classified as shoulder surfing, it does not matter if the motivation to shoulder surf is simply curiosity or a deliberate attempt to steal information [9]. In fact, shoulder surfing mainly occurs in an opportunistic, non-malicious way [14]. Nonetheless, failing to prevent bystanders from observing sensitive information can lead to negative consequences such as financial loss, public exposure, and embarrassment [3]. An example of a shoulder surfing attack is shown in Fig. 1.

In the following, we provide an overview of different dimensions that help describe and classify shoulder surfing. The goal is not to present a complete overview of all dimensions relevant to shoulder surfing but rather to discuss different aspects that should be considered:

Motivation of Attack: Shoulder surfing attacks can be either *intentional* or *unintentional*, whereas unintentional means in an opportunistic, non-malicious way [9]. In most cases, shoulder surfing is unintentional and does not have serious consequences [14]. Nonetheless, it can evoke negative feelings for both parties and result in various coping strategies.

Attack Pattern: Shoulder surfing attacks can follow different attack patterns. Abdrabou et al. [2] found three different patterns: *continuous attacks*, *cautious attacks*, and *repeated attack*. While continuous attacks are characterized by bystanders looking at the target device for an extended period with few or no gaze shifts, cautious and repeated attacks alternate between observing the target device and looking away. For the latter two, the difference is the victim's behavior, who either looks up from the target device (from time to time) or shows high engagement. Friends, family, or colleagues at work may repeatedly observe their peers and thereby combine multiple partial observations to form a hypothesis of a target device's secret [37, 57].

Number of Attackers: In theory, a shoulder surfing attack can be performed by *multiple attackers*. While some research considers threat models with more than *one attacker* [24], many studies simplify this aspect and study 1:1 relationships between victim and attacker.

Relationship Between Victim and Attacker: Besides the number of attackers, the *type of relationship* (family, friend, colleague, stranger) is important as well. Muslukhov et al. [37] conducted surveys and interviews to investigate users' concerns about unauthorized access by insiders and strangers. They concluded that observing unlock attempts, memorizing it, and thus gaining unauthorized access by insiders are highly likely to occur. That is directly linked to insiders' ability to observe interactions closely and repeatedly. Farzand et al. [16] showed that the type of relationship impacts the choice of mitigation behavior. Moreover, depending on the relationship with the attacker, victims often do not want them to know they were caught.

Victim–Attacker Relative Pose: To successfully shoulder surf, the content on the target device must be directly visible to the attacker (unless we reconstruct the screen content from visual reflections with machine learning [60]). Thus, the relative *pose between victim and attacker* is important, as the used term shoulder surfing illustrates. A sitting pose, for example, enables shoulder surfing more than a standing pose [46]. Furthermore, *viewing angle* and *distance* play an important role as well [6]. However, tilting the device away from the observer, a widely adopted defense strategy, provides limited protection from shoulder surfing attacks [25].

Type of Device: Different devices can be the target of a shoulder surfing attack, including but not limited to notebooks, tablets, smartphones, and smartwatches. However, shoulder surfing can also occur when using shared devices or accessing private information on public devices [9]. The main prerequisite for shoulder surfing is that a bystander can observe the user's screen. Hence, smartglasses are unaffected and can be used as a mitigation strategy [58].

Type of Content: Mainly, two different types exist: (1) authentication-based and (2) content-based shoulder surfing [18]. The primary focus of many shoulder surfing studies is to investigate secure password or PIN entry [8]. While authentication is, of course, important and prone to observational attacks, other types of content can also be observed. Moreover, content-based shoulder surfing is more frequently experienced than authentication-based shoulder surfing [18]. Previous work has examined different content types such as notifications, texts, photos, social media, and gaming [6, 46]. Nevertheless, while different types of content are affected by shoulder surfing, there are differences in their perceived sensitivity [17].

Type of Environment: Shoulder surfing can take place in different environments such as buses, trains, airports, coffee shops, and restaurants [3]. These environments can be classified in two different ways. One can either distinguish private, semi-public (work), or public contexts [45] or differentiate between personal and professional contexts [62]. Independent of the classification choice, the location cannot be neglected when studying shoulder surfing attacks as it influences victim and attacker behavior [48].

2.2 Research Methods

As outlined in the chapter "Empirical Research Methods in Usable Privacy and Security" , privacy and security research has applied various methods. In this section, we highlight the methods that were previously used to study shoulder surfing. In summary, we classify these methods into four categories: (1) surveys and interviews, (2) lab studies, (3) field/in-the-wild studies, and (4) studies in extended reality. The following subsection describes the different methods and highlights their advantages and disadvantages. Our goal is to provide an overview of the different methods to support researchers and practitioners (new to the field) in deciding which method to apply in their research.

Surveys and Interviews Surveys and interviews are helpful tools for privacy researchers to gather valuable insights into a broader population or specific user groups [36]. The difference between surveys and interviews is that in interviews, a researcher takes an active role and directs questions to the interviewee (cf., Lazar et al.[27, 28]), while in surveys, a set of predefined questions is presented to the participants. With surveys and interviews, it is possible to achieve various objectives. On one side, researchers can use them to gather evidence for shoulder surfing attacks in the real world and get insights into personal experiences with the phenomenon from both victims and attackers of shoulder surfing incidents (e.g., [14]). On the other side, they help to understand preliminary performance metrics of authentication techniques against observation attacks (e.g., robustness [4]) and can even be used to quantify which parameters of these techniques help to make them less observable (e.g., [54]). Different approaches to constructing surveys exist. Noticeable is the inclusion of video material to present recreations of shoulder surfing attacks to participants [4]. Aviv et al. [5] show that these videos embedded in surveys can achieve results comparable to user studies in the lab.

Compared to other research methods, surveys allow larger sample sizes as researchers can reach and recruit more participants. Nevertheless, sample sizes vary enormously for shoulder surfing research. Previous work has reported studies with more than 1000 participants ($n = 1173$) [4] to smaller numbers that remain in the hundreds (e.g., $n = 298$ [54] or $n = 174$ [14]). Compared with other research methods, surveys often report higher numbers of participants. Recently, crowdsourcing platforms have entered the stage of privacy research and provide researchers with access to different user groups (that can be specified concerning various dimensions) [23]. Nowadays, researchers can more easily recruit a diverse set of participants.

In addition to surveys, in-depth interviews can be a sensible next step that allows scientists to understand the reasons behind the observed data [14]. Nonetheless, interviews can also be applied as a standalone method. For interviews, the more active participation of a researcher asking questions can lead to more detailed responses [28]. Moreover, interviews allow the live demonstration of specific techniques under controlled conditions. For example, the interviewer can present different shoulder surfing mitigation strategies to participants during the interview [16].

Finally, there has been a recent study that explored shoulder surfing through a longitudinal investigation, meaning they performed a diary study with 23 participants over one month [18]. They found that content-based shoulder surfing takes place more frequently than authentication-based shoulder surfing.

While we presented different methods in this part, they all have in common that they rely on self-reporting. While self-reporting is frequently deployed in privacy research, it has a few noteworthy drawbacks. As researchers do not directly observe a phenomenon, factor, or effect, they rely on the subjective perception of the participant, which can include a recall bias [43]. Moreover, not every type of information can be gathered with self-reporting; however, asking indirect and anonymity-preserving questions can minimize social desirability bias [33, 53].

Lab Studies Scientists often conduct experiments to answer their research questions concerning shoulder surfing. In experiments, it is often necessary that researchers can observe a shoulder surfing situation taking place. Due to the challenges of researching the phenomenon during field or in-the-wild studies (see below), these studies are primarily carried out in the lab. Moreover, compared to surveys and interviews, recruiting participants is more difficult, and conducting the experiment is often more workload-intense. As a result, experiments generally report smaller sample sizes. Nevertheless, a lab study also has certain advantages, for example, compared to field or in-the-wild studies. The most significant benefit (compared to other study types) is the high degree of control over the experimental conditions. Moreover, a lab study allows gathering consent from all involved parties before the experiment.

When conducting a lab study to research different dimensions of a shoulder surfing attack (e.g., the resilience of authentication techniques against human shoulder surfers), a challenge is to replicate these attacks for the study [56]. In lab studies, participants often take over the role of the attacker (e.g., [46]). Nevertheless, it remains challenging to replicate realistic attacks, as often they are performed out of boredom in opportunistic moments [14]. Simply instructing participants to perform a shoulder surfing attack would broadly differ from the behavior observable during an actual attack. To overcome this challenge, researchers have designed studies that inform participants about the study's goals toward the end (e.g., [46]). These studies partially deceive participants by leaving out specific study details not to influence their behavior. However, it should be noted that deceiving participants in a user study can be problematic and not justified. Hence, it is strongly encouraged to balance ethical implications and knowledge gain and act cautiously when deceiving participants.

A different approach is to research factors and effects that are not related to the timing, occurrence, or behavior of shoulder surfing attacks but instead focus on aspects that can be researched with the research goal out in the open. For example, a previous study has investigated the effect viewing angle and distance have on the success of shoulder surfing attacks [6]. Here, a lab study can offer control to isolate research factors from others that would introduce too much complexity to the experiment.

In-the-Wild or Field Studies Researching the phenomenon of shoulder surfing with in-the-wild or field studies sheds more light on the contexts in which these attacks take place and could provide insights into the behavior of attackers and victims. However, performing these studies is very challenging and, thus, rarely conducted. One of these studies was a two-week in-the-wild study conducted by Schneegass et al. [48], where they investigated the likelihood of shoulder surfing attacks occurrence during unlock events. Nonetheless, shoulder surfing is socially unacceptable and privacy-invasive. Hence, observing these attacks requires consent, potentially biasing participants and making it very difficult to observe authentic interactions. Moreover, outside the lab, bystanders get involved quickly; when that happens, their consent is also necessary (e.g., when recording video for eye tracking). In the past, researchers have primarily relied on surveys and interviews to assess in-the-wild experiences [14], relying on self-assessment as the most frequent research method. To encompass both the benefits of a study in the lab (such as its associated high degree of control) and to enable researching more realistic (in situ) shoulder surfing scenarios, researchers have applied eXtended Reality as a study method.

Studies in Extended Realities Recently, eXtended Reality (XR) [42] entered human–computer interaction (HCI) as a means to conduct user studies that are not directly related to XR but use XR as a modality to conduct user studies instead (e.g., [31]). This is particularly the case for user studies that are taking place in virtual reality (VR) in a virtual environment (VE), whereas XR could implicate "augmented reality" (AR) or "mixed reality" (MR) as well. The trend of using XR as a research method got amplified with the ongoing Covid-19 pandemic as different frameworks appeared [19, 40].

Using VR to research the shoulder surfing phenomenon has several inherent benefits. First, a virtual environment allows a more believable recreation of a real-life situation, which would otherwise be hard to recreate in the lab (e.g., a bus stop or office environment with different people present [2]; see Fig. 2). In addition to the realistic recreated scenes, VR allows maintaining the consistency among study participants, avoiding external uncontrolled situations. With eye trackers embedded in the head-mounted displays (HMD), researchers are able to capture and analyze the gaze of the participants. Accordingly, they are able to profoundly understand the observation attacks cycles and expect what triggers the observers' attention. As VR is associated with a high degree of immersion, it allows placing the subject in a simulated, virtual environment, where they can experience the situation as intended

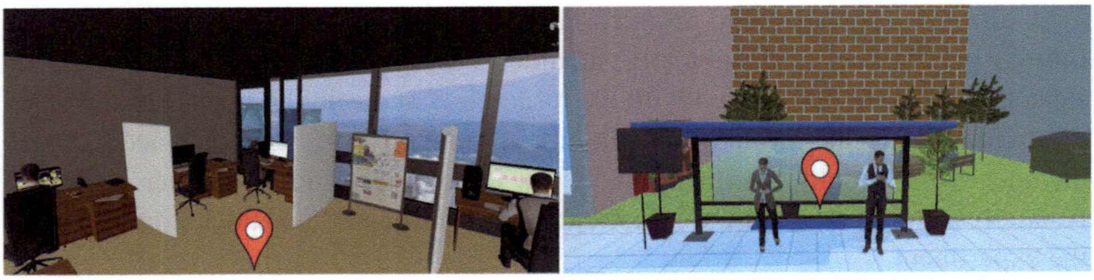

Fig. 2 Example taken from a previous paper that studied shoulder surfing in virtual reality [2]. The figure shows two virtual scenes that were used to investigate observing others' displays in an open office space (left) and a bus stop (right). The read markers indicate the participants' initial position

by the researchers. Here, the degree of presence can be assessed through the usage of presence questionnaires [50, 51, 59].

Potentially, such studies can also run outside the lab on HMDs owned by participants [40], and they were validated for usable security evaluations [35]. Additionally, user studies in XR allow fulfilling particular requirements specific for shoulder surfing studies. One is *privacy*, as conducting a user study in a real-world environment with real victims can be considered ethically challenging, whereas shoulder surfing a virtual avatar in a virtual environment (VE) is less likely an issue. Furthermore, conducting a user study in a VE allows for a very high degree of control since the environment is simulated by a computer, often exceeding the capability of control that an experimenter has over a real-world situation, even if it takes place in a lab. The high degree of control allows for replicability of such user studies between participants, as the experienced situation can be made to be precisely always the same.

2.3 Key Findings on Shoulder Surfing Behavior

With the growing number of studies investigating shoulder surfing events, we highlight the key findings on observers behaviors that we believe are of high relevance.

Observations Are Often More Random Than Planned In the survey by Eiband et al. [14], the main findings showed that despite the fact that observations are frequently conducted on an opportunistic basis, they go beyond exposing the authentication. Several participants reported negative feelings when other content such as personal photos or texts are exposed.

Victim–Attacker Pose Relationships Are Unalike In 2021, Saad et al. [46] explored the tendency of bystanders to shoulder surf in a scenario within an underground train. To that end, they varied the point of view of the attacker (standing vs. sitting) and the position of the victim (again standing vs. sitting) and used a

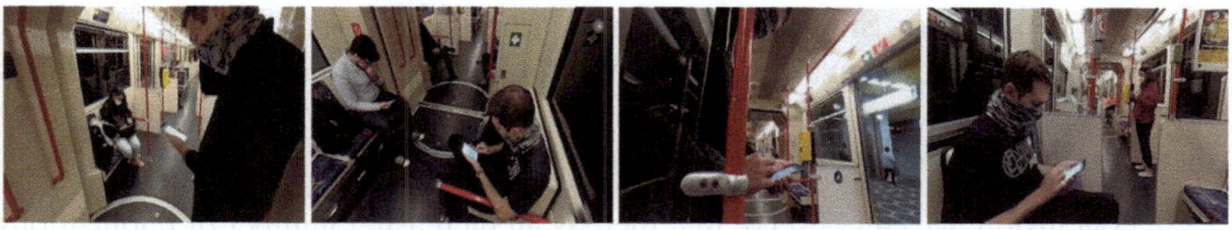

Fig. 3 User study conducted in virtual reality to investigate shoulder surfing attacks with prerecorded 360° videos [46]. Left to right: viewpoints of the participants with four different relative poses to the (virtual) victim: standing to standing, standing to sitting, sitting to standing, and sitting to sitting

360° camera to obtain a photorealistic recording of this setting, where several actors played either the role of the victim or became extras to simulate other people on the train. This recording then was played back to participants in a user study on an HMD that was equipped with an eye tracker in a lab study, and the point of view of the participants is seen in Fig. 3. Through the eye-tracking data, it was apparent that participants gazed at the object of interest, a smartphone held by the victim, and 11.16% of the time they were nearby.

VR Reflects Genuine Behavior... In 2022, Abdrabou et al. [2] conducted another project on the understanding of shoulder surfer behavior and the associated attack patterns. Here, they created a simulation in virtual reality with virtual, human-like avatars who were either located at a bus stop scene or within an office. The human participant of this study then was placed inside this VE through a VR HMD, which was again equipped with an eye tracker. The experimenters then recorded the participants' gaze and their walking patterns in VR and found that participants looked at several objects of interest (e.g., smartphones in the bus stop scene or monitors in the office scene) 5.7 times on average, whereas the average eye contact duration was 1.61 s.

...but Immersion Is Needed. Also in 2022, Mathis et al. [34] considered the differences between non-immersive and immersive VR for shoulder surfing research and conducted a user study to explore the characteristics of both settings. They considered shoulder surfing attacks on automated teller machines, smartphone personal identification numbers (PIN), and smartphone pattern unlock mechanisms. They compare three scenarios, 2D video observations, 3D observations, and VR observations. The first scenario, 2D video observations, consists of the study participants watching a video of the shoulder surfing situation that they cannot influence on a traditional computer monitor, whereas in 3D observations, they could use the keyboard and mouse to walk around. These two conditions then were compared against each other and VR observations, where participants were wearing a VR headset and could freely move around and adjust their observation perspective. The authors found that VR observations lead to a significantly higher sense of presence and involvement and that VR observations also lead to the most accurate shoulder surfing observations.

There Is More than Smartphones There are other devices that are becoming more ubiquitous nowadays, smartwatches for instance. Recently, more studies are proposing authentication approaches for smartwatches, with resilience against shoulder surfing as a key metric for robustness [38, 39].

In conclusion, we can observe that there is an increasing number of publications that utilize XR, particularly VR, as a research method for shoulder surfing research. The high degree of immersion lets the participants of a user study easily take the role of the attacker, while such a lab study setting allows for an efficient resolution of the problematic aspects connected to ethics in this kind of research. Furthermore, VR allows the study to be exactly the same for each subject, as the computer-driven simulation creates an easily repeatable environment. Thereby, realistic scenarios can effectively be replicated in the lab.

3 Mitigating Shoulder Surfing Attacks

For the mitigation of shoulder surfing attacks, it is important to note that not every shoulder surfing incident is equally problematic. One important aspect to consider is the type of content visible. For content-based shoulder surfing, we need to understand what is considered sensitive content as it plays an important role in selecting a suitable mitigation strategy. To tackle this challenge, Farzand et al. [17] present a typology of perceived sensitivity that can help to understand the content sensitivity. Furthermore, one needs to take into account that the perception of shoulder surfing is different between cultures [47]. As a consequence, it also differs what is considered sensitive content.

In the following section, we look at research that aims to find solutions to mitigate shoulder surfing attacks. Therefore, we start by looking at different threat models against which researchers and practitioners can evaluate their mitigation strategies. After that, we briefly describe technical approaches to detect shoulder surfing and their current limitations. Finally, we present an overview of different mitigation strategies.

3.1 Threat Models

Threat models provide a systematic approach to investigate potential weaknesses to privacy and security [32]. For shoulder surfing, different threat models have been considered in the literature. Below, we provide a selection of these models and describe them briefly. It should be noted that also mixes of these are possible (e.g., a repeated attack that is technology-supported [7]):

Weak Attacks: A shoulder surfing attack is considered a weak attack if it is performed by a human observer without the help of any technology and with only limited practice [11].

Trained Shoulder Surfers. Compared to weak attacks, trained shoulder surfers are more effective by training themselves. They often employ cognitive strategies that help to reach higher success rates [26]. Please note that trained shoulder surfers manage to be more effective without using recording devices.

Repeated Attacks: The repeated attacks threat model assumes that an attacker can repeatedly observe the target device of the victim. Moreover, this threat model often considers the attacker to be at close range—the attacker quite literally looks over the victims' shoulder [7].

Insider Attacks: Quite similar to the repeated attacks threat model are the insider attacks. The main difference is that for this type of attack, family, friends, or colleagues perform them. They may repeatedly observe the victim, and by combining these partial observations, it is easier to form a hypothesis on the victim's secret [57].

Multiple Attackers: The shoulder surfing attacks become more threatening when multiple attackers try to observe the target device. In this case, attackers can coordinate by either focusing on specific parts or organizing distraction and information stealing roles between attackers [24].

Technology-Supported Attacks: The probably strongest form of shoulder surfing attacks are technology-supported ones. In these cases, an attacker is recording the victim's interactions, for example, when drawing money from an ATM [10]. With recent technology advances, camera-based sensors can be manufactured in very tiny proportions, allowing attackers to seamlessly integrate them in their clothing or accessories. When analyzing the recorded data with machine learning, breaches of privacy are possible even when the attacker is not direct line of sight because reflections on glasses are sufficient for reconstruction of screen content [60].

3.2 *Algorithmic Detection of Attacks*

To mitigate shoulder surfing attacks, they first need to be detected. In previous research, detecting shoulder surfing attacks is primarily achieved by focusing on the human attackers. Here, algorithmic approaches oftentimes rely on visual sensor data (i.e., monochrome and RGB cameras). As shoulder surfing is frequently researched for mobile devices, the built-in camera is a good source for visual information to detect attackers. For example, Ali et al. [3] investigated the use of the built-in camera on mobile devices to detect if an unauthorized person tries to gain access to the device. Here, to detect an observer, face detection is applied to the incoming video feed. Interestingly, popular operating systems such as Android come with real-time face detection capabilities that can be used for detecting

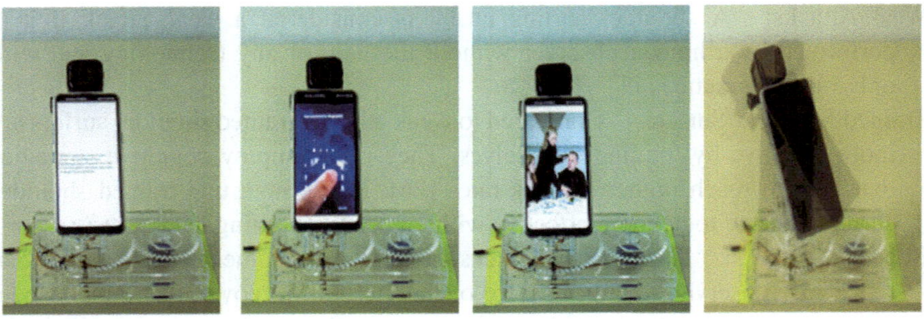

Fig. 4 Study apparatus to investigate the influence of distance and viewing angle on shoulder surfing success rate, figure taken from Bâce et al. [6] licensed under CC BY-NC-ND 4.0. The subfigures show examples of different content types on the phone display: (left to right): text, PIN, photo, and no content visible. The mechanical prototype visible rotated the smartphone between 0, 30, and 60°

shoulder surfers [7]. Nonetheless, not every detected face is necessarily a potential attacker as other factors play an important role as well, such as gaze direction and context, among others. In a recent study, different angles and distances have been investigated to understand which of them are most critical as they provide a good position for shoulder surfing [6]. The threat model was also based on evaluating people's perception on the displayed content that varied between visual, textual, and authentication, as seen in Fig. 4.

Nevertheless, visual detection of potential shoulder surfing also comes with a few downsides. First, they require the camera to be active and to record the scene. This scene likely involves the users of the device as well and, thereby, introduces another privacy risk. Furthermore, not only the privacy of a user may be violated, but also that of bystanders (as it continuously records the scene). Another issue is that the continuous recording and processing of the video feed drains the battery more quickly [7]. Hence, researchers have explored other options as well. For example, Lian et al. [29] used "multiple sensors, i.e., video camera module, ultrasonic distance module, light sensor module, to detect screen peeping, user distance and environmental lightness." Here, future studies should compare the different sensor technologies and develop adaptive strategies that take the context into consideration. For example, when a user is logged in to their wireless network at home and no other Bluetooth signatures are around, continuous monitoring via the built-in camera to detect shoulder surfing may not be necessary.

3.3 Prevention Strategies

Oftentimes, a detection algorithm proposed by researchers goes hand in hand with an implementation of a mitigation strategy (cf. [44]). In the following, we discuss two different strategy types into which proposed systems can be classified.

On one side, there are strategies that try to be generalizable toward every kind of content, and on the other side, there are strategies that focus on mitigating attacks against specific types of contents. These two strategies are in line with how we categorize shoulder surfing attacks into authentication-based and content-based shoulder surfing. Here, it is important to note that while authentication-based shoulder surfing is perceived as more problematic, content-based shoulder surfing is occurring more frequently [18].

Strategies Independent of Content Often times, researchers propose systems that mitigate shoulder surfing attacks independent of the content shown by the target device. Different systems have been proposed that try to create awareness for an actively ongoing shoulder surfing attack. For example, Ali et al. [3] proposed a system that informs users whether text on the screen could be read by an attacker. To better understand, in which way users want to be alerted, researchers have conducted a user study to compare four different methods: vibro-tactile, front LED, on-screen icons, and video feedback, finding that vibro-tactile feedback works best, as seen in Fig. 5. Their findings showed that vibration feedback allowed for a faster response time, in comparison to the other three methods [44]. Moreover, it has been examined how additional parameters such as distance and orientation can benefit victims in applying appropriate actions [62].

While awareness-based systems leave it to the user to decide on how they want to react, researchers have proposed different strategies that help users in their actions [9] or automatically react to shoulder surfing attacks [29]. Here, users can either move or hide information presented on the screen by performing explicit interactions [9] or information is automatically masked [9, 29] (e.g., with the help of eye tracking [41]). Lian et al. [29] found that with limited brightness or contrast, only the user could read the screen, while others have trouble reading it [29].

Furthermore, different strategies have been proposed that do not rely upon detecting a shoulder surfer at first, but rather are applied constantly. For example, Chen et al. [12] developed Hide Screen, which utilizes human vision characteristics to preserve privacy. Simplified, the approach allows changing the readability of information based on the viewing angle. Instead of hiding the information from an attacker, Watanabe et al. [55] suggest adding additional information that is designed to throw an attacker off. They suggest showing multiple cursors on the screen

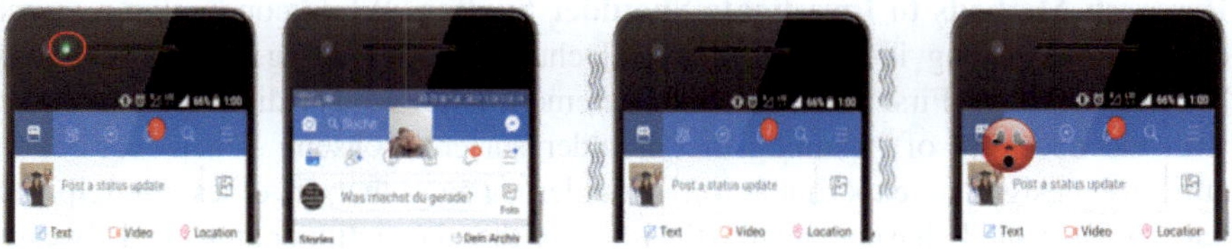

Fig. 5 Different feedback conditions to communicate a shoulder surfing incident investigated in previous work [44]. The different feedback conditions are (from left to right): (1) front LED, (2) video preview, (3) vibro-tactile, and (4) on-screen icon. The authors found that vibro-tactile feedback results in the lowest reaction time

and, thereby, effectively hiding the real cursor for an observer. Finally, it has been proposed to extend an observable screen with a second screen that is not observable and can be used to show private information. For example, Winkler et al. [58] are using smartglasses to show private information that would have otherwise be shown on the smartphone display.

Strategies Focused on Specific Types of Content Because not every type of content requires the same level of protection, many proposed strategies that are highly dependent on the type of content that they protect. In particular, authentication approaches need high protection against shoulder surfing attacks. Hence, researchers have suggested a variety of authentication techniques that are more resilient against observational attacks.

Bianchi et al. [7] proposed to use a composition of non-visual cues (i.e., audio and haptic cues) to enter a password. As a result, an observational attack cannot rely on visual information only to decipher the password. Furthermore, others have suggested to use gaze as an input modality in combination with graphical passwords [10]. Thereby, an attacker would need to observe the eye gaze of the victim additionally to the phone screen, making it very challenging to reconstruct the password. Another strategy is to extend the input surface for the authentication scheme toward the backside of the smartphone, which is more difficult to observe [13].

Besides authentication approaches, researchers have focused on other types of content. For example, Eiband et al. [15] have investigated how text can be presented in a way that is readable to the user but unreadable to an observer. In essence, they propose to display text in the user's own handwriting. While this does not prevent an attacker from reading the text, it significantly slows them down.

4 Challenges and Future Research Directions

In the following, we present challenges and research directions concerning the methodology of researching shoulder surfing and the phenomenon itself. These are particularly related to the methodology of shoulder surfing research and the attacker's behavior.

Research Methods to Investigate Shoulder Surfing While conducting research on shoulder surfing in the wild, several challenges regarding the methodology became apparent. First of all, a central element is an ethical dilemma associated with the necessity of obtaining the shoulder surfer's consent. When researchers ethically design an experiment on shoulder surfing that involves participants, participants usually have to get into the role of either the victim or attacker. However, shoulder surfing usually is an interaction that is very affective by its nature [14], hence instructing participants on the roles that they should get into highly inflects their behavior, and thus, results elicited from the study. Consequently, there is a dichotomy between asking for consent and subjects' unchanged behavior that needs to be weighed individually for each study, taking the objectives of the study into account.

Another argument on shoulder surfing studies is to *simultaneously* consider both roles of the attacker and the victim. Considering only the role of the observer and not the victim could leave out vital parts of the shoulder surfing incident, such as the occlusion of the phone display by the victim [6].

Virtual Reality for User Studies To overcome some of the challenges related to this ethical dichotomy, several research projects utilized virtual reality to simulate the shoulder surfing interaction with virtual avatars [2, 34, 46]. Although it is not necessary to obtain consent from a virtual avatar that has the role of the victim, it, however, still is necessary to obtain consent from a participant that gets into the role of the attacker. Furthermore, virtual reality allows for a simulation of the environment; hence, the interaction can be explored in different settings that would be hard to replicate in a physical lab.

However, virtual reality is also only a limited solution, as there are certain aspects impacted by the simulation of the environment. For example, today's head-mounted displays can influence people's behavior such as their movement [20] or also their social comfort distance that is less in virtual reality than in reality [22]. They can, however, help in recreating scenarios from the real-world by simulating them in a lab, as conducting field studies or in-the-wild experiments is particularly challenging due to the ethical aspects, particularly, when uninvolved third parties become part of the investigation. The same applies to other methodologies such as the usage of recording videos outside the lab, the so-called "lifelogs", as using cameras impacts the protection of private information of both the wearer and potential bystanders [21].

Identifying Sensitive Content In general, two types of shoulder surfing are distinguished: authentication-based and content-based shoulder surfing. While authentication-based shoulder surfing is inherently problematic as it exposes sensitive information (e.g., PIN or password), it is more complicated for content-based shoulder surfing that happens more frequently [18]. Privacy is an individual concept. Hence, what one person considers sensitive information may not be considered sensitive by someone else. This makes it very difficult to have an overall solution that equally protects all users. As a consequence, we need to investigate what content is considered sensitive (e.g., [17]). Furthermore, we need to examine different factors that can influence the perception of what is considered sensitive content such as cultural differences [47].

Understanding the Attacks and Behavior Another open research direction is to create an understanding of the shoulder surfing interaction itself, by, for instance, creating models of it. Here, Abdrabou et al. have created one of the first works in creating a model of attack patterns [2]. Their study took place in virtual reality; hence, creating a model-based understanding of the phenomenon, in reality, is still an open research opportunity nowadays. It is therefore necessary to conduct further studies to determine more substance to derive models about behavior within more contexts of the interaction. This includes, but is not limited to, in-the-wild studies as well as long-term studies to understand, whether the behavior changes over time.

Additionally, recent studies focus on password attacks but do not have a strong focus on understanding shoulder surfing behavior in general [8]. However, when considering only the attacks on passwords, such as android pattern locks, models were already created that predict the grade of observability [54]. This also opens up the opportunity to further explore the type of content that is particularly attracting shoulder surfing attacks, which partly has been covered by recent studies [2, 46].

5 Conclusion

In this chapter, we presented lessons learned from research on the shoulder surfing phenomenon and attack mitigation strategies. We started with a definition of shoulder surfing and an introduction of different types of attacks. After that, we present different research methods that have been applied in the past and discussed key findings related to shoulder surfing behavior. Next, we gave an overview of different threat models for shoulder surfing and discussed algorithmic detection of these attacks and different mitigation strategies. We concluded the chapter with an outlook on persistent challenges and future research directions. We believe that this book chapter offers a great starting point for new researchers and practitioners in the field. Moreover, we see great potential for eXtended Reality to overcome the limitations that field and in-the-wild studies introduce.

References

1. Abdelrahman, Y., Khamis, M., Schneegass, S., & Alt, F. (2017). Stay cool! Understanding thermal attacks on mobile-based user authentication. In *Proceedings of the 2017 CHI Conference on Human Factors in Computing Systems* (pp. 3751–3763).
2. Abdrabou, Y., Rivu, S. R., Ammar, T., Liebers, J., Saad, A., Liebers, C., Gruenefeld, U., Knierim, P., Khamis, M., Makela, V., Schneegass, S., & Alt, F. (2022). Understanding shoulder surfer behavior and attack patterns using virtual reality. In P. Bottoni & E. Panizzi, (Eds.), *Proceedings of the 2022 International Conference on Advanced Visual Interfaces* (pp. 1–9). ACM.
3. Ali, M. E., Anwar, A., Ahmed, I., Hashem, T., Kulik, L., & Tanin, E. (2014). Protecting mobile users from visual privacy attacks. In *Proceedings of the 2014 ACM International Joint Conference on Pervasive and Ubiquitous Computing: Adjunct Publication*, UbiComp '14 Adjunct (pp. 1–4). Association for Computing Machinery.
4. Aviv, A. J., Davin, J. T., Wolf, F., & Kuber, R. (2017). Towards baselines for shoulder surfing on mobile authentication. In *Proceedings of the 33rd Annual Computer Security Applications Conference* (pp. 486–498).
5. Aviv, A. J., Wolf, F., & Kuber, R. (2018). Comparing video based shoulder surfing with live simulation. In *Proceedings of the 34th Annual Computer Security Applications Conference*, ACSAC '18 (pp. 453–466). Association for Computing Machinery.

6. Bâce, M., Saad, A., Khamis, M., Schneegass, S., & Bulling, A. (2022). PrivacyScout: Assessing vulnerability to shoulder surfing on mobile devices. *Proceedings on Privacy Enhancing Technologies, 1*, 21.
7. Bianchi, A., Oakley, I., Kostakos, V., & Kwon, D. S. (2010). The phone lock: Audio and haptic shoulder-surfing resistant PIN entry methods for mobile devices. In *Proceedings of the Fifth International Conference on Tangible, Embedded, and Embodied Interaction*, TEI '11 (pp. 197–200). Association for Computing Machinery.
8. Bošnjak, L., & Brumen, B. (2020). Shoulder surfing experiments: A systematic literature review. *Computers & Security, 99*, 102023.
9. Brudy, F., Ledo, D., Greenberg, S., & Butz, A. (2014). Is anyone looking? Mitigating shoulder surfing on public displays through awareness and protection. In *Proceedings of The International Symposium on Pervasive Displays*, PerDis '14 (pp. 1–6). Association for Computing Machinery.
10. Bulling, A., Alt, F., & Schmidt, A. (2012). Increasing the security of gaze-based cued-recall graphical passwords using saliency masks. In *Proceedings of the SIGCHI Conference on Human Factors in Computing Systems*, CHI '12 (pp. 3011–3020). Association for Computing Machinery.
11. Chakraborty, N., & Mondal, S. (2014). An improved methodology towards providing immunity against weak shoulder surfing attack. In A. Prakash & R. Shyamasundar (Eds.), *Information Systems Security* (pp. 298–317). Springer International Publishing.
12. (Daniel) Chen, C.-Y., Lin, B.-Y., Wang, J., & Shin, K. G. (2019). Keep others from peeking at your mobile device screen! In *The 25th Annual International Conference on Mobile Computing and Networking*, MobiCom '19. Association for Computing Machinery.
13. De Luca, A., Harbach, M., von Zezschwitz, E., Maurer, M.-E., Slawik, B. E., Hussmann, H., & Smith, M. (2014). Now you see me, now you don't: Protecting smartphone authentication from shoulder surfers. In *Proceedings of the SIGCHI Conference on Human Factors in Computing Systems*, CHI '14 (pp. 2937–2946). Association for Computing Machinery.
14. Eiband, M., Khamis, M., Von Zezschwitz, E., Hussmann, H., & Alt, F. (2017). Understanding shoulder surfing in the wild: Stories from users and observers. In *Proceedings of the 2017 CHI Conference on Human Factors in Computing Systems* (pp. 4254–4265).
15. Eiband, M., von Zezschwitz, E., Buschek, D., & Hußmann, H. (2016). My scrawl hides it all: Protecting text messages against shoulder surfing with handwritten fonts. In *Proceedings of the 2016 CHI Conference Extended Abstracts on Human Factors in Computing Systems*, CHI EA '16 (pp. 2041–2048). Association for Computing Machinery.
16. Farzand, H., Bhardwaj, K., Marky, K., & Khamis, M. (2021). The interplay between personal relationships & shoulder surfing mitigation. In *Mensch Und Computer 2021*, MuC '21 (pp. 338–343). Association for Computing Machinery.
17. Farzand, H., Marky, K., & Khamis, M. (2022). "I hate when people do this; there's a lot of sensitive content for me": A typology of perceived privacy-sensitive content in shoulder surfing scenarios. In *Proceedings of the Eighteenth USENIX Conference on Usable Privacy and Security*. USENIX Association.
18. Farzand, H., Marky, K., & Khamis, M. (2022). Shoulder surfing through the social lens: A longitudinal investigation & insights from an exploratory diary study. In *2022 European Symposium on Usable Security* (pp. 85–97).
19. Gruenefeld, U., Auda, J., Mathis, F., Schneegass, S., Khamis, M., Gugenheimer, J., & Mayer, S. (2022). VRception: Rapid prototyping of cross-reality systems in virtual reality. In *Proceedings of the 2022 CHI Conference on Human Factors in Computing Systems*, CHI '22. Association for Computing Machinery.
20. Hollman, J. H., Brey, R. H., Robb, R. A., Bang, T. J., & Kaufman, K. R. (2006). Spatiotemporal gait deviations in a virtual reality environment. *Gait & Posture, 23*(4), 441–444.
21. Hoyle, R., Templeman, R., Anthony, D., Crandall, D., & Kapadia, A. (2015). Sensitive lifelogs: A privacy analysis of photos from wearable cameras. In *Proceedings of the 33rd Annual ACM Conference on Human Factors in Computing Systems*, CHI '15 (pp. 1645–1648). Association for Computing Machinery.

22. Iachini, T., Coello, Y., Frassinetti, F., Senese, V. P., Galante, F., & Ruggiero, G. (2016). Peripersonal and interpersonal space in virtual and real environments: Effects of gender and age. *Journal of Environmental Psychology, 45*, 154–164.
23. Jin, H., Shen, H., Jain, M., Kumar, S., & Hong, J. I. (2021). Lean privacy review: Collecting users' privacy concerns of data practices at a low cost. *ACM Transactions on Computer-Human Interaction, 28*(5), 1–55.
24. Khamis, M., Bandelow, L., Schick, S., Casadevall, D., Bulling, A., & Alt, F. (2017). They are all after you: Investigating the viability of a threat model that involves multiple shoulder surfers. In *Proceedings of the 16th International Conference on Mobile and Ubiquitous Multimedia*, MUM '17 (pp. 31–35). Association for Computing Machinery.
25. Khan, H., Hengartner, U., & Vogel, D. (2018). Evaluating attack and defense strategies for smartphone PIN shoulder surfing. In *Proceedings of the 2018 CHI Conference on Human Factors in Computing Systems*, CHI '18 (pp. 1–10). Association for Computing Machinery.
26. Kwon, T., Shin, S., & Na, S. (2014). Covert attentional shoulder surfing: Human adversaries are more powerful than expected. *IEEE Transactions on Systems, Man, and Cybernetics: Systems, 44*(6), 716–727.
27. Lazar, J., Feng, J. H., & Hochheiser, H. (2017). Chapter 5: Surveys. In J. Lazar, J. H. Feng, & H. Hochheiser (Eds.), *Research methods in human computer interaction* (2nd ed., pp. 105–133). Morgan Kaufmann.
28. Lazar, J., Feng, J. H., & Hochheiser, H. (2017). Chapter 8: Interviews and focus groups. In J. Lazar, J. H. Feng, & H. Hochheiser (Eds.), *Research methods in human computer interaction* (2nd ed., pp. 187–228). Morgan Kaufmann.
29. Lian, S., Hu, W., Song, X., & Liu, Z. (2013). Smart privacy-preserving screen based on multiple sensor fusion. *IEEE Transactions on Consumer Electronics, 59*(1), 136–143.
30. Maggi, F., Volpatto, A., Gasparini, S., Boracchi, G., & Zanero, S. (2011). Poster: Fast, automatic iPhone shoulder surfing. In *Proceedings of the 18th ACM Conference on Computer and Communications Security* (pp. 805–808).
31. Mäkelä, V., Radiah, R., Alsherif, S., Khamis, M., Xiao, C., Borchert, L., Schmidt, A., & Alt, F. (2020). Virtual field studies: Conducting studies on public displays in virtual reality. In R. Bernhaupt, F. F. Mueller, D. Verweij, J. Andres, J. McGrenere, A. Cockburn, I. Avellino, A. Goguey, P. Bjørn, S. Zhao, B. P. Samson, & R. Kocielnik (Eds.), *Proceedings of the 2020 CHI Conference on Human Factors in Computing Systems* (pp. 1–15). ACM.
32. Marback, A., Do, H., He, K., Kondamarri, S., & Xu, D. (2013). A threat model-based approach to security testing. *Software: Practice and Experience, 43*(2), 241–258.
33. Marques, D., Guerreiro, T., & Carriço, L. (2014). Measuring snooping behavior with surveys: It's how you ask it. In *CHI '14 Extended Abstracts on Human Factors in Computing Systems*, CHI EA '14 (pp. 2479–2484). Association for Computing Machinery.
34. Mathis, F., O'Hagan, J., Khamis, M., & Vaniea, K. (2022). Virtual reality observations: Using virtual reality to augment lab-based shoulder surfing research. In *2022 IEEE Conference on Virtual Reality and 3D User Interfaces (VR)* (pp. 291–300). IEEE.
35. Mathis, F., Vaniea, K., & Khamis, M. (2021). RepliCueAuth: Validating the use of a lab-based virtual reality setup for evaluating authentication systems. In Y. Kitamura, A. Quigley, K. Isbister, T. Igarashi, P. Bjørn, & S. Drucker (Eds.), *Proceedings of the 2021 CHI Conference on Human Factors in Computing Systems* (pp. 1–18). ACM.
36. Müller, H., Sedley, A., & Ferrall-Nunge, E. (2014). Survey research in HCI. In *Ways of knowing in HCI* (pp. 229–266). Springer.
37. Muslukhov, I., Boshmaf, Y., Kuo, C., Lester, J., & Beznosov, K. (2013). Know your enemy: The risk of unauthorized access in smartphones by insiders. In *Proceedings of the 15th International Conference on Human-Computer Interaction with Mobile Devices and Services* (pp. 271–280).
38. Nagatomo, M., Watanabe, K., Aburada, K., Okazaki, N., & Park, M. (2019). Proposal and evaluation of authentication method having shoulder-surfing resistance for smartwatches using shift rule. In *International Conference on Network-Based Information Systems* (pp. 560–569). Springer.

39. Park, M., Aburada, K., & Okazaki, N. (2021). Proposal and evaluation of a gesture authentication method with peep resistance for smartwatches. In *2021 Ninth International Symposium on Computing and Networking Workshops (CANDARW)* (pp. 359–364). IEEE.
40. Radiah, R., Mäkelä, V., Prange, S., Rodriguez, S. D., Piening, R., Zhou, Y., Köhle, K., Pfeuffer, K., Abdelrahman, Y., Hoppe, M., Schmidt, A., & Alt, F. (2021). Remote VR studies: A framework for running virtual reality studies remotely via participant-owned HMDs. *ACM Transactions on Computer-Human Interaction, 28*(6), 1–36.
41. Ragozin, K., Pai, Y. S., Augereau, O., Kise, K., Kerdels, J., & Kunze, K. (2019). Private reader: Using eye tracking to improve reading privacy in public spaces. In *Proceedings of the 21st International Conference on Human-Computer Interaction with Mobile Devices and Services*, MobileHCI '19. Association for Computing Machinery.
42. Rauschnabel, P. A., Felix, R., Hinsch, C., Shahab, H., & Alt, F. (2022). What is XR? Towards a framework for augmented and virtual reality. *Computers in Human Behavior, 133*, 107289.
43. Robins, R. W., Fraley, R. C., & Krueger, R. F. (2009). *Handbook of research methods in personality psychology*. Guilford Press.
44. Saad, A., Chukwu, M., & Schneegass, S. (2018). Communicating shoulder surfing attacks to users. In *Proceedings of the 17th International Conference on Mobile and Ubiquitous Multimedia*, MUM 2018 (pp. 147–152). Association for Computing Machinery.
45. Saad, A., Gruenefeld, U., Mecke, L., Koelle, M., Alt, F., & Schneegass, S. (2022). Mask removal isn't always convenient in public!—The impact of the Covid-19 pandemic on device usage and user authentication. In *Extended Abstracts of the 2022 CHI Conference on Human Factors in Computing Systems*, CHI EA '22. Association for Computing Machinery.
46. Saad, A., Liebers, J., Gruenefeld, U., Alt, F., & Schneegass, S. (2021). Understanding bystanders' tendency to shoulder surf smartphones using 360-degree videos in virtual reality. In *Proceedings of the 23rd International Conference on Mobile Human-Computer Interaction* (pp. 1–8). ACM.
47. Saleh, M., Khamis, M., & Sturm, C. (2019). What about my privacy, Habibi?. In D. Lamas, F. Loizides, L. Nacke, H. Petrie, M. Winckler, & P. Zaphiris (Eds.), *Human-computer interaction —INTERACT 2019* (pp. 67–87). Springer International Publishing.
48. Schneegass, S., Saad, A., Heger, R., Delgado, S., Poguntke, R., & Alt, F. (2022). An investigation of shoulder surfing attacks on touch-based unlock events. In *Proceedings of the 24th International Conference on Human-Computer Interaction with Mobile Devices and Services*, MobileHCI '22. Association for Computing Machinery. To Appear.
49. Schneegass, S., Steimle, F., Bulling, A., Alt, F., & Schmidt, A. (2014). SmudgeSafe: Geometric image transformations for smudge-resistant user authentication. In *Proceedings of the 2014 ACM International Joint Conference on Pervasive and Ubiquitous Computing*, UbiComp '14 (pp. 775–786). Association for Computing Machinery.
50. Schubert, T. W. (2003). The sense of presence in virtual environments: A three-component scale measuring spatial presence, involvement, and realness. *Zeitschrift für Medienpsychologie, 15*(2), 69–71.
51. Schwind, V., Knierim, P., Haas, N., & Henze, N. (2019). Using presence questionnaires in virtual reality. In *Proceedings of the 2019 CHI Conference on Human Factors in Computing Systems, volume 2019 of CHI '19* (pp. 1–12). Association for Computing Machinery.
52. Shin, H., Sim, S., Kwon, H., Hwang, S., & Lee, Y. (2022). A new smart smudge attack using CNN. *International Journal of Information Security, 21*(1), 25–36.
53. Tourangeau, R., & Yan, T. (2007). Sensitive questions in surveys. *Psychological Bulletin, 133*(5), 859.
54. von Zezschwitz, E., De Luca, A., Janssen, P., & Hussmann, H. (2015). Easy to draw, but hard to trace? On the observability of grid-based (un)lock patterns. In *Proceedings of the 33rd Annual ACM Conference on Human Factors in Computing Systems*, CHI '15 (pp. 2339–2342). Association for Computing Machinery.
55. Watanabe, K., Higuchi, F., Inami, M., & Igarashi, T. (2012). CursorCamouflage: Multiple dummy cursors as a defense against shoulder surfing. In *SIGGRAPH Asia 2012 Emerging Technologies*, SA '12 (pp. 1–2). Association for Computing Machinery.

56. Wiese, O., & Roth, V. (2015). Pitfalls of shoulder surfing studies. In *In NDSS Workshop on Usable Security 2015 (USEC'15)* (pp. 1–6). Internet Society.
57. Wiese, O., & Roth, V. (2016). See you next time: A model for modern shoulder surfers. In *Proceedings of the 18th International Conference on Human-Computer Interaction with Mobile Devices and Services*, MobileHCI '16 (pp. 453–464). Association for Computing Machinery.
58. Winkler, C., Gugenheimer, J., De Luca, A., Haas, G., Speidel, P., Dobbelstein, D., & Rukzio, E. (2015). Glass unlock: Enhancing security of smartphone unlocking through leveraging a private near-eye display. In *Proceedings of the 33rd Annual ACM Conference on Human Factors in Computing Systems*, CHI '15 (pp. 1407–1410). Association for Computing Machinery.
59. Witmer, B. G., & Singer, M. J. (1998). Measuring presence in virtual environments: A presence questionnaire. *Presence: Teleoperators and Virtual Environments, 7*(3), 225–240.
60. Xu, Y., Heinly, J., White, A. M., Monrose, F., & Frahm, J.-M. (2013). Seeing double: Reconstructing obscured typed input from repeated compromising reflections. In *Proceedings of the 2013 ACM SIGSAC Conference on Computer & Communications Security*, CCS '13 (pp. 1063–1074). Association for Computing Machinery.
61. Ye, G., Tang, Z., Fang, D., Chen, X., Wolff, W., Aviv, A. J., & Wang, Z. (2018). A video-based attack for Android pattern lock. *ACM Transactions on Privacy and Security, 21*(4), 1–31.
62. Zhou, H., Ferreira, V., Alves, T., Hawkey, K., & Reilly, D. (2015). Somebody is peeking! A proximity and privacy aware tablet interface. In *Proceedings of the 33rd Annual ACM Conference Extended Abstracts on Human Factors in Computing Systems*, CHI EA '15 (pp. 1971–1976). Association for Computing Machinery.

Privacy Research on the Pulse of Time: COVID-19 Contact-Tracing Apps

Eva Gerlitz and Maximilian Häring

1 Introduction

In 2020, COVID-19 hit the World, and with it came the desire for a well-functioning and a fast-working possibility to trace contacts of those people who tested positive for the virus, a method called *contact tracing*.

> **Definition: Contact Tracing**
> Following the Cambridge Dictionary, contact tracing is "the process of finding any other people that an infected person has met or had close contact with, usually in order to control the spread of an infectious disease" [8]. Similar definitions are used elsewhere, e.g., by the World Health Organization (WHO) [11] and the European Centre for Disease Prevention and Control (ECDC) [9].

Early on, digital contact tracing was seen as a tool to interrupt chains of infection. This led to a discussion about apps to automatically trace and store with whom a user had been in contact with and, as a result, would warn those who might have become infected. Digital contact tracing was even advertised as a "key" in fighting the pandemic [12]. It has several advantages compared to a manual approach done

Authors are listed in alphabetical order and contributed equally

E. Gerlitz
Fraunhofer FKIE, Bonn, Germany
e-mail: gerlitz@cs.uni-bonn.de

M. Häring (✉)
Universität Bonn, Institut für Informatik, Bonn, Germany
e-mail: haering@cs.uni-bonn.de

N. Gerber et al. (eds.), *Human Factors in Privacy Research*,
https://doi.org/10.1007/978-3-031-28643-8_11

by health workers, e.g., that it enables to warn more people who else could not have been notified due to incomplete memory or knowledge about contacts of an infected person. Digital contact tracing also supports the authorities by notifying contacts of positive tested persons: Instead of calling each person one by one, the information can be transferred immediately to all persons at once.

Most of the digital contact tracing approaches were realized through smartphone apps. The idea of using apps that help fight a disease was not new in 2020. In Africa, e.g., an app supported contact tracing personnel in faster submitting the information to help combat Ebola in 2019 [33].

One of the first COVID-19 focusing apps was launched in February 2020 by the Chinese government. It was specifically designed to warn its users about a contact with someone who is infected with the virus [7]. Many other governments followed, and a lot of those contact tracing apps (CTA) based their tracing on Bluetooth or the users' location. As of March 2021, the MIT Technology Review lists 49 contact tracing apps in 48 countries from around the World [1] and an overview from Google lists 60 apps that make use of their provided framework [27].

Depending on how automated tracing is implemented, it is necessary to capture and store sensitive information about the user, such as where the user has been, who they were in contact with, and their health status. All of this entails the potential of mission creep and surveillance. Based on the possibility of misuse, a lot of public discussions in 2020 revolved around the architecture of such tracing apps. Many experts and organizations worldwide made a strong case for apps that should technically prevent abuse [10].

Researchers from the University of Oxford estimated what percentage of the population would need to install a contact tracing app for it to be effective, depending on further measures that were taken throughout the country. Their results indicate that adoption of 60% could stop the pandemic, but already smaller installation numbers would reduce the number of infections and deaths [17]. In public discussions, this number of 60% was often misreported to be the threshold that needs to be achieved in order to fight COVID-19 [24].

Taken together, these requirements (privacy preserving and the need to reach a large part of the population) were able to influence political decisions, e.g., in Germany [6], where the government switched to a more privacy-preserving app after another one had already been planned.

The concerns for misuse of the captured data were, in fact, not unfounded: Later, in at least one case in Germany, data of a private app that was used to check-in into restaurants and that stored the data centrally were used by the criminal investigation department to find witnesses of an accident. This happened even though it is illegal to use these data for law enforcement purposes, according to the Infection Protection Act for reasons of data protection [22].

In Singapore, data captured through the widespread contact tracing app "Trace-Together," about which it was claimed after its release that the data would only be accessed if a user tests positive for COVID-19, were used in a murder investigation [32].

So, privacy has been a big topic in the development and the public discussions centered around contact tracing apps. But how big of a role does and did privacy actually play in the mind of potential users when they needed to decide whether or not to install a contact tracing app? And what can privacy research learn from that?

This chapter is a starting point for every reader interested in these questions. In this chapter, we:

- Give a brief outline of the tracing technologies and their implications for the users' data and therefore privacy.
- Look at scientific studies with end users and how their privacy concerns impacted their decision to install a contact tracing app.
- Set the study results in the context of the used methodology (e.g., the time the study was conducted or who was asked).

After reading this chapter, the reader will have an overview of the general privacy discussion on contact tracing apps in the context of COVID-19 and hints on where to find further information.

2 Tracing Technologies

This section gives a brief overview of technical possibilities to automatically warn people who had been in contact with someone who later tests positive for COVID-19. Worldwide, different versions of contact tracing apps were proposed, discussed, and rolled out. The task of apps in this context ranged from simply informing users about their contact and asking them to start a voluntary quarantine (e.g., in Germany [25]) to functioning as access control (e.g., in China [23]).

Obviously, it is (currently) not feasible to technically directly track whether a person met another person; therefore, many solutions use the personal smartphone as a proxy. The apps capture whether a device was in proximity to another device, therefore also called proximity tracing. For simplicity, we assume in the following that people always carry their smartphones with them, and we will use the ideas of "Who met whom" and "Which device encountered which device" interchangeably.

The following two sections detail the steps of such a digital contact tracing: The tracing itself and the details of when and how a user is informed about meeting someone who tested positive. Our goal is to give enough detail about the essential technology for the reader to have a general overview and can follow the debates around the different apps, their approaches, and possible implications for the users. Please note that this is not a complete list of technologies.

2.1 Proximity Tracing

For a contact tracing app to work, first and foremost, it must be logged who was in contact with whom. There are different approaches to accomplish this and different ways to categorize them: Huan et al. [18], for example, used a categorization where approaches are separated based on the data collection method: *cell phone base station data*, *location history*, and *Bluetooth proximity data*. Another possible taxonomy could be built based upon the interaction and setup needed (e.g., device-to-device communication directly via Bluetooth), indirect via participation tracking (e.g., at an event through QR codes [15]), or the not-so-common usage of already existing data (e.g., cell phone base station data).

To understand a lot of the research focusing on privacy in the contact tracing context, one has to look at the storage location of the logged contact data and the usage of Bluetooth Low Energy (LE). It works as follows: devices broadcast IDs via Bluetooth LE. The received IDs are stored together with the sent ones, and some information is added/derived, such as a distance and time metric. Those stored IDs are later matched with a list of IDs representing infected persons. If a device keeps the gathered IDs stored locally and compares them locally to a public list of IDs representing an infected person, the approach is called *decentral*. On the other hand, *central* means that the devices upload at least the seen and gathered IDs to a central entity/server.

Both approaches have their disadvantages, but the threat model differs. In the centralized approach, parties hosting or having access to the service (e.g., the government) could gain access to the data [28]. In this case, the third party could, for example, learn about the users' social graph. Compared to this, in the decentralized approach, an attacker needs to be in close vicinity to gain knowledge, as explained by Baumgärtner et al. [5].

Independent of how the approaches are categorized, tracing was discussed in many different ways, and for further research in this area, we suggest further literature and projects (e.g., [4, 5, 14, 26, 29]).

2.2 Risk Calculation and Informing Those at Risk

For efficient contact tracing, it is not only necessary to trace contacts, but also to inform those who had been in close contact with infected people (and possibly also give advice or instructions on how they should behave). This can be divided into the following three problem spaces:

Medical Basis for Risk Calculation The fundamental question is who should be informed and under what circumstances. For this, requirements from epidemiologists and virologists need to be implemented, concerning, for example, the distance and time after which an infection becomes more likely.

Technical Implementation of Risk Calculation There are different possibilities for where the actual risk calculation can occur. Research and politics in the EU favored mainly the previously outlined decentralized approach. In this approach, the assessment of whether the user is at risk is calculated on the phones directly. In the centralized approach, this calculation happens on a central server. Independent of the approach is the fact that the risk calculation can only be an estimation of what actually happened. False positives and true negatives have to be balanced. On either side, it can result in a negative effect on the adoption and effectiveness of the app.

How to Inform Those at Risk In the decentralized approach, no central entity knows the contacts of an infected person and therefore cannot inform them. Each device itself is "responsible" to inform its user. In a centralized setting, the server knows who is at risk. Therefore, even out-of-band contact, e.g., via phone, is possible depending on what data are available.

3 Privacy and Contact Tracing Apps—User Studies

The previous sections concerned technical circumstances to give the reader an overview of the situation. This section now focuses on the end user, thus the person owning a smartphone, and who is the potential user of an app. We give insight into what the studied participants think about contact tracing apps in terms of privacy, and how privacy considerations impact the willingness to use such apps.

For this, we conducted a literature review. In 2020, the topic of contact tracing apps was highly relevant and design decisions needed to be made urgently, so many researchers around the world examined the effect of different app properties and their general acceptance in the public population: The ACM Digital Library [2], for example, as of September 2022, lists around 32K publications published since 2020 when searching for "contact tracing."

We thus specified our search term such that the terms "contact," "trac*," and "priv*" had to be found in either the title or the abstract. Our full search comprised the databases ACM Digital Library [2], IEEE Xplore [19], and Web of Science [38]. We also analyzed the Google Scholar top twenty security conferences and journals if their names included "privacy" and the A* and A CORE-ranked privacy conferences and journals. Only those that were not already included in the previous database search underwent a manual title search. This included the Symposium On Usable Privacy and Security (SOUPS) and the International Conference on Security and Privacy for Communication Networks (SecureComm).

After this search, we ended up with 245 papers. We manually reviewed all abstracts and only picked those that fit our requirements. Articles were excluded if they matched the following criteria:

- Not related to contact tracing technology to combat COVID-19.
- No user study was conducted. (This included all studies that looked at user feedback from the App stores of Apple or Google.)
- The user study did not look at sentiments of users concerning the privacy aspects of contact tracing apps.

We ended up with 13 papers that are covered in this chapter. Table 1 gives a brief overview of the included studies.

It must be noted that because of the urgency and its possible high relevance to ongoing discussions, many studies were not only published in a peer-reviewed conference or journal but faster published, e.g., by uploading on arXiv. Those are not necessarily of bad quality but have to be read more carefully than work that was

Table 1 Brief overview of the presented studies. If a specific contact tracing app was investigated, this information is included in brackets

Authors	Country	*n*	Purpose of the app (CT = Contact tracing)	Used standardized questionnaire?
Huang et al. [18]	USA	44	CT, Home quarantine, Epidemiological investigation support system, Information tracking of dine-in customers, E-permit service	No
Häring et al. [16]	Germany	744	CT (CWA)	No
Utz et al. [37]	Germany, USA, China	1003, 1003, 1019	CT, Symptom Check, Quarantine Enf., Information, Health Certificate	IUIPC, 2004
Redmiles et al. [28]	USA	1000	CT, Information	No
Xie et al. [39]	Ireland	286	CT (COVID Tracker)	Westin's privacy segmentation index (PSI), privacy attitude questionnaire (PAQ)
Trestian et al. [36]	Ireland	258	CT (COVID Tracker)	Westin's privacy segmentation index (PSI)
Lu et al. [21]	USA	291	CT (identifying and notifying close contacts) + monitoring symptoms	No
Dooley et al. [13]	USA	7,010,271 impres-sions	CT	–
Zampedri et al. [40]	Belgium	15	CT	No
Sharma et al. [30]	27 different countries	261	CT, information, self-assessment	No
Trestian et al. [35]	Ireland	1001	CT (COVID Tracker)	Westin's privacy segmentation index (PSI)
Jamieson et al. [20]	USA	290	CT	UTAUT
Aji et al. [3]	Malaysia	505	CT (MySejahtera)	No

already peer-reviewed. For this reason, we only include peer-reviewed work in this chapter but would like to point out that many (in our sample of papers 9) of those cite such publications. Also, we want to point out to the reader that the studies were not conducted in the same setting: Some asked about a hypothetical app, others studied an existing app, and others an app that was about to be published. Additionally, the design of the presented apps differed, making the comparison additionally hard.

When presenting the results of a paper, we will clarify what app was examined throughout the study.

3.1 Results from User Studies—Privacy Concerns

This section presents the privacy concerns participants had or mentioned in user studies. Following our goal, we will only describe the privacy-relevant questions for each paper. Please note that the studies use a concept "privacy" that not always means the same thing, e.g., the PSI (privacy segmentation index) or a direct question about privacy concerns.

Although timed differently and with different local effects, there were some very similar progression steps of the pandemic worldwide. The following publications are thus sorted by the date the reported studies were started, so that the reader is able to set them into context of the situation at that time.

Some of the presented studies did not solely concentrate on contact tracing but also included other purposes of COVID-19 apps, such as symptom checks or providing information about the current COVID-19 situation. An overview of the apps that were included is given in Table 1. We want to note that we will not report statistical results in detail, as this would come with the need of a detailed description of the used tests. Instead, if the publication mentions a statistically significant test, we report that.

Huang et al. [18] conducted 44 interviews concerning six made-up information-tracking solutions that were based on existing apps. Those were not just contact tracing apps but also apps to, e.g., monitor quarantine. The participants were asked about their perceptions of the different solutions. The interviews were conducted between May 12, 2020, and January 4, 2021. Regarding privacy, participants expressed concerns about the data used (e.g., selfies and location data), data that might be collected undocumented, the long-term misuse of personal data, and further usage, such as unauthorized sharing. Among others, threats like identifying theft or data breach were mentioned.

Häring et al. [16] conducted a survey study in June 2020 in Germany, right before the official German contact tracing app, the Corona-Warn-App (CWA), was published. The 744 participants were asked what attributes of the soon-to-be-released app were true, and whether they would use the app. The authors also asked the participants to rate the influence of potential properties. They found that over a fourth of the participants believed the app would threaten their privacy. Six of the potential properties could be attributed to a centralized approach. The authors saw that those properties that would be beneficial to the user or society in general (e.g., allowing a better assessment of the situation or allowing the official health institute to see contacts in order to warn contacts) statistically positively impacted the intention to install the app, whereas those that focus on the potential disadvantages (e.g., Health officials seeing distance violations) statistically impacted the intention negatively.

Utz et al. [37] surveyed people in Germany, the USA, and China. The surveys were conducted between June and August 2020. They presented their participants' ten different hypothetical COVID-19 apps and then asked them to rate the apps based on different criteria. The apps were built using the following properties (among others): data collected, user anonymity, data receiver, and data transmission. The participants were also asked for general negative and positive reasons why they would or would not install a contact tracing app. After that, they filled the Internet Users' Information Privacy Concerns (IUIPC, 2004) constructs (see also the chapter "Toward Valid and Reliable Privacy Concern Scales: The Example of IUIPC-8"). The authors found that 40% of the participants reported being generally concerned about their privacy regarding contact tracing apps. Participants from Germany who had high concerns regarding data collection (IUIPC: Collection) were significantly less likely to use any app.

Redmiles et al. [28] surveyed 1000 US Americans in June 2020 about specific privacy concerns of COVID-19 apps. Forty-eight percent were concerned about someone being able to learn their location information, and 31% feared someone could find out who they have been in contact with. Thirty-six percent were not concerned about any of the presented possible concerns.

Sharma et al. [30] distributed a survey from July to August 2020 on social media and community groups. The survey was conducted in English, and they gathered 261 complete responses from 27 countries. They found that the participants' privacy concerns were about data privacy and data practices.

Xie et al. [39] and Trestian et al. [36] report the results from a survey pilot study ($n = 286$ [39], $n = 258$ [36]) that was open for one month from August 27, 2020, on. The full study that was conducted between November 2020 and January 2021 with 1001 participants is reported by Trestian et al. [35]. They made use of the PSI (privacy segmentation index) to check for connections between privacy attitudes and installation willingness. All three papers found that participants who identified as privacy fundamentalists, a category of the PSI, were least willing to share their personal data with a contact tracing app.

Lu et al. [21] did not only investigate contact tracing apps but compared them to human contact tracers. For this, they surveyed 291 Americans in August 2020. They asked how comfortable they would be with an app or a human to identify close contacts, be notified as a contact, and share a daily health status. The percentage of participants who reported being very comfortable or comfortable with each approach was in the range of 50.1–69.5%. There was no overall difference between human and digital contact tracing. However, the authors found an interaction effect: Participants were significantly more comfortable using digital tracing for monitoring their daily health status. In an open-ended question, the authors also wanted to know about benefits and risks of digital and human contact tracing. The results indicate that digital contact tracing could, in fact, also have perceived privacy benefits: Participants mentioned that technology could allow for anonymity, avoid being judged, and might also not bring up social anxiety when dealing with sensitive health topics. Finally, the authors asked for the participants' willingness to share different types of personal information typically collected by one or both contact tracing approaches. For this question, they saw that participants were still significantly more comfortable sharing data with human contact tracers. The concern to share data with digital contact tracing was often related to concerns related to data security.

Dooley et al. [13] conducted an experiment in Louisiana in February 2021. In this, they tested different advertisements for Louisiana's COVID-19 exposure-notification app. In the advertisements, they included a collective or individual benefit of the app and different nuances of privacy and data collection transparency. The authors then analyzed the proportion of people who clicked on the advertisement. They found that data collection and privacy transparency have different impacts, depending on the appeal (collective or individual). Ads with a collective appeal perform better when paired with privacy transparency statements but worse with data transparency statements. A data transparency statement increased the number of clicks for ads that point to an individual benefit, whereas a technical privacy statement decreased this number. In their discussion, the authors assume that combining a collective benefit with information about individual data that is collected might conflict with peoples' sense of collectivist purpose.

3.2 *Influence of Privacy on Using a CTA*

In this section, we summarize findings that bring privacy considerations and the users' intention to use a contact tracing app together, answering the question how they relate and whether concerns affect the intention to install.

First of all, it was reported that **only a few participants have concerns** about their privacy in the context of contact tracing in general: Lu et al. [21] found 6.5% of their participants not to be comfortable with contact tracing in general, no matter if done by a human or digitally. Many of them (4% of all participants) explicitly mentioned privacy as the reason for this.

However, when specifically asking about contact tracing apps, the feelings seem to shift. Thus, studies found an **influence of privacy concerns on the willingness to install** a contact tracing app: Häring et al. [16] saw that the general concern that the app threatens one's privacy and the belief that the government would be able to see distance violations significantly influenced the willingness to use the app negatively. Utz et al. [37] saw that German participants were less likely to use an app if the data would be transferred to private companies, law enforcement, or the general public. They also found that participants with "higher privacy concern with regard to data collection practices (IUIPC (2004): Collection)" were significantly less likely to install an app. In an interview study with 15 participants, Zampedri [40] found that many of those who did not download the Belgian contact tracing app worry about privacy violations and lack of data transparency. Even though the Belgian app followed the decentralized approach, the participants believed the app to be privacy-invasive and that the government had access to the data. Huang et al. [18] found that privacy concerns are associated "with participants' unwillingness to adopt the solutions". In a study by Sharma et al. [30], 25% of the participants mentioned that a privacy breach would be a reason to uninstall the app. Yet, if a government forces their citizens to use an app, privacy concerns seem to be overruled: Aji et al. [3] conducted a survey in Malaysia with 505 participants who were users of the Malaysian official contact tracing app. This app asks for personal information, such as the telephone number and the name, and can be used to check-in to places the user is visiting. The authors were interested in the participants' data usage and privacy awareness about the app. They saw that in general, most participants were aware of issues the app had, e.g., that the government has access to the user's location and personal information when using the app.

So while privacy concerns seem to be an influencing factor for not installing a contact tracing app, **the opposite does not seem to apply**. Having no privacy concerns did not necessarily lead people to install an app, as Jamieson et al. [20] found. They saw that being unconcerned about privacy or data leakage was not enough to actually motivate people to install a contact tracing app. However, other motivations were needed, such as providing evidence of the app's effectiveness in protecting members of one's community. The implementation of an app thus seems to be just one part of the story.

4 Privacy: A Matter of Asking? Looking at Different Methods

In this section, we want to give an overview of how the different aspects of the chosen methodology could have impacted the results. While some of these aspects cannot always be prevented, they should be kept in mind when evaluating the results.

We would like to point out that most of the following methodological aspects apply to almost any study conducted with humans but might take on different dimensions, depending on the subject that is studied. We discuss aspects in the context of contact tracing apps that also apply to other research areas.

4.1 Timing and Context

Since the topic of contact tracing apps was urgent and relatively new, all studies presented in this chapter are cross-sectional studies and not longitudinal studies, which means that users' sentiments and concerns were only captured at one specific time. Some of the presented studies are more than two years old, and replications are missing in this set of publications.

With this, overall societal attitudes can have more impact on the data than what would be seen if looking at a topic at several times. For contact tracing apps in particular, there could be a difference in attitudes depending on whether an app is already published for some time without any major issues reported or whether an app is still in the implementation phase and several details are not known yet. Apart from that, public discussion could influence the feeling toward technology. Some of these topics could also be seen in the publications:

Privacy Concerns Might Be Overruled by Extreme Situations Trestian et al. [35] investigated the privacy paradox of contact tracing apps. This paradox refers to the discrepancy between expressed privacy concerns and actual behavior. The authors surveyed Irish citizens and classified them into three privacy groups according to the Privacy Segmentation Index (PSI) (privacy fundamentalists, pragmatics, and unconcerned). To analyze the privacy paradox, they asked the participants whether they would be willing to share their mobile data (a) to help defeat COVID-19 and (b) in normal circumstances. They found that in all privacy groups, more participants are willing to share their data in the context of COVID-19: Looking at all participants combined, the percentage rose from 14% for those who would normally share their data to 61% for fighting COVID-19. Still, the numbers depend on the PSI: Participants classified as "privacy fundamentalists" were less willing to share compared to those who were classified as "unconcerned" [39]. The authors also found that of those who use the app (55% of 258), 18% mentioned privacy concerns, and 26% thought it could be used for surveillance [36]. Sharma et al. [30] reported as a typical response to why the participants installed the app that they were not concerned about their privacy and that other aspects, such as "a sense of responsibility," were more important.

Political Enforcement's or the Provider's Role in Decision-Making Aji et al. [3] conducted their survey in Malaysia and found that most participants were aware of issues the app had (e.g., access of the government to user's location and personal information). Yet, in Malaysia, citizens could be punished if not using the app and the authors conclude that privacy concerns seem to be overruled by the enforcement issued by the government.

Sharma et al. [30] report that trust in the provider plays an important role in the adoption and that privacy benchmarks and transparency in data policy are not enough, if the data are not handled by a trusted entity. For the global North, the authors found that over 50% of the participants believed that university research groups and healthcare providers would protect the collected data. On the other hand, over 50% did not put this amount of trust into industry startups and large corporations. The importance of the provider was also emphasized by Huang et al. [18]. They found that "most participants were very comfortable with the health authorities and the government as the solution provider."

Intention–Behavior Gap Third, people do not always follow what they intend to do, known as intention–behavior gap [31] (see also the chapter "From the Privacy Calculus to Crossing the Rubicon: An Introduction to Theoretical Models of User Privacy Behavior"). Jamieson et al. [20] examined this in the context of contact tracing apps by conducting a study among 290 Americans who were presented with a hypothetical app. Depending on the state the participants lived in, they were separated into those who had access to a contact tracing app and those who did not. They found that while privacy concerns influenced people's stated intention to install a contact tracing app, privacy was no longer an influential factor for installing an app. It seems that other considerations became more important.

4.2 Who Is Asked?

The participants and thus the recruitment can have a strong influence on the results. In the following, we will summarize what we found in the literature to have an effect on the results in the context of CTA.

Personal views on topics can influence how properties related to them are seen: Häring et al. [16] presented potential properties the CWA could have. One of them proposed that the RKI,[1] would see that users are not keeping a minimal distance to each other. If this were true, it would, in fact, be a problem for the users' privacy. When the participants were asked how this property would influence their decision to install or not install the app, the authors saw that 25% of the people who were very certain about their installation decision ("I will definitely install the app") would reconsider if this property were true. When looking at the group of participants who were not so certain ("I will probably install the app"), the number of participants who indicated not to like this property went up to 35%. This general tendency should be considered when recruiting participants or interpreting the results.

Cultural Differences Utz et al. [37], who conducted the same survey in Germany, USA, and China, found that participants from China were much more open to installing a contact tracing app in general, compared to the other countries, even when those have real consequences for the users' freedom, such as quarantine enforcement. When asked for general negative aspects of contact tracing apps, 37.5% of the Chinese participants mentioned either something positive or stated that they do not see any issues (compared to 11% in the USA and 12.6% in Germany).

Sharma et al. [30] conducted a survey from July 13, 2020, to August 13 with 261 participants from 27 countries. While having similar motivations, they found differences in the willingness to share personal information and with whom between the "Global North" and the "Global South," e.g., people from the North reported more often discomfort about sharing tracing data with large corporations.

Political Debates Häring et al. [16] asked for the participants' preferred political party and hypothesized that this could affect the willingness to install the German contact tracing app. The results do not indicate the preferred political party to be a factor; however, trust toward the government significantly influenced whether participants indicated they would use the app. Utz et al. [37] found something

[1] "The government's central scientific institution in the field of biomedicine"[34].

similar: A favorable rating of the state government and health authorities had a significant positive effect on the decision of whether the participants wanted to use an app. An unfavorable rating of the federal government had a negative influence.

For topics that are highly discussed in politics and where political stakeholderrs influence the outcome, it might thus be necessary to both ask participants about their general rating of these stakeholders, but also to understand and discuss the political and the societal situation in the country at the time of the study.

4.3 Privacy Concerns != Privacy Concerns

The literature shows that for many research questions, it is essential to know the participants' understanding of different concepts. Privacy concerns are likely based on participants' mental model of how things work, but that does not mean that the technical model actually has these problems; worse, it may even already mitigate the issues. Additionally, participants can have different understandings of what exactly would be privacy-invasive.

As an example of the latter, Häring et al. [16] reported that around 27% had concerns about their privacy in general. At the same time, many (around 58%) assumed that the app shows infected persons in the vicinity. It seems that many of the participants did not see this to be a problem for their privacy.

Utz et al. [37] asked for general negative aspects of contact tracing apps and also wanted to know why the participants did not use such an app at the time of the survey. Both questions were open-ended. While 40% of the German participants had privacy concerns in general, only around 10.5% mentioned privacy to be the reason why they currently use no such app. However, the unavailability of an app and the app being unnecessary were mentioned by 31% and 23% of the participants. It has to be noted that the official German app was released shortly after the survey. So while privacy concerns exist in general, this does not necessarily mean those privacy concerns also impact the rating of one specific app.

Lu et al. [21] further point out that in the context of contact tracing, there is a difference between informational privacy (e.g., control over personal information), social privacy (e.g., impression management), and interactional privacy (e.g., control of who to interact with). This might lead to seemingly inconsistent results: While participants, for example, mentioned that digital contact tracing would allow for anonymity, they were, at the same time, more willing to share data with a human contact tracer.

While the studies use privacy as a concept the way they ask about it is not always clearly defined. Two studies that used standardized approached were from Utz et al. [37] and Xie et al./Trestian et al. [35, 36, 39]. Utz et al. [37] used the IUIPC (2004) and found that, maybe counterintuitively, participants from China with higher privacy concerns (IUIPC: Control and Awareness) were more likely to use corona apps (not only tracing apps, but also apps with the purpose of symptom checks or quarantine enforcement). Xie et al. [39] found that, for Irish participants, the group of privacy fundamentalists (according to the PSI) was linked to the lowest adoption rate.

These are not per se contradicting results. One possible reason for that is that while the PSI and IUIPC have a similar goal, they cover slightly different aspects. The PSI concerns general privacy concerns, while the IUIPC focuses on online privacy. Also, both studies are conducted in different countries. Although the studies were conducted at similar times, the local situation may have influenced the results, e.g., Xie et al. report that the participants of their study report a change in their privacy concerns during the pandemic.

These different studies highlight again that it is crucial to be specific when talking about and conducting privacy research. For some of the studies, it is not clear what participants understood under the umbrella of "privacy." They provide valuable insights, but the different, sometimes unspecified, notions of privacy are hard to link to standardized tests, such as the IUIPC.

5 Conclusion

This chapter looked at contact tracing apps designed to combat COVID-19 and gave a brief overview of used techniques. Several studies looked at how the population perceived the aspect of privacy and how privacy concerns impacted the intention to install a contact tracing app. Summarized, it can be said that the public discussion was not completely detached from users as studies indeed reported privacy concerns, some of which are well-founded. Other concerns, however, cannot be traced back to actual technology. One example of this is the belief that the government has access to the data, even if the app followed a decentralized approach. Privacy concerns (in general or related to the app) were often associated with a lower willingness to install and use a CTA. However, when participants had the feeling that the app is unnecessary, they would also not install the app, even without privacy concerns. Another aspect of privacy research in the contact tracing context were methodological aspects that might have an influence on the results, such as the timing of the study, the context (e.g., political discussions), and the recruitment of participants. While the topic of contact tracing apps may vanish from the public radar as the need shrinks, there are still open research questions. It is still unclear whether and how the insights and opinions of the people shift over time. Additionally, it is to be seen if and how the findings can be applied to other areas as well, e.g., for enhanced monitoring of the spread of other diseases.

References

1. A flood of coronavirus apps are tracking us. Now it's time to keep track of them. Retrieved July 22, 2022, https://www.technologyreview.com/2020/05/07/1000961/launching-mittr-covid-tracing-tracker/
2. ACM Digital Library. Retrieved July 22, 2022, https://dl.acm.org/
3. Aji, Z. M., Mohd Salleh, N. S., Zakaria, N. H., & Mohd Khalid, A. H. (2021). Are you aware of your data privacy? The case of digital contact tracing applications (MySejahtera). In *2021 7th International Conference on Research and Innovation in Information Systems (ICRIIS)* (pp. 1–6).
4. Apple and Google partner on COVID-19 contact tracing technology. Retrieved 20, 2022, https://www.apple.com/newsroom/2020/04/apple-and-google-partner-on-covid-19-contact-tracing-technology/
5. Baumgärtner, L., Dmitrienko, A., Freisleben, B., Gruler, A., Höchst, J., Kühlberg, J., Mezini, M., Mitev, R., Miettinen, M., Muhamedagic, A., Nguyen, T. D., Penning, A., Pustelnik, D., Roos, F., Sadeghi, A.-R., Schwarz, M., & Uhl, C. (2020). Mind the GAP: security & privacy risks of contact tracing apps. In *2020 IEEE 19th International Conference on Trust, Security and Privacy in Computing and Communications (TrustCom)* (pp. 458–467).
6. Bundesregierung denkt bei App um. Retrieved July 22, 2022, https://www.tagesschau.de/inland/coronavirus-app-107.html
7. China launches coronavirus 'close contact detector' app. Retrieved July 22, 2022, https://www.bbc.com/news/technology-51439401
8. Contact tracing. Retrieved June 8, 2022, https://dictionary.cambridge.org/de/worterbuch/englisch/contact-tracing
9. Contact tracing in the European Union: Public health management of persons, including healthcare workers, who have had contact with COVID-19 cases—fourth update. Retrieved June 8, 2022, https://www.ecdc.europa.eu/en/covid-19-contact-tracing-public-health-management.
10. Contact Tracing Joint Statement. Retrieved July 22, 2022, https://www.esat.kuleuven.be/cosic/sites/contact-tracing-joint-statement/
11. Coronavirus disease (COVID-19): Contact tracing. Retrieved June 28, 2022, https://www.who.int/news-room/questions-and-answers/item/coronavirus-disease-covid-19-contact-tracing
12. COVID-19 and Your Health. Retrieved June 8, 2022 https://www.cdc.gov/coronavirus/2019-ncov/daily-life-coping/contact-tracing.html
13. Dooley, S., Turjeman, D., Dickerson, J. P., & Redmiles, E. M. (2020). Field evidence of the effects of privacy, data transparency, and pro-social appeals on COVID-19 app attractiveness. In *CHI Conference on Human Factors in Computing Systems*, CHI '22. Association for Computing Machinery.
14. DP3T - Decentralized Privacy-Preserving Proximity Tracing. Retrieved July 22, 2022, https://github.com/DP-3T/documents
15. Eventregistrierung in der Corona-Warn-App. Retrieved July 22, 2022, https://www.bundesregierung.de/breg-de/themen/coronavirus/corona-warn-app-version-2-0-1889868
16. Häring, M., Gerlitz, E., Tiefenau, C., Smith, M., Wermke, D., Fahl, S., & Acar, Y. (2021). Never ever or no matter what: Investigating adoption intentions and misconceptions about the Corona-Warn-App in Germany. In *Seventeenth Symposium on Usable Privacy and Security (SOUPS 2021)* (pp. 77–98). USENIX Association.
17. Hinch, R., Probert, W., Nurtay, A., Kendall, M., Wymant, C., Hall, M., Lythgoe, K., Cruz, A. B., Zhao, L., Stewart, A., Ferretti, L., Parker, M., Meroueh, A., Mathias, B., Stevenson, S., Montero, D., Warren, J., Mather, N. K., Finkelstein, A., Bonsall, D., and Fraser, C. (2020). Effective configurations of a digital contact tracing app: A report to NHSX (p. 29).
18. Huang, Y., Obada-Obieh, B., Redmiles, E. M., Lokam, S., and Beznosov, K. (2022). COVID-19 information-tracking solutions: A qualitative investigation of the factors influencing people's adoption intention. In *ACM SIGIR Conference on Human Information Interaction and Retrieval*, CHIIR '22 (pp. 12–24). Association for Computing Machinery.

19. IEEE Xplore. Retrieved July 22, 2022, https://ieeexplore.ieee.org/Xplore/home.jsp
20. Jamieson, J., Epstein, D. A., Chen, Y., & Yamashita, N. (2022). Unpacking intention and behavior: Explaining contact tracing app adoption and hesitancy in the United States. In *CHI Conference on Human Factors in Computing Systems*, CHI '22. Association for Computing Machinery.
21. Lu, X., L. Reynolds, T., Jo, E., Hong, H., Page, X., Chen, Y., & A. Epstein, D. (2021). Comparing perspectives around human and technology support for contact tracing. In *Proceedings of the 2021 CHI Conference on Human Factors in Computing Systems*, CHI '21. Association for Computing Machinery.
22. Mainzer Polizei nutzte Daten aus Luca-App ohne Rechtsgrundlage. Retrieved July 22, 2022, https://www.swr.de/swraktuell/rheinland-pfalz/mainz/polizei-ermittelt-ohne-rechtsgrundlage-mit-daten-aus-luca-app-100.html
23. Mozur, P., Zhong, R., and Krolik, A. (2020). In *Coronavirus fight, China gives citizens a color code, with red flags*. Retrieved September 26, 2022, https://www.nytimes.com/2020/03/01/business/china-coronavirus-surveillance.html
24. No, coronavirus apps don't need 60% adoption to be effective. Retrieved July 22, 2022, https://www.technologyreview.com/2020/06/05/1002775/covid-apps-effective-at-less-than-60-percent-download/
25. Open-Source-Projekt Corona-Warn-App – FAQ. Retrieved June 8, 2022, https://www.coronawarn.app
26. Pan-European Privacy-Preserving Proximity Tracing. Retrieved July 22, 2022, https://web.archive.org/web/20200409221119/https://www.pepp-pt.org/
27. Publicly-available Exposure Notifications apps. Retrieved July 29, 2022, https://developers.google.com/android/exposure-notifications/apps
28. Redmiles, E. M. (2020). User concerns & tradeoffs in technology-facilitated COVID-19 response. *Digital Government: Research and Practice, 2*(1), 1–12.
29. Shahroz, M., Ahmad, F., Younis, M. S., Ahmad, N., Boulos, M. N. K., Vinuesa, R., & Qadir, J. (2021). COVID-19 digital contact tracing applications and techniques: A review post initial deployments. *Transportation Engineering, 5*, 100072.
30. Sharma, T., Islam, M. M., Das, A., Haque, S. M. T., & Ahmed, S. I. (2021). Privacy during pandemic: A global view of privacy practices around COVID-19 apps. In *ACM SIGCAS Conference on Computing and Sustainable Societies*, COMPASS '21 (pp. 215–229). Association for Computing Machinery.
31. Sheeran, P. (2002). Intention–behavior relations: A conceptual and empirical review. *European Review of Social Psychology, 12*(1), 1–36.
32. Singapore reveals Covid privacy data available to police. Retrieved July 22, 2022, https://www.bbc.co.uk/news/world-asia-55541001
33. Speeding up detection to slow down Ebola: Smartphone app is game-changer for contact tracing in hotspots in the Democratic Republic of the Congo. Retrieved July 22, 2022, https://www.afro.who.int/news/speeding-detection-slow-down-ebola-smartphone-app-game-changer-contact-tracing-hotspots
34. The Robert Koch Institute. Retrieved July 22, 2022, https://www.rki.de/
35. Trestian, R., Celeste, E., Xie, G., Lohar, P., Bendechache, M., Brennan, R., & Ta, I. (2022). The privacy paradox—investigating people's attitude towards privacy in a time of COVID-19. In *2022 14th International Conference on Communications (COMM)* (pp. 1–6).
36. Trestian, R., Xie, G., Lohar, P., Celeste, E., Bendechache, M., Brennan, R., & Tal, I. (2021). PRIVATT—a closer look at people's data privacy attitudes in times of COVID-19. In *2021 IEEE International Mediterranean Conference on Communications and Networking (MeditCom)* (pp. 174–179).
37. Utz, C., Becker, S., Schnitzler, T., Farke, F. M., Herbert, F., Schaewitz, L., Degeling, M., & Dürmuth, M. (2021). Apps against the spread: Privacy implications and user acceptance of COVID-19-Related smartphone apps on three continents. In *Proceedings of the 2021 CHI Conference on Human Factors in Computing Systems*, CHI '21. Association for Computing Machinery.

38. Web of Science. Retrieved July 22, 2022, https://www.webofscience.com/
39. Xie, G., Lohar, P., Florea, C., Bendechache, M., Trestian, R., Brennan, R., Connolly, R., & Tal, I. (2021). Privacy in times of COVID-19: A pilot study in the republic of Ireland. In *The 16th International Conference on Availability, Reliability and Security*, ARES 2021. Association for Computing Machinery.
40. Zampedri, G. (2021). To download, or not to download, that is the question: Investigating Belgian residents' motivation to download or not download the COVID-19 contact-tracing app Coronalert. In *2021 14th CMI International Conference—Critical ICT Infrastructures and Platforms (CMI)* (pp. 1–5).

Privacy Perception and Behavior in Safety-Critical Environments

Enno Steinbrink, Tom Biselli, Sebastian Linsner, Franziska Herbert, and Christian Reuter

1 Introduction

Privacy and security are becoming increasingly important due to the developments in the area of Big Data and the growing social importance of online services. Since the perception of missing privacy or security by users can be reflected in changes in behavior (or maybe not, as can be seen in the privacy paradox, see also the chapter "From the Privacy Calculus to Crossing the Rubicon: An Introduction to Theoretical Models of User Privacy Behavior"), part of the research is concerned with contextual factors that can moderate the relationship between perception and behavior. Especially in areas where data leaks or security breaches could lead to serious consequences, it is worth taking a closer look. In these critical environments, whether in personally sensitive contexts, in the context of crises, or in the area of critical infrastructure, it becomes apparent that the frequently cited privacy paradox does not have general validity.

Helen Nissenbaum's [37] theory of Privacy as Contextual Integrity already describes the idea that the adequacy of privacy is linked to specific contexts that have to be considered from the perspective of governance. According to this view, norms of appropriateness and information flows, among others, form the context in which the realm of acceptable disclosure can be conceptualized. Relevant parameters, which determine the acceptability of information disclosures, thereby include information type, involved actors, and transmission principles. From this

E. Steinbrink (✉) · T. Biselli · S. Linsner · C. Reuter
Technical University of Darmstadt, Darmstadt, Germany
e-mail: steinbrink@peasec.tu-darmstadt.de; biselli@peasec.tu-darmstadt.de; linsner@peasec.tu-darmstadt.de; reuter@peasec.tu-darmstadt.de

F. Herbert
Ruhr-University Bochum, Bochum, Germany
e-mail: franziska.herbert@rub.de

N. Gerber et al. (eds.), *Human Factors in Privacy Research*,
https://doi.org/10.1007/978-3-031-28643-8_12

point of view, it becomes understandable that attitudes and behaviors toward privacy protection are not static, but strongly situation-dependent. In the following chapter, these aspects will be highlighted from the perspective of the users by presenting examples of relevant research on these topics. Since in safety-critical environments the aspect of security behavior is often the more salient aspect in research, the first question to be addressed is to what extent security and privacy behavior are interrelated and how these terms can best be conceptualized. In addition, we will specifically look at the issue of context-dependency, namely what kind of data is perceived as private and under which circumstances and with whom users are willing to share this data.

Subsequently, we will look at how privacy is perceived in safety-critical environments. For this purpose, on the one hand, a study will be presented that deals with the question to what extent privacy is a relevant aspect in the smartphone and Internet use of asylum seekers during their flight and which strategies emerge to protect digital privacy during the journey. Afterward, a study is presented that deals with privacy within agriculture, as part of the critical infrastructure. This will shed light on how privacy concerns impact the adoption of digital technologies.

2 On the Relationship Between Cyber Privacy and Security Behavior

How Are Cyber Privacy and Security Behavior Interrelated? In today's digital world, appropriate privacy and security behavior are more imperative than ever. Yet, the precise interrelationship between privacy and security behavior is still unclear, as it is rarely addressed in the relevant literature. To date, when it comes to the precise nature or characteristics of the relationship between privacy and security, there is no general consensus. However, against the background of effectively improving both privacy and security behavior in society, it is essential to gain an accurate understanding of how the two are interrelated in different contexts, where conceptual similarities or differences exist, as well as whether privacy and security behavior are similarly influenced by certain factors. Building on a better understanding of these interrelationships is the only way to ensure the development of adequate interventions and software that provide support to users in enhancing their privacy and security.

In general, privacy refers to the prevention of exposure of sensitive information about individuals or groups. This includes, among other things, the nondisclosure of behavior, communications, and descriptive personal data [41]. The general notion of the term privacy today is still quite close to Westin's early definition, which described privacy as "the claim of individuals, groups, or institutions to determine for themselves when, how, and to what extent information about them is communicated to others" [51]. However, maintaining privacy in a rapidly changing digital environment is much more difficult today. This may be one of the reasons why there is still no general consensus on the exact scope of the concept of privacy.

IT security, on the other hand, refers to the protection of computer systems from theft and damage of hardware, software, and information as well as the disruption of services they are supposed to provide [34]. A good conceptualization of this protection is provided by the so-called CIA triad: secure IT systems should therefore maintain confidentiality, integrity, and availability [40]. Confidentiality hereby refers to the prevention of unauthorized viewing, integrity to the unauthorized modification, and availability to the preservation of access [40]. These definitions suggest that security may, but may not completely, encompass the privacy domain. There is a particular overlap in the factor confidentiality since unauthorized viewing is associated with both unauthorized access as a security breach and the possible exposure of sensitive information about individuals as a privacy breach. On the other hand, integrity and availability define characteristics that may be distinguished from privacy more clearly.

In order to illuminate the relationship between privacy and security, in a previous study we examined privacy and security behavior in connection with certain socio-demographic factors (gender, age, educational background, political ideology). Within a representative survey, people in Germany ($N = 1219$) were asked to report about their privacy and security behavior regarding the private use of digital devices [5]. The survey evaluation indicates that there is only a low correlation between self-reported privacy and security behavior. Furthermore, the two concepts are differently influenced by certain socio-demographic factors. For example, age and education have a significant impact on security behavior in that older and more educated people show higher security behavior. Such correlations, however, could not be established for privacy behavior. Moreover, for the factor political ideology, there was no relation found with either privacy or security behavior.

On this basis, the often-presumed inherent connection and interrelationship between privacy and security behavior must be put into question. With view to the overall research landscape in this subject area, our study is in stark contrast to the results of many other studies, which generally and interchangeably speak of privacy and security [23, 28, 42]). With view to these studies, there is the risk that false connections are inferred and thereby, e.g., security aspects are wrongly attributed to privacy improvements, when in fact they are only appropriate for enhancing security. This understanding of the relevance of the appropriate distinction between the two concepts can potentially be relevant for both the education of individuals regarding private privacy and security behavior and software developers in terms of the proper protection of either privacy or security.

The results of this current study contribute to the existing body of literature, with particular reference to the investigation of privacy and security behaviors in contrast to people's corresponding attitudes toward these concepts. Our findings are consistent with previous research indicating that personality traits have different influences on attitudes toward privacy and security and that in this regard, the correlation between respective privacy and security attitudes is marginal [14, 15].

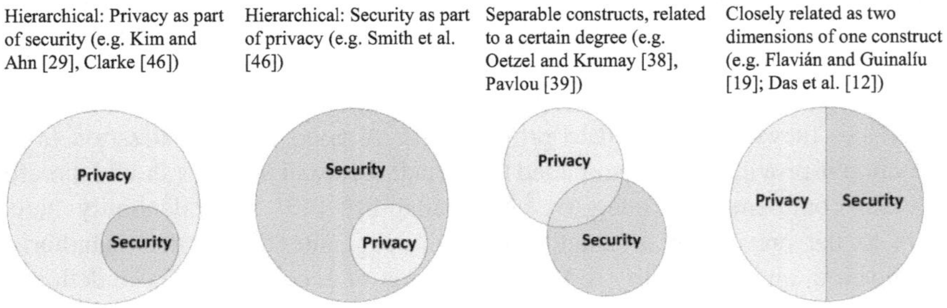

Fig. 1 Conceptualizations of the relationship between privacy and security proposed in the literature (as depicted in [5], inspired by Hurlburt [26])

Since a general consensus about the relation between privacy and security is lacking, different studies on the subject implicitly or explicitly suggest hierarchical relationships in which security is either part of privacy [46], privacy is part of security [6, 7, 29], the two are separate domains that gradually overlap [38, 39], or both are just different but related dimensions of the same construct [12, 19] (see Fig. 1). In the empirical study described earlier, the observable correlation between privacy and security was low, but not totally absent. However, this could not be attributed to certain demographic factors. Therefore, it is unclear at which level privacy and security are connected and where exactly the common ground between the two concepts might lie. In order to address this issue of comprehensively contrasting both concepts, we fall back on the technology threat avoidance theory (TTAT). TTAT examines cognitive processes related to threat assessment—including the perception of susceptibility and severity—and coping evaluation that significantly influence subsequent behavior regarding IT threats [32]. TTAT does not differentiate between IT threats associated with privacy and those associated with security. However, we suspect that perceived security could be a dimension on the basis of which different influences on privacy and security could be observed. In other words, the observation of high privacy behavior can give a direct indication of the existence of a high level of perceived severity, as only a person who is concerned about the collection of personal data would show corresponding behavior. With view to security on the other hand, the consequences of insufficient behavior are much more immediately noticeable, e.g., with regard to computer viruses, than with regard to the more abstract risks of inadequate privacy protection. For this reason, a person's security behavior might be high and their privacy behavior low at the same time. Building on this, TTAT suggests that a common factor for both might lie in the complete avoidance of certain threats associated with technology. While this would only explain a low correlation, there may be differences in privacy and security behavior depending on a person's underlying beliefs for particular aspects of this common factor, such as accurately assessing an IT threat by evaluating the corresponding perceived severity. In this context, it would be very plausible that age and educational background have an influence on privacy and security behavior—which was confirmed by our empirical study.

Based on these findings so far and given the marginal correlation between the two concepts and the ambiguous influence of certain demographic factors, the relation cannot yet be clearly determined. While it is not obvious whether privacy and security represent two hierarchical, merely overlapping, or even interrelated approaches, against the background of TTAT, we propose that the link between privacy and security might be best illustrated by the picture of different but related dimensions of the same construct. According to TTAT, this link is probably best visible by looking at the behavior of avoiding IT threats. In summary, the aspects outlined demonstrate the urgent need for a fine-grained differentiation of privacy and security in order to effectively influence corresponding behavior in the future.

3 Awareness on Data Sharing Functionalities and Acceptance of Private Data Sharing

Which kind of data is perceived as private data? When and with whom are people willing to share their private data?

Today, ever greater amounts of data are being produced, stored, and shared. While in some cases people share their data intentionally, e.g., by the use of social media or messengers, beyond that, much more data are shared by smartphones if certain data sharing functionalities, such as GPS, Bluetooth, or Wi-Fi, are activated. This raises several questions, including whether people are generally aware of the data sharing in public spaces, whether they actively switch on and off certain data sharing functionalities, and which kind of data people want to share with whom. By conducting a representative online survey ($n = 980$) and face-to-face-interviews ($N = 58$) with smartphone users in Germany, we investigated self-reports as well as actual data sharing practices (see [25]). The results provide insights into the circumstances in which private data are shared voluntarily, conditionally, or involuntarily.

Many study participants classified all examined data types (name, address, date of birth, bank details, identity card number, personal files, personal location data, personal communication data) as private, whereby bank data in particular were considered as private by nearly all participants. Related research suggest that women are generally more cautious with regard to their private data than men [17, 35, 44, 45]. Moreover, people who care about data protection might also have preferences for certain software. Hence, prior to the study, we presumed a correlation between the classification of data as private and the installed operating system as well as the gender of the study participants. Based on the data collected, only the latter could be confirmed, as female respondents were more likely to consider data as private than male participants, for all data types. With view to the factor of socio-cultural characteristics, other studies have shown that countries differ

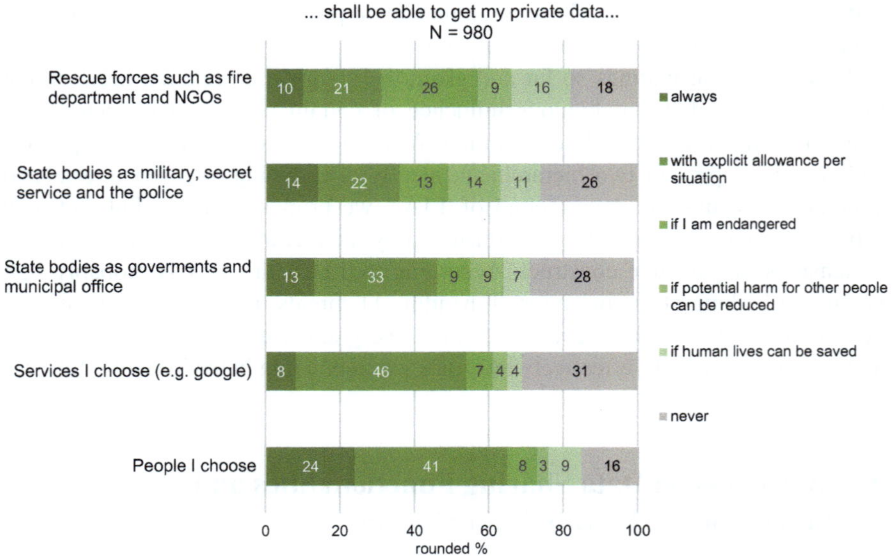

Fig. 2 Answers to "who shall be able to view my private data in which cases? Online survey from [25]

in terms of the use of social media during emergencies [43], privacy concerns [4], trust in social network sites [30], and the openness in using COVID-19 related apps [48] (see also the chapter "Privacy Research on the Pulse of Time: COVID-19 Contact-Tracing Apps"). In comparison to respondents asked in other European countries [43], the participants in our German sample showed a lower use of social media. Compared to study participants in the USA [30], Germans seem to have a lower level of trust in social media. However, Germans show higher acceptance regarding COVID-19 related apps than US participants, but lower acceptance than Chinese respondents [48]. Regarding the willingness to share private data, we found that this depends on the type of data, the data recipients, and the specific circumstances (see Fig. 2). This is consistent with Helen Nissenbaum's theory of contextual integrity [36] stating that the flow of private data is dependent on the specific attributes (i.e., data types), actors (i.e., data receiver), and transmission principles. This indicates that attitudes and behavior toward data sharing are highly volatile and difficult to generalize across all types of data and actors involved. With regard to our findings, in terms of data sharing with services or acquaintances, many study participants (46%, 41%) stated that they would want to decide this based on the specific situational context. Prior to the study, we assumed that some people would frequently share their private data and keep Wi-Fi, GPS, and Bluetooth activated simply for convenience. However, while we found this was true for more than 10% of respondents who would always share their data, more than 16% of study participants said they would never agree to share their private data, in each case regardless of the recipient. The services with the highest percentage of people not wanting to share their data with were Google and similar services, which could be due to a lack of incentives or a perceived deficit of transparency and trust. Our results are in line with other studies that show that users care about both the recipient and transparency about the data shared [1, 10].

According to the study participant's self-assessment, most would be more likely to share personal data if they themselves were at risk than if other people were at risk. Regarding the latter aspect, 39% of the respondents stated that their data sharing behavior would not change if someone they knew were in danger, and however, more than 50% of the people asked would be willing to disclose more data in this case if these were shared with government agencies or emergency responders. In accordance with previous studies ***(e.g., [48, 49], this highlights that users want to retain control over their data and are generally more willing to disclose private data if they know who the recipients of this data will be and for what purpose it will be used.

Generally, most people are not inclined to always share their data, which however is assumed by some studies [21, 24]. In fact, our results indicate that between 16% and 31% (deviations with regard to certain recipients) of users would never be willing to share private data. Since some people generally see no value at all in data sharing [3], lack of transparency or control over personal data might therefore lead to limited or no use of certain applications [49]. Especially with regard to the disclosure of sensitive medical data, it was shown that this would only be considered acceptable if the societal benefits were extremely high [50].

With view to the activation status of data sharing functionalities, we also compared respondents' statements with the actual settings on their smartphones. Regarding the settings of Wi-Fi and GPS, we found only minor discrepancies (4%). However, about 17% of study participants were not aware that their Bluetooth function was activated. Building on this, we want to underline the importance of raising users' awareness about the potential benefits and risks involved in activating these functionalities and how personal data can be adequately protected.

4 Critical Environment I: Digital Privacy Perceptions of Asylum Seekers in Germany

How relevant is digital privacy for asylum seekers? How is privacy related knowledge acquired during the flight? What strategies emerge among asylum seekers to protect their digital privacy during the journey?

After examining how privacy and security behavior are related and what data is typically considered private, we want to shift the focus on privacy considerations in safety-critical environments. An example for such a safety-critical environment for individual users is the context of flight and displacement.

In light of the continuing and increasing migration movements within which numerous refugees try to come to Europe to seek asylum, the role of digital

Table 1 Identified strategies of asylum seekers to protect their digital privacy from Steinbrink et al. [47]. These strategies are characterized by specific protection behaviors that could be identified within the interviews

(1) Anonymity efforts	(2) Adaption of communication	(3) Adaption of user behavior	(4) Renouncement
• Anonymous purchase of a SIM card • Anonymous purchase of a smart-phone • Use of pseudonyms • VPN connection	• Minimization of communication • Selective communication • Code language	• Selective usage • Variety of apps • Delete data • Use of external storage media	• Disposal of smart-phone or SIM card • Hiding the smart-phone • Use of non-technical solutions (e.g., face-to-face communication)

infrastructures such as smartphones has gained on importance, since many asylum seekers frequently use these in order to get information or communicate with friends, family, or acquaintances [13, 47]. To be able to reach their target countries, asylum seekers rely heavily on the access to mobile Internet and online services. While the use of such technologies offers many possibilities, little is known about associated risks perceived by asylum seekers. We addressed this research gap by examining how asylum seekers use mobile information technologies during their flight to Europe, focusing particularly on potential privacy concerns faced by these users (see [47]). By conducting 14 qualitative interviews with asylum seekers in Germany, we especially wanted to investigate digital privacy perceptions and corresponding privacy protection behavior. We found that most asylum seekers are aware of several risks of using digital technologies during their journey, such as surveillance and related possible prosecution by states or other actors. Against this background, it can be observed that there are several strategies to deal with these risks, which is mainly shown in avoidance behavior, but also mitigation strategies (see Table 1). Since such behavior is caused by perceived lack of privacy and trust in certain applications and online services, the design of assistance apps and collaboration platforms should be specifically targeted at these needs.

Most of our results conducted from the interviews were in accordance with research in the field. With regard to the possession of phones, 11 out of 14 study participants stated to have owned a smartphone at least for some time, while the three remaining owned a simple mobile phone. These results are in line with insights stated by Emmer et al. [16]. The interview respondents reported of several incidents during their flights in which their smartphones were confiscated by smugglers or police officers. According to the respondents, they sometimes disposed of the smartphones themselves in order to prevent someone controlling it.

When asked for what purpose they mostly needed their smartphones, the respondents stated (1) GPS applications and (2) communication with relatives, friends, other asylum seekers, or smugglers as the most frequently used smartphone applications, which is in line with previous research [13, 16]. Specifically, online and offline maps were of great importance for the asylum seekers since the access to map applications was essential for their autonomy. These results are confirmed by the findings of Zijlstra and van Liempt [52], who state that map applications contribute to refugees' mobility and ability to cross borders since they are less reliant on smugglers. The crucial significance of smartphones, e.g., for planning and orientation, was also confirmed by several other studies, including [2, 13, 22].

The results of our study indicate that the challenges and threats associated with border controls may have an effect on the way smartphones are used. Here different user behavior is partly dependent on the origin and reason for fleeing of the respective person. This is partly supported by the findings of Gillespie et al. [22], who find a shift in refugees' digital practices in that they require online (in)visibility to protect themselves from detection, arrest, or deportation. We find that it is primarily asylum seekers who themselves experienced the negative impact of government surveillance and persecution in their countries of origin who have acquired a profound awareness of the importance of digital privacy. One interview respondent stated his digital privacy to be directly related to his family's well-being as one must always fear surveillance and arrest due to critical opinions or actions. This finding is supported by the International Rescue Committee [9] and Latonero and Kift [31]. Coles-Kemp and Jensen [8] found that asylum seekers trying to adapt to a new country primarily want to use the advantages of digital services and are less concerned with their data privacy. However, we discovered a direct link between the precarity of a situation and drastic user behavior which can lead up to abandonment of the smartphone or the digital services. This especially applies if data privacy is directly linked to glaring threats, such as the risk of detention or possibly life-threatening consequences. With regard to our interviewees, we could observe that their awareness for data privacy often relied on a specific experience or event. Beyond that, we suspect preconceptions and technological literacy as important drivers behind smartphone use. Taking these aspects more into account during the development of digital tools or the conceptualization of digital help offers could support asylum seekers in using ICT securely.

5 Critical Environment II: The Role of Privacy in Digitalization—Analyzing Perspectives of German Farmers

How relevant is digital privacy for small and medium enterprises? How does privacy affect the adoption of digital technology in agriculture?

While advances in technology bring many conveniences and benefits, they can disrupt entire sectors of society and reshape the fundamental ways in which we interact and work together. Especially changes in domains of critical infrastructure, such as food supply, therefore need to be carefully considered. In doing so and in light of current privacy research, we wanted to examine the effect of the advancing digitalization in the area of agriculture. For this purpose, we conducted 52 qualitative interviews with farmers in Germany (see [33]). Against the background of the introduction of digital tools and services in the sector of agriculture, great challenges arise, particularly for small- and medium-sized enterprises (SMEs) [11]. On the one hand, businesses have to follow the requirements of consumers and retailers, who demand transparency and information about products and respective supply chains [27]. On the other hand, sharing this kind of data with multiple actors along the supply chain also involves a privacy risk that should not be neglected [18, 20]. Especially in light of the hard competition in agriculture, in which small farms have to assert themselves against large players driving technological innovation, data becomes an important resource [20, 33]. Therefore, they have to be careful about who they give insight into their operational data and how much. While too much digital privacy could lead to a further weakening of the market position of small farms and serious consequences for the whole domain, rejection of digital tools will do the same. Here, we found that privacy behavior and the adoption of digital tools are mutually interrelated [33]. In the following, we outline the challenges associated with this. With view to the protection of privacy, digital data have to be adequately managed in order to achieve an appropriate balance between data dissemination and data protection [20, 33]. For this purpose, adequate infrastructure has to be provided which takes these requirements into account from the outset. The introduction of digital processes not only affects individual work steps but also permanently changes the entire profession of a farmer [11, 33]. The smaller the business, the more difficult and expensive it is to introduce automation and digitalization processes. Specifically, this concerns, e.g., the size of the farmland or the increased bureaucratic workload [20, 33].

The traditional farming job mainly consisted of manual work including field or barn work and the maintenance of farming equipment and machinery and only marginally included the planning of work steps and corresponding agreements with other actors. Today, however, with view to the large number of stakeholders and specialists in the field of modern agriculture, the latter aspect takes up a large part of the work. In the conducted interviews, some respondents even indicated that digitalization has not resulted in an easing of workload but has only increased office work. Furthermore, farmers can no longer manage the work of their business themselves but are heavily dependent on third parties for maintenance or data management, whereby sensitive data is potentially exposed to several actors. A further challenge concerns the comprehensive and time-consuming processes of data collection for automated machines. Here, one interviewee expressed his displeasure about the fact that this work step only brings little benefit for the own business compared to the large disclosure of sensitive data. The aspect of increased dependencies is also visible regarding customer retention in relation to manufacturers of digital tools. As farmers are often contractually bound to purchase all their digital tools from only one provider, this often leads to a so-called vendor lock-in effect. Yet another new challenge of modern agriculture concerns the dependencies—not only on the weather as in the past but also on quality network connections, mobile phone coverage, and satellite availability. Based on

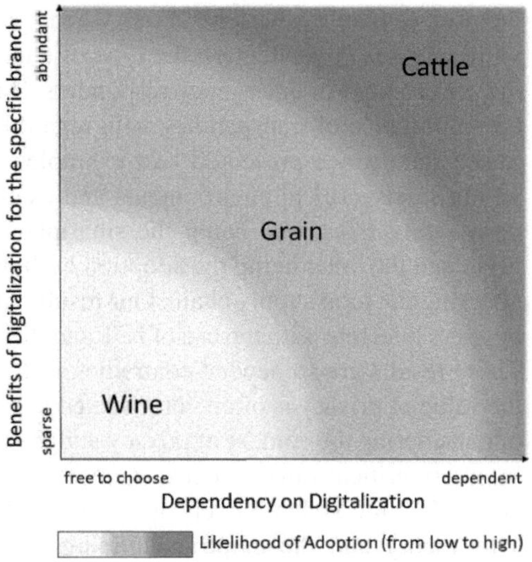

Fig. 3 Different agricultural subsectors and how they are affected by digitalization, influencing attitudes toward privacy and the likelihood of adopting new technologies as depicted in [33]

the outlined challenges, there is nothing surprising in the fact that digital processes are implemented by farms to very different extents. Here, our findings underline that both likelihood and extent of the integration of digitalization deviate with view to different subsectors in the domain (see Fig. 3) and are highly dependent on the anticipated benefit and the necessity of introduction based on competitive pressure or legal requirements. With regard to the impacts on privacy behavior, we found that farmers with advanced experience with the technology have also less reservations about it, which is reflected in the respective data management behavior. At the same time, the reliance on certain technologies may mean that farmers have no choice but to share their data. Interestingly, both factors seem to lower the inhibition threshold regarding privacy concerns, which again highlights the importance of privacy as decisive factor with view to the implementation of digital tools. What became clear through this study is that the advantageous adoption of digital processes by agricultural businesses not only involves financial and physical resources but also to a large extent time flexibility as well as willingness to share sensitive operational data.

6 Conclusion

In this chapter, we explored how privacy perceptions and behaviors are affected in safety-critical environments. We investigated the relationship between security and privacy behaviors and the results of our study suggested privacy and security behavior are distinct, but possibly two different dimensions of the same construct. Then, we considered the perception of sensitivity of data by end-users in Germany. Our study showed that users are more willing to provide data if they are at risk themselves than if others are at risk when it contributes to their safety. However, this willingness is dependent on the type of data and the party receiving the data (e.g., the government or emergency responders). Furthermore, the results pointed toward the importance of transparency with regard to what purposes the data are used for. Subsequently, we presented two examples of privacy perceptions and behaviors within safety-critical environments and examined the extent to which digital tools pose safety risks, one being the smartphone use of asylum seekers during their flight and the other being the adoption of digital tools within agriculture, potentially affecting the food supply chain. Our results suggest that the perception of these risks by users interfere with the use of ICT and the potential advantages associated with it. These results are somewhat contradictory to the privacy paradox, which states that the value of privacy is often not reflected in behavior. This highlights the importance of considering the context of privacy and the corresponding user behavior, especially in safety-critical environments. Consequently, the results presented can be well embedded into Helen Nissenbaum's theory [37] of Privacy as Contextual Integrity. This theory and the empirical findings highlighted suggest that privacy cannot be viewed as a static value, but that conformity to individual norms about what constitutes appropriate disclosure is variable and can alter how privacy is valued. Hence, it is crucial for future research to take contextual factors into account, when regarding privacy protection behaviors.

Acknowledgments This chapter builds upon own previously published work. Section 2 summarizes and is based on Biselli and Reuter [5], Section 3 on Herbert et al. 2021 [25], Section 4 on Steinbrink et al. [47], and Section 5 on Linsner et al. [33]. The research is funded by the Deutsche Forschungsgemeinschaft (DFG, German Research Foundation)—251805230/RTG 2050 and 236615297/SFB 1119 (CROSSING)—and by the German Federal Ministry of Education and Research and the Hessian Ministry of Higher Education, Research, Science, and the Arts within their joint support of the National Research Center for Applied Cybersecurity ATHENE.

References

1. Aldehoff, L., Dankenbring, M., & Reuter, C. (2019). Renouncing privacy in crisis management? People's view on social media monitoring and surveillance. In *Proceedings of the Information Systems for Crisis Response and Management (ISCRAM)*, València, Spain. ISCRAM Association.
2. Alencar, A. (2020). Mobile communication and refugees: An analytical review of academic literature. *Sociology Compass, 14*, e12802.
3. Balebako, R., Jung, J., Lu, W., Cranor, L. F., & Nguyen, C. (2013). "little brothers watching you": Raising awareness of data leaks on smartphones. In *Proceedings of the Ninth Symposium on Usable Privacy and Security. SOUPS '13*. Association for Computing Machinery.
4. Bellman, S., Johnson, E. J., Kobrin, S. J., & Lohse, G. L. (200). International differences in information privacy concerns: A global survey of consumers. *The Information Society, 20*, 313–324.
5. Biselli, T., & Reuter, C. (2021). On the relationship between it privacy and security behavior: A survey among German private users. In F. Ahlemann, R. Schütte, & S. Stieglitz (Eds.), *Innovation through information systems* (pp. 388–404). Springer.

6. Bubaš, G., Orehovaíki, T., & Konecki, M. (2008). Factors and predictors of online security and privacy behavior. *Journal of Information and Organizational Sciences, 32*, 79–98.
7. Clarke, R. (2009). Privacy impact assessment: Its origins and development. *Computer Law and Security Review, 25*, 123–135.
8. Coles-Kemp, L., & Jensen, R. B. (2019). Accessing a new land: Designing for a social conceptualisation of access. In *Proceedings of the 2019 CHI Conference on Human Factors in Computing Systems, 1–12. CHI '19*. Association for Computing Machinery.
9. International Rescue Committee. (2017). *Using ICT to facilitate access to information and accountability to affected populations in urban areas: A review of the serviceinfo and refugee.info platforms*. International Rescue Committee.
10. Consolvo, S., Smith, I. E., Matthews, T., LaMarca, A., Tabert, J., & Powledge, P. (2005). Location disclosure to social relations: Why, when, & what people want to share. In *Proceedings of the SIGCHI Conference on Human Factors in Computing Systems, 81–90. CHI '05*. Association for Computing Machinery.
11. Cravotta, S., & Grottke, M. (2019). Digitalization in German family firms—some preliminary insights. *Journal of Evolutionary Studies in Business, 4*(1), 1–25.
12. Das, S., Kim, T. H.-J., Dabbish, L. A., & Hong, J. I. (2014). The effect of social influence on security sensitivity. In *SOUPS '14: Proceedings of the Tenth USENIX Conference on Usable Privacy and Security* (pp. 143–157).
13. Dekker, R., Engbersen, G., Klaver, J., & Vonk, H. (208). Smart refugees: How Syrian asylum migrants use social media information in migration decision-making. *Social Media + Society, 4*, 2056305118764439.
14. Egelman, S., & Peer, E. (2015). Predicting privacy and security attitudes. In *ACM SIGCAS computers and society* (Vol. 45, pp. 22–28).
15. Egelman, S., & Peer, E. (2015). Scaling the security wall: Developing a security behavior intentions scale (SeBIS). In *Conference on Human Factors in Computing Systems–Proceedings* (pp. 2873–2882).
16. Emmer, M., Richter, C., & Kunst, M. (2016). Flucht 2.0: Mediennutzung durch flüchtlinge vor, während undnach der flucht.
17. Felt, A. P., Egelman, S., & Wagner, D. (2012). I've got 99 problems, but vibration ain't one: A survey of smartphone users' concerns. In *Proceedings of the Second ACM Workshop on Security and Privacy in Smartphones and Mobile Devices, 33–44. SPSM '12*. Association for Computing Machinery.
18. Ferris, J. L. (2017). Data privacy and protection in the agriculture industry: Is federal data privacy and protection in the agriculture industry: Is federal regulation necessary? Regulation necessary? *Science & Technology Minnesota Journal of Law*, 18, 309.
19. Flavián, C., & Guinalíu, M. (2006). Consumer trust, perceived security and privacy policy: Three basic elements of loyalty to a web site. *Industrial Management & Data Systems, 106*, 601–620.
20. Gandorfer, M., Schleicher, S., Heuser, S., Pfeiffer, J., & Demmel, M. (2017). Landwirtschaft 4.0—digitalisierung und ihre herausforderungen. *Ackerbau-technische Lösungen für die Zukunft* 9–19.
21. Gao, H., Liu, C. H., Wang, W., Zhao, J., Song, Z., Su, X., Crowcroft, J., & Leung, K. K. (2015). A survey of incentive mechanisms for participatory sensing. *IEEE Communications Surveys & Tutorials, 17*, 918–943.
22. Gillespie, M., Osseiran, S., & Cheesman, M. (2018). Syrian refugees and the digital passage to Europe: Smartphone infrastructures and affordances. *Social Media + Society, 4*, 2056305118764440.
23. Halevi, T., Lewis, J., & Memon, N. (2013). A pilot study of cyber security and privacy related behavior and personality traits. In *WWW '13 Companion—Proceedings of the 22nd International Conference on World Wide Web* (pp. 737–744).
24. Hann, I.-H., Hui, K. L., Lee, S.-Y. T., & Png, I. P. (2002). Online information privacy: Measuring the cost-benefit trade-off. In *ICIS*.

25. Herbert, F., Schmidbauer-Wolf, G. M., & Reuter, C. (2021). Who should get my private data in which case? Evidence in the wild. In *Mensch Und Computer 2021, 281–293. MuC '21*. Association for Computing Machinery.
26. Hurlburt, G. F., Miller, K. W., Voas, J. M., & Day, J. M. (2009). Privacy and/or security: Take your pick. *IT Professional, 11*, 52–55.
27. Kamath, R. (2018). Food traceability on blockchain: Walmart's pork and mango pilots with IBM. *The Journal of the British Blockchain Association 1. The British Blockchain Association* 1–12.
28. Kang, R., Dabbish, L., Fruchter, N., & Kiesler, S. (2015). "my data just goes everywhere:" User mental models of the Internet and implications for privacy and security. In *SOUPS 2015—Proceedings of the 11th Symposium on Usable Privacy and Security* (pp. 39–52).
29. Kim, M. S., & Ahn, J. H. (2006). Comparison of trust sources of an online market-maker in the e-marketplace: Buyer's and seller's perspectives. *Journal of Computer Information Systems, 47*, 84–94.
30. Krasnova, H., & Veltri, N. F. (2010). Privacy calculus on social networking sites: Explorative evidence from Germany and USA. In *43rd Hawaii International Conference on System Sciences* (pp. 1–10).
31. Latonero, M., & Kift, P. (2018). On digital passages and borders: Refugees and the new infrastructure for movement and control. *Social Media + Society, 4*, 2056305118764432.
32. Liang, H., & Xue, Y. (2009). Avoidance of information technology threats: A theoretical perspective. *MIS Quarterly, 33*, 71–90.
33. Linsner, S., Kuntke, F., Steinbrink, E., Franken, J., & Reuter, C. (2021). The role of privacy in digitalization—analysing the German farmers' perspective. In *Proceedings on Privacy Enhancing Technologies (PoPETs)* (Vol. 2021).
34. Mihajlov, M., Josimovski, S., & Jerman, B. (2011). A conceptual framework for evaluating usable security in authentication mechanisms—usability perspectives. In *Proceedings—2011 5th International Conference on Network and System Security, NSS* (pp. 332–336).
35. Moscardelli, D. M., & Divine, R. (2007). Adolescents' concern for privacy when using the Internet: An empirical analysis of predictors and relationships with privacy-protecting behaviors. *Family and Consumer Sciences Research Journal, 35*, 232–252.
36. Nissenbaum, H. (2009). *Privacy in context: Technology, policy, and the integrity of social life*. Stanford University Press.
37. Nissenbaum, H. (2004). Washington law review: Privacy as contextual integrity. *Washington Law Review, 79*, 119–157.
38. Oetzel, M. C., & Krumay, B. (2011). Differentiating privacy and security: A content analysis of B2C websites. In *17th Americas Conference on Information Systems 2011, AMCIS 2011* (Vol. 3, pp. 1891–1900).
39. Pavlou, P. A. (2011). State of the information privacy literature: Where are we now and where should we go? *MIS Quarterly: Management Information Systems*.
40. Pfleeger, C. P., Pfleeger, S. L., & Margulies, J. (2015). *Security in computing* (5th ed.). Pearson.
41. Pfleeger, S. L., & Pfleeger, C. P. (2009). Harmonizing privacy with security principles and practices. *IBM Journal of Research and Development, 6*, 1–6.
42. Redmiles, E. M., Kross, S., & Mazurek, M. L. (2019). How well do my results generalize? Comparing security and privacy survey results from MTurk, web, and telephone samples. In *IEEE Symposium on Security and Privacy, 2019-May:1326–1343*. Institute of Electrical and Electronics Engineers.
43. Reuter, C., Kaufhold, M.-A., Schmid, S., Spielhofer, T., & Hahne, A. S. (2019). The impact of risk cultures: Citizens' perception of social media use in emergencies across Europe. *Technological Forecasting and Social Change, 148*, 119724.
44. Rowan, M., & Dehlinger, J. (2014). Observed gender differences in privacy concerns and behaviors of mobile device end users. *Procedia Computer Science, 37*, 340–347.
45. Sheehan Bartel, K. (1999). An investigation of gender differences in online privacy concerns and resultant behaviors. *Journal of Interactive Marketing, 13*, 24–38.

46. Smith, H. J., Milberg, S. J., & Burke, S. J. (1996). Information privacy: Measuring individuals' concerns about organizational practices. *MIS Quarterly: Management Information Systems, 20*, 167–195.
47. Steinbrink, E., Reichert, L., Mende, M., & Reuter, C. (2021). Digital privacy perceptions of asylum seekers in Germany: An empirical study about smartphone usage during the flight. In *Proceedings of the ACM on Human-Computer Interaction 5* (pp. 1–24). Association for Computing Machinery
48. Utz, C., Becker, S., Schnitzler, T., Farke, F. M., Herbert, F., Schaewitz, L., Degeling, M., & Dürmuth, M. (2021). Apps against the spread: Privacy implications and user acceptance of covid-19-related smartphone apps on three continents. In *Proceedings of the 2021 CHI Conference on Human Factors in Computing Systems. CHI '21*. Association for Computing Machinery.
49. Van Kleek, M., Liccardi, I., Binns, R., Zhao, J., Weitzner, D. J., & Shadbolt, N. (2017). Better the devil you know: Exposing the data sharing practices of smartphone apps. In *Proceedings of the 2017 CHI Conference on Human Factors in Computing Systems*, CHI '17 (pp. 5208–5220). Association for Computing Machinery.
50. Watson, C., & Smeddinck, J. D. (2020). Unconsented data transfusions: Attitudes towards extracting personal device data for public health emergencies. In *Proceedings of the Conference on Mensch Und Computer, 205–209. MuC '20*. Association for Computing Machinery.
51. Westin, A. F. (1967). *Privacy and freedom*. Atheneum.
52. Zijlstra, J., & Liempt, I. (2017). Smart(phone) travelling: Understanding the use and impact of mobile technology on irregular migration journeys. *International Journal of Migration and Border Studies (IJMBS), 3*, 174–191.

Part IV
Solutions

Generic Consents in Digital Ecosystems: Legal, Psychological, and Technical Perspectives

Bianca Steffes, Simone Salemi, Denis Feth, and Eduard C. Groen

1 Challenge and Vision

Since the rise of digitization and the widespread use of digital services in everyday life, user interactions with technology have become increasingly intricate. The simple design of services used only by one client and one server at a time is long outdated. Modern digital services are typically divided into sub-services that perform very specific tasks: for example, online purchases are made via a central marketplace, i.e., an online platform that allows multiple (independent) asset providers to offer assets, such as goods or services; payments are processed via dedicated payment providers; and physical assets are delivered by an external (transport) service provider. Furthermore, platforms may include additional third-party services like RSS feeds or location services to enhance the user experience. The result is a digital ecosystem where every player benefits from participating. Koch et al. [25] provide deeper insights into the interplay between aspects and actors of a digital ecosystem. We define a digital ecosystem as a socio-technical system that brings together various independent providers and consumers of digital goods. A digital ecosystems service could be a website or a web service or even a locally installed software provided by a third party.

B. Steffes (✉)
Saarland University, Saarbrücken, Germany
e-mail: bianca.steffes@uni-saarland.de

S. Salemi
Saarbrücker Zentrum für Recht und Digitalisierung, Saarbrücken, Germany
e-mail: simone.salemi@zrd-saar.de

D. Feth · E. C. Groen
Fraunhofer IESE, Kaiserslautern, Germany
e-mail: denis.feth@iese.fraunhofer.de; eduard.groen@iese.fraunhofer.de

N. Gerber et al. (eds.), *Human Factors in Privacy Research*,
https://doi.org/10.1007/978-3-031-28643-8_13

In almost every digital ecosystem service, personal data of the users are processed. In the European Union, personal data is protected by the General Data Protection Regulation (GDPR; [15]) and may only be processed under the conditions such as stipulated by Article 6. A common legal basis for the processing of personal data in digital services is the *consent* of the data subject to the processing of their personal data. Consent is defined as "any *freely given*, *specific*, *informed* and *unambiguous* indication of the data subject's wishes" (Article 4 (11) GDPR). Because of the requirement that the consent be *specific*, it is pivotal from a legal perspective that users consent to each individual instance of data processing.

One example that shows the consequences of constantly repeated requests for consent are cookie banners used on websites. As obtaining consent concerning cookies is addressed in Art. 5 (3) ePrivacy Directive of the European Union and not in the GDPR, this only serves as an example to show how data subjects usually act when they are overwhelmed by the number of requests asking for their consent in the processing of their data. Unlike the GDPR, the ePrivacy Directive (just as the upcoming ePrivacy Regulation) applies to the processing of any electronic communication data arising from the provision or use of electronic communication services, as well as information related to the end-users terminal equipment (Art. 2 (1) ePrivacy Regulation (draft)). As cookies do not always relate to the processing of personal data, the consent concerning cookies is regulated in the ePrivacy Directive (or soon in the ePrivacy Regulation). In order to get rid of cookie banners, users often just accept everything instead of making an informed decision. This phenomenon is known as *cookie fatigue* [21]. The consents that are collected on websites in a digital ecosystem, for example to finish an ordering process, may lead to a problem comparable to the cookie fatigue. In a digital ecosystem, there are usually many different actors involved in one process, so there might be more than one consent necessary to complete one process. It is therefore conceivable that data subjects will act the same way as they act when too many cookie banners show up: they will just accept everything in order to proceed as quickly as possible. The way such requests are usually made (i.e., by cookie banners) cannot be considered to be an *informed* consent as demanded by Article 4 (11) GDPR [26, p. 407]. Besides cookies, websites or digital ecosystem services might also have other features for which explicit consent must be obtained. For example, weather forecast services frequently make use of the user's current location obtained from their device. In this case, no other legal basis than a consent is possible.

The more specific consents are, the more elaborate the requests for consent are, causing greater informational and cognitive load on the user, which negatively affects usability [34]. In practice, this causes the number of consents to quickly become unmanageable for users. Research in this field has provided a number of solutions on how to request and receive consent in a consolidated and simplified way, yet these approaches have failed to fulfill their purpose as they were never widely adopted and integrated into services. We argue that digital ecosystems as self-contained systems need only a smaller and more manageable number of services to implement a possible solution for communicating consent. Therefore, we propose (Sect. 2) and legally assess (Sect. 3) *generic consents* as a possible solution for requesting and handling user consents in the scope of digital ecosystems. Focusing on the consumers' needs, we also present our ideas of a trial period (Sect. 4) that allows the users to test and gain trust in a digital ecosystem service before giving further consent. Next, we discuss implementation options that allow us to demonstrate the general technical feasibility of our proposed solutions (Sect. 5), and we conclude with a discussion on the practicality and advantages of our solutions compared to previous work (Sect. 6).

2 Generic Consents

Consents form one of the bases for processing referred to in Article 6 GDPR, besides contracts, legal obligations, vital interests, public interests, and legitimate interest. In the following, we will primarily address two of the required characteristics of consents, namely, being specific and unambiguous. In a digital ecosystem, a consent relates to the interplay between an asset provider, a digital ecosystem service, a data category, a processing type, and a purpose.

For a consent granted by a user to be considered an **explicit consent**, four conditions have to be met: (1) it is granted to one service or one provider, (2) it is one specific processing permission, (3) it pertains to one concrete data item, and (4) it applies to one specific purpose. To counteract the problem of users getting overloaded by the plethora of explicit consents to cover the variability of these aspects, we propose the use of **generic consents** that can apply to several or selected groups of providers, services, data categories, processing types, or purposes (cf. Fig. 1). Note: The term *explicit consent* does not impose any restrictions on the way consent may be given. Both generic and explicit consents may be declared explicitly (e.g., in a written declaration of intent) or implicitly (e.g., through conduct implying an intent).

In general, the use of generic consents cannot be sustained because a consent that is too broad is not lawful. In the context of digital ecosystems, however, we argue that when properly implemented, generic consents can be used to express the users' data protection demands in an abstract way while still being specific

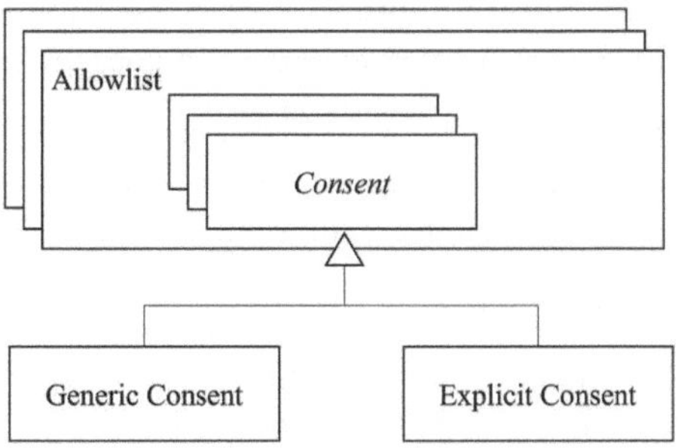

Fig. 1 Concept of generic consents and allowlists

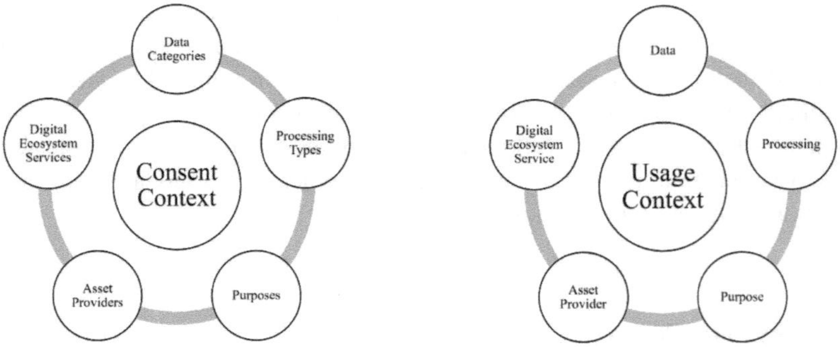

I grant my to consent to *all marketplace sellers* for the *disclosure of my location data* to *delivery services rated 4.5+ stars* for the purpose of *asset delivery*.

SuperShop24 wants to *transfer* your *address to DHL for delivering ordered goods*.

Fig. 2 Example: consent context vs. usage context

and unambiguous *enough* to be GDPR-compliant (cf. Sect. 3) and still being manageable. It is also conceivable that a whole set of consents is proposed to the user, for example, by neutral bodies or by the platform itself. These are essentially **allowlists**, which of course must always be compiled in the interests of the user. Therefore, this approach is not suitable for all digital ecosystems, but only for those where the platform provider is particularly trustworthy (e.g., data trustee) or where trustworthy interest groups exist to take care of this.

Determining whether a consent exists for a specific processing purpose is not trivial when using generic consents. In particular, one must always consider the digital ecosystem's state, which we call *context*. To determine whether consent does or does not exist, the consent must be interpreted with respect to two temporally disjunct circumstances: the context when it was given—"consent context"—and the context of the current activity—"usage context"—(cf. Fig. 2). Then the fundamental question is whether the usage context is covered by the consent context. For each aspect, we will argue in the following how the characteristics of digital ecosystems can be used to achieve compliant generic consents.

Data First, we must answer the question of whether the data to be processed (usage context) is covered by the given consent (consent context).

Design Challenges Different asset providers may use different terms for the same data category (e.g., geo-data vs. location data). In a digital ecosystem, this could be resolved through central standardization, which would also boost comprehensibility. Another problem is that categories are often related to each other. In the simplest case, this results in a hierarchy. For example, consent for the super-category "location data" should also apply to the subcategories "GPS-based location data" and "network-based location data." The resulting taxonomy defines the consent context for the data.

However, it might not be possible to establish a clean, overlap-free hierarchy in all digital ecosystems. In practice, one and the same data item can be assigned to several categories. Accordingly, this might give rise to the problem that generic consents and objections may contradict each other. If a data item is in categories A and B and there is only consent for A, the consent can be regarded as given. However, if there is an objection for B, the objection takes precedence over the consent given in A.

Asset Provider The next aspect to be assessed is whether the provider that wants to process a data item (usage context) is covered by the given consent (consent context). Asset providers within the same digital ecosystem can be categorized easily in most cases. For example, consent could apply to all "payment providers" or all "shipping service providers". The categorization of providers could be performed by the operator of the digital ecosystem as part of the on-boarding process, which should make this a relatively simple exercise compared to the data categories. The resulting taxonomy thus defines the consent context in a structured manner.

Design Challenges This notion relies on the assumption that asset providers in a given category are so similar that a user would always treat them in the same way when consenting. However, this decision could go beyond a simple categorization. For example, users could base their consent on the provider's reputation (e.g., "consent only for companies rated 4.5 stars or higher"). The set of providers that fulfill this criterion is not fixed and must thus be reflected by the usage context. Since there are also various non-rational and non-measurable criteria that we cannot formally cover, there should be the possibility to exclude a provider explicitly from the generic consent ("all except provider x"). Another concern is whether the consent also applies for providers that joined the ecosystem after the generic consent was given. One could argue that these new players are not covered because they were not part of the consent context. On the other hand, it is likely in the interest of the users, and the basic idea of our approach is that they do not have to reconsider their consent every time a new asset provider joins the digital ecosystem. This also contradicts the flexibility and openness that are central to digital ecosystems. Naturally, when in doubt, users could also be given a choice of whether they want new asset providers to automatically be included in their generic consent. Another viable compromise would be to ask the users to reconfirm a previously given consent. If a user decides to give consent anew, the consent context would then get updated.

Digital Ecosystem Service Digital ecosystem services can strongly vary between digital ecosystems, but they can usually be classified by their characteristics. The resulting taxonomy thus defines the consent context for the providers.

Design Challenges Some aspects of a service offering can be rather dynamic, which will need to be taken into account in the consent. For example, consent could be related to the service level offered to specific users (e.g., "24/7 phone support") or to temporary offers (e.g., "free returns only this weekend"). The digital ecosystem services that fulfill these criteria are not fixed and must thus be reflected by the usage context.

Processing Type Both generic and explicit consents may specify the allowed processing type(s). According to Article 4 (2) GDPR, processing types includes "collection, recording, organisation, structuring, storage, adaptation or alteration, retrieval, consultation, use, disclosure by transmission, dissemination or otherwise making available, alignment or combination, restriction, erasure or destruction".

Design Challenges Even though these processing types could be used, it must be checked whether the users understand them and whether the users' mental models (i.e., their own individual understanding, see also the chapter "Achieving Usable Security and Privacy Through Human-Centered Design") fit the actual meaning of the processing type. Thus, a clear, understandable, and ecosystem-wide definition of each processing type is essential.

Purpose Finally, the core of any consent is the intended purpose of the processing. Limiting data processing to only those purposes that are defined in advance is a consequence of the so-called *purpose limitation principle*. As demanded in Art. 5 (1) (b) GDPR, personal data shall only be collected for specified, explicit, and legitimate purposes. Changes of the primary purpose are only lawful if they comply with the prerequisite of Art. 6 (4). The purpose typically relates to specific business processes, such as ordering, payment, or advertising.

Design Challenges These can vary between asset providers. However, the number of purposes in a digital ecosystem aimed at the users is actually quite limited. The operator of the digital ecosystem should therefore define the list of purposes for which generic consent can be obtained. The resulting taxonomy thus defines the consent context for purposes. Here, we do not assume a usage context. If the definition of a purpose changes, either the consent becomes invalid or the old definition (i.e., the consent context) has to be applied.

To summarize, it is primarily the task of the operator of the digital ecosystem to define precisely how the "context" of a consent is defined in their digital ecosystem. This task might sound like a lot of work, but we suggest to not over-engineer this distinction. Because users will not be giving their generic consent for special cases, it should be sufficient to cover the most common cases in the categories. However, these cases should be clearly defined in order to avoid (unintentional or intentional) assignment of consents to categories for which they were not intended, which would legally invalidate these consents.

3 Legal Assessment

As described above, the GDPR provides a well-defined set of requirements for lawful requests for consent: the basic requirements of Art. 4 (11) GDPR, as well as their modifications described in Arts. 6 (1) (a), 7, and 9 (2) (a) GDPR.

Among other things, consent may only be given by a data subject when they have knowledge of the full facts and circumstances. This results from the demand for the consent to be informed, specific, and unambiguous. A data subject's consent must be given in respect to a specific data processing. In particular, it may not be derived from another expression of intent, not even if they are of comparable subject matter [32, para. 38]. Furthermore, informed consent can only be given if the controller provides the data subject with the information demanded by the GDPR (described in more detail in Sect. 2) in clear and plain language and in an easily accessible form [32, para. 40]. A data processing controller is the natural or legal person, public authority, agency, or other body that, alone or jointly with others, determines the purposes and means of the processing of personal data; cf. Art. 4 (7) GDPR. Subjects must be able to foresee the precise consequences of their consent. However, these strict requirements lead to the *cookie fatigue* problem described in Sects. 1 and 4.1: to fulfill the demands for consent to be informed, specific, and unambiguous, a multitude of consents is obtained, which causes users to experience overload and take on a dismissive stance, rather than leading to a more detailed understanding of the matter (which Sect. 4.1 explores in greater depth). With digital ecosystems encompassing a variety of different services, it makes them more susceptible to this effect.

To counteract this phenomenon, the legal literature has, in recent years, proposed and discussed different proposals to improve data protection and consent managements. One of the suggested Privacy Enhancing Technologies (PETs, see also the chapter "Acceptance Factors of Privacy-Enhancing Technologies on the Basis of Tor and JonDonym") is a *Personal Information Management System* (PIMS) [6, p. 946]. The goal of PIMS is to enable users to manage their personal data in one place [17, p. 2241]. As a consent management system, a PIMS could be utilized to request and obtain the generic consents described in Sect. 2 in a lawful way. The basic idea of providing users with a central place to manage their data protection preferences is not new: back in 2002 already, the W3C recommended adopting the *Platform for Privacy Preferences Project* (P3P) [46]. The goal of P3P was to allow websites to present their data collection practices in a standardized, machine-readable, and easy-to-locate manner, thereby enabling web users to understand what kind of data will be collected, how it will be used, and what they can change about that [46]. Thus, P3P was a mechanism to support the protection of data and privacy [18, p. 157], just as PIMS are, and required users to indicate their data protection preferences beforehand [18, p.159], [46]. Unfortunately, P3P rarely got adopted in practice. Perhaps PIMS will be able to achieve the original goals of P3P. One factor that will allow PIMS to be adopted more widely than P3P is the introduction of new legislation recommending their use in practice. The enactment of the *Telekommunikations-Telemedien-Datenschutz-Gesetz, TTDSG* (Telecommunications-Telemedia Data Protection Act) in Germany in December 2021 marked the first introduction of a regulation regarding PIMS (§ 26 TTDSG). The TTDSG is partly based on the Directive 2002/58/EC of the European Parliament and of the Council concerning the processing of personal data and the protection of privacy in the electronic communications sector (Directive on privacy and electronic communications, or "ePrivacy Directive"). Art. 10 ff. of the *Data Governance Act* of the European Union further provides regulations concerning Data Sharing Services, which serve similar purposes as PIMS. Even though this demonstrates that regulations regarding PIMS are being drafted, their implementation into digital ecosystems is still a long way off. The regulations of the TTDSG regarding consent management systems (§ 26 TTDSG) correspond only to consent in terms of § 25 TTDSG. The prerequisite determining when this regulation applies is not governed by the processing of personal data, but by the storage of data on a user's personal devices [43, *Ettig*, § 25 TTDSG para. 3]. Further restrictions by the German legislator regarding consent were not possible: regulations that deviate from the GDPR would be unlawful because of the precedence that laws of the European Union have over national legislation [23, Ambrock, Teil A, II. Rechtliche Grundlagen, para. 52 ff.]. Moreover, the Data Governance Act is not yet applicable in the Member States of the European Union. Thus, although these regulations may also be of relevance, the implementation of a PIMS to obtain generic consents in a digital ecosystem should be examined particularly in terms of its compliance with the demands of the GDPR. Because there is a vivid discussion about PIMS in Germany since the TTDSG was discussed in the German Parliament, our particular focus will be on the German jurisdiction and literature on this topic.

3.1 Personal Information Management Systems in Digital Ecosystems

A PIMS could be used as a central consent management system in digital ecosystems. A possible implementation for the central management of consents could take the form of a dashboard or cockpit: such a system would encompass both easy access to overviews on consents already given and management (e.g., reviewing, tracking, and withdrawal) of the consents themselves [6, p. 947]. To be precise, users should be enabled to define privacy settings only *once* and in an *abstract* way, so that these can be a basis on which future requests for consent can be answered automatically [6, p. 947]. Our idea assumes the use of a privacy cockpit combined with a central platform as an intermediation service between data subject and controller. This privacy cockpit would then be responsible for obtaining the user's consents, which the individual services would then technically enforce in a given context. The privacy cockpit would then forward the given consents to the ecosystem services, which would implement them in their context. Should a consent cover the specific context only partially, the user should be presented with options to be notified of missing required consents in a non-intrusive way and should be supported in making adjustments. This is in line with the abstract model of a PIMS used as a consent management system.

3.2 Obtaining Consent via a PIMS

The legal community's view on PIMS is in parts quite skeptical of PIMS because when consents are given via such a system, these consents are given in advance and without detailed knowledge of the specific data processing involved. Giving consents in an automated way based on settings unrelated to individual cases fundamentally contradicts the aforementioned principle of consent for personal data processing having to be specific. It harbors the danger of providing unlawful blanket consents. If a consent is insufficiently specified regarding its content, purpose, or consequences, it will be legally void for being too unspecific and ambiguous [43, *Arning/Rothkegel*; cf. Art. 4 GDPR para. 329]. Yet, as it is the goal of PIMS to decrease the effort and the sheer endless number of consents to be given by a user, one could then question whether it is even possible to obtain consents via a PIMS in a lawful way and resolve problems like overload and fatigue.

3.3 Using Allowlists in Digital Ecosystems

To ensure that a consent is specific, unambiguous, and informed enough on the one hand and to eliminate the *cookie fatigue* problem on the other hand, a compromise might be needed. The question of how lawful such abstract ex ante consents are was central in a research assessment of § 26 TTDSG, in which the authors proposed the implementation of so-called *whitelists*[1] as a solution to this problem [42, p. 42 ff.]. Allowlists are intended to improve usability: users should be enabled to give consents for fine-grained processing purposes based only on knowledge about groups (i.e., categories of controllers) [42, p. 6 para. 5]. Using these allowlists, one could considerably reduce the number of consents to be given if a whole group of controllers could be accepted with one click. At the same time, such a list can be highly informative, specific, and unambiguous and eliminate the need to repeatedly have to obtain unique consents, by providing the user with an overview of all controllers and their corresponding processing purposes to which they grant consent.

The concept of allowlisting is certainly not new in the legal community. In the domain of competition and copyright law, various courts already addressed the issue of ad blockers implementing allowlists [8, 29]. They considered the question of whether for websites financed through advertisements, buying a placement on a allowlist that allows them to display advertisements even while an ad blocker is active complies with competition law [29]. The notion of using allowlists to simplify data processing consents has not been discussed in detail yet. Note that it is not our goal to establish a similar practice of granting a position on a allowlist

[1] This source used the term *whitelists*; for cultural sensitivity reasons, we use the term *allowlists*.

in return for a fee because that might lead to allowlists becoming only available to services with sufficient financial backing and without the assurance that the controllers comply with data protection regulations. Thus, a different approach to creating such allowlists is needed, which can take either of two forms:

Solution 1: Organizational Allowlists

The first possibility is to create allowlists as suggested in the aforementioned research assessment concerning § 26 TTDSG [42]. The authors propose different approaches to allowlisting by organizations, such as using a neutral third party (the research assessment proposed NGOs) to curate a listing of trustworthy controllers [42, p. 5, para. 4], or by the operator of a PIMS. They propose the following procedure: first, the controllers register in the PIMS. Next, the PIMS operator creates allowlists containing all applicable controllers and their processing purposes, which can be tailored to the users. The allowlists would be adapted to a user's specified preferences, but the user could still choose to accept or reject them [42, p. 43].

Solution 2: User-Defined Allowlists

The other possibility is to have the users generate their own allowlists in the PIMS. After registering, they could be prompted to set up their preferences and then create their own allowlist by specifying the controllers they deem trustworthy and to whom they would like to grant consent for the specified processing purposes. To implement this solution, the PIMS operator would have to invest some additional effort. They would have to determine how to classify services (e.g., payment services or shipping services) and processing purposes in the system so that the users can define their choices for these classes. The advantage of this solution is that it offers increased flexibility and empowers the users to define a allowlist that is completely attuned to their own preferences.

3.4 Legal Conclusion

We can conclude that, from a legal point of view, a dashboard or cockpit as a PIMS specialized in managing consents is an appropriate solution for requesting and managing consents in a centralized and lawful way within the context of digital ecosystems. Skeptics might argue that giving consent in advance without knowing about the intended data processing in detail fundamentally contradicts the regulations of the GDPR. Though our idea of using generic consents entails exactly that, just as systems like PIMS do in general, our approach should not be rejected right away when seeking a solution for the serious challenges that exist with consent handling. The requirements for consents to be informed, specific, and unambiguous were intended to enable data subjects to completely grasp the consequences and implications of their granted consents. It is indisputable that detailed information of the users is necessary to fulfill these demands, but whether the level of detail that the GDPR demands is truly needed is up for debate. For instance, if a user consents to a delivery service provider processing their address for the purpose of delivering a shipment, we can assume that the user has a clear understanding of the consequences of their consent. To them, it makes little difference whether this processing is done by delivery service provider A or delivery service provider B—as long as both are trustworthy. Granting consent to a group of controllers thus does not contradict the spirit and purpose of the GDPR. The reason why the GDPR requires consent to be given in an informed way is because it aims to protect the data subjects. They should only give their consent if they are completely aware of and agree with the consequences of this action [43, *Taeger*, Art. 6 GDPR para. 37]. This intended protection is best achieved when data subjects take note of and seriously consider the provided information. Overloading the user with information or repeatedly asking for consent does not fulfill this aim [43, *Taeger*, Art. 6 GDPR para. 40], as also demonstrated by the *cookie fatigue* problem. It is more likely that data subjects will examine the information provided in a PIMS if they have to specify their privacy settings and preferences regarding data processing by specific data processors only once [5, p. 10]. When applying a teleological interpretation (an interpretation in the spirit and purpose of the law) of the regulations of the GDPR, one will come to the conclusion that requesting generic consents is indeed lawful as long as the information demanded by the GDPR is at least provided for categories or groups (see also [42, p. 6, para. 5]).

4 User-Oriented Redesign of Consent Handling

In Sect. 3, we learned about the legal basis and the logic behind obtaining consents. However, we also established that from a legal perspective, it can be challenging to obtain these consents in a practical way that fits the purpose and is free from dispute. But what if we approach this problem from another perspective: that of the human actors involved in the transaction of data processors requesting consent and data subjects giving consent? What challenges do they face with the current implementation of the legal requirements, and how can they benefit from the concept of generic consents? In this chapter, we will take a psychological perspective on the consent handling process.

When data protection regulations started to mandate consents, little guidance was given on how this should be implemented. Processors of personal data had to quickly find ways to get consent and make sure to do so in a legally effective way, but without agreeing on a technical privacy standard. This resulted in a proliferation of consent handling tools [12, 27]. Dentists and other doctors had their patients fill out hastily created consent forms, and associations scurried to obtain explicit consent from their members—typically by email—to store their data and opt into their newsletter.[2] However, there is one type of consent that people continue to be confronted with most frequently: website cookies, which we briefly mentioned in Sect. 1. Because nearly all B2C digital ecosystems use a website as their primary front-end, this is an important topic for operators of digital ecosystems. As a result, cookie banners are a very compelling and tangible example of a current problem with consents, which allows us to explore the challenges and possible solutions in more depth. We understand that this is just one example instance of a problem area, and we believe that the solutions we propose can be transferred to other challenges related to consents.

In Sect. 4.1, we will review what problems exist with cookies and how they affect end-users from a psychological perspective. In Sect. 4.2, we will propose solutions for these problems. In essence, we will propose a solution that promotes privacy-driven human-centered design (see also the chapter "Achieving Usable Security and Privacy Through Human-Centered Design") and emphasizes a positive user experience through an approach in which a website—which may be the front-end for a digital ecosystem—aligns with its users over time to find the optimal level of consent for them. Drawing from psychological concepts of how we build relationships as humans, this approach assumes that users must first be given an opportunity to build up *trust* [3, 9, 24, 39]. During this time, they do not have to concern themselves with consents. In this context, we specifically consider *online trust*, which Shankar et al. [39] define as "a reliance on a firm by its stakeholders with regard to its business activities in the electronic medium, and in particular, its Web site". Empowering users to build up trust can foster their perception of self-determination, boost their loyalty, and ultimately provide them with a better user experience [2]. Although our focus lies on websites (i.e., front-ends of digital ecosystems), these solutions could also be transferred to digital ecosystems where services take the place of websites.

4.1 Psychological Effects of Cookie Banners

Whether or not they are aware of it, a user and the provider of a digital ecosystem enter into a mutual relationship; not only a contractual relationship but also an actual relationship that involves dependencies and feelings [16]. The users take the role of *asset consumers*, who depend on the digital ecosystem to be the *asset broker* that helps them find and obtain these assets, while the operator of the digital ecosystem is also the *data processor*, who depends on the users to be willing *data subjects* [25]. This relationship is obviously most harmonious if both parties feel comfortable. Particularly the fact that a user has trust in the operator of the website has been found to have a strong impact on that user's willingness to provide personal

[2] Note that a transitional arrangement did not demand this action for existing patients, members, or other persons already in the database, but it was often done to be on the safe side.

information [50] and thus on the likelihood that they will give consent. But as we will see, in current practice, this relationship is often distorted or even unhealthy. This is unfortunate because this relationship is not just defined over the user's personal data; the operator of a digital ecosystem does not merely want to store cookies, but they also want to be liked by the users in hopes that they will become loyal customers.[3] On the other hand, they must respect that the users do not want to feel too exposed on the one hand; on the other hand, however, they do not like to feel limited in their use of the website due to their privacy settings [9, 24]. In this section, we will explore six closely related problems related to cognitive aspects.

Problem 1: Upfront Consents

Obtaining consent from persons like patients or association members is markedly different from asking website visitors to consent to cookies. In the former case, there is always a basic level of trust that has been able to grow over some period of time. Before a patient seeks medical attention, they surmise that the medical professional will be able to treat them, often based on another doctor's referral or recommendations from friends or online reviewers. By the time the patient is asked to fill out a consent form in the waiting room, they have already gone through these preparations, and during their visit they can determine whether the facilities are inviting and the staff is friendly. Similarly, one typically only joins an association that suits oneself. With websites, this is entirely different. Before one can go on to explore, the first thing one has to do is to provide consent. This is strange because in most cases, the visitor has no relation to whoever is behind that website and has not had the opportunity to gain (online) trust [3, 24]. Even if the website is the portal to a well-respected bank or news outlet, the visitor who has not had dealings with them before will not be able to decide for themselves whether or not they should entrust them with their personal data at this point. Thus, especially when a user has never visited that particular website before, it is too early to reasonably expect them to make a well-informed decision. They might be unsure as to whether the website is even suited for them or whether they can trust the website's operator, especially if the latter is shrouded in anonymity. Consequently, in many cases, consent is provided indiscriminately, making this action—except in the case of the rare user who reads the Terms & Conditions and the Privacy Policy—more a case of *blind consent* rather than actual *informed consent* [11, 45].

[3] We would like to emphasize that data processors are rarely greedy data collectors with ill intent; they are stakeholders that often have a genuine need for the data, which may include offering data-driven services.

Problem 2: Coerced Consents

So far, we have assumed the normal case, where a user still has the freedom to not share their personal data, even though this procedure is still somewhat intrusive. It might be compared to a real-life situation where one asks a random person on the street for their phone number as soon as they make eye contact. Unfortunately, the reality can be less idyllic. Sometimes users are downright coerced into consenting to (all) cookies before they can continue to use the website; this practice is employed, for example, by various prominent newspapers [40]. The lawfulness of this can be disputed [12, 38] because the user is no longer free to make this choice. They are rather pushed toward a decision because the website uses the psychological concept of *fear of missing out (FOMO)* [36] and because they know that without granting consent, they will not be able to access everything [49]. As a practical example, visitors of Healthline.com seeking medical advice can only choose between allowing and disallowing *all* purposes at the same time. In the latter case, they are redirected to the ad-free and tracking-free portion of the website,[4] which offers nearly no functionality. Users can thus either decide to give in and allow all purposes so they can access the content or consult other websites instead.

Particularly the example of newspaper websites (further explored by Soe et al. [40]) perfectly illustrates just how bad things have become with cookies, which in real life would be the equivalent of medical practices employing bouncers forcing patients to sign a release form before they can enter or associations requiring potential members to sign a non-disclosure agreement before they can even get in touch. Needless to say, that is not a great start for a mutual and trustful relationship to develop because it disrespects the autonomy that data subjects are entitled to have. Other websites use the strategy of pleading or *confirmshaming* [28] in a follow-up dialog. Such emotional appeals might not only push uncertain users in the direction intended by the designer but can also spark a feeling of discomfort, aversion, and mistrust both among users who reluctantly accept this as well as users who double down on their dismissal. Here, too, the basis for the relationship with the user is unhealthy and unbalanced.

Problem 3: Poor User Experience

So far, we have analyzed how users fundamentally and subconsciously approach their relationship to operators of a digital ecosystem. These aspects can be difficult to measure, but they play a central role in driving the user's actions and perceptions. However, an aspect that is far more obvious to the user is how they consciously perceive their interaction with the website, i.e., their *user experience* [22]. Being confronted with a cookie pop-up (and possibly other pop-ups for subscribing to a newsletter, activating notifications, and accepting advertisements) before they

[4] Located at https://anon.healthline.com/.

can access a website's contents can be a nuisance.[5] Earlier, we discussed that the mandate for obtaining consents was not complemented with guidance on how to implement this. Moreover, as explained in Sect. 5, the industry has so far been unable to agree on a shared standard. This has obvious consequences for cookie banners. Instead of being based on a reference architecture or system to meet the legal requirements in a structural manner, each website operator has essentially been left to their devices in terms of finding and implementing their own solution [12, 27]. This is why cookie banners come in all shapes and sizes, resulting in a "Frankenstein implementation" that has become a user experience nightmare.

Problem 4: Unclear Utility

A user browsing the Internet typically accesses a website for a reason, like accessing resources (e.g., texts, multimedia) or online functions (e.g., web shops, instant messaging). This experience can be typified as instant gratification [41, 51], meaning the user clicks on a link in order to immediately have access to content that entertains, provides information, or in another way helps them achieve their goal. In this cognitive process, a cookie banner and other pop-ups are an undesirable hurdle that stands in the user's way of receiving their gratification and adds to the time and effort needed for them to achieve their goal. This is why they experience it as a nuisance that negatively impacts the user experience. It is further exacerbated by the fact that it is not obvious to them how these additional—seemingly unnecessary—actions will benefit them. For example, on most websites, a user will not be able to directly see the effect of accepting or declining functional cookies. The consequences of the additional actions are often neither desired nor do they appear to be beneficial, so they have a low perceived value to the user. This, in turn, lowers the user's motivation to spend their cognitive resources on deliberating about them (cf. [40, 45]). The user's new sub-goal thus becomes overcoming the annoyance (a) with as little cognitive load as possible and (b) as quickly or efficiently as possible. Put more plainly, the user wants to think as little as possible about this activity and ideally just wants to click somewhere to be done with it [19]. This is what leads to the phenomenon known as *cookie fatigue* [21], which we discussed in Sect. 1.

Problem 5: Dark Patterns

Unfortunately, designers of cookie banners are aware of the *mental offloading* that takes place in users and in some cases have chosen to exploit it. This explains the success of so-called *dark patterns* [28, 40] (see also the chapter "The Hows and

[5] Recommended viewing on this topic: Tom Scott—*Why The Web Is Such A Mess*, https://www.youtube.com/watch?v=OFRjZtYs3wY.

Whys of Dark Patterns: Categorizations and Privacy") being successfully applied because they guide an inattentive user to the result desired by the data processor and not the result that is necessarily in the user's best interest. Through a *misdirection* pattern, providing consent becomes as easy as pressing the highlighted button, while objecting will, in the best scenario, involve the cognitive effort of reading what is on the highlighted and non-highlighted buttons and pressing the latter. Typically, this also involves a tedious and demanding process of finding a link in a paragraph of text or pressing an unlikely button (e.g., "more information") to find the settings. In the worst-case scenario, it forces users to individually revoke their consent for legitimate interest processing purposes, after which they must still be careful not to click on the highlighted "Accept all" button [40]. This is not congruent with the careful deliberation a user is supposed to enter into when deciding what cookies to accept but instead is more about jumping through the right hoops; by this time, the cognitive load and the time that must be invested are so high that even privacy-aware users may already have given up and just consent to all [19]. Moreover, users rarely revisit their cookie settings on their own initiative—assuming they are even able to find the location where they can update their settings, provided this option is offered at all [12]. This actually makes it quite rewarding to lure users into giving their consent through dark patterns. Some websites even repeatedly ask for consents anew from users who have not consented to all before, in which case the ability to process more personal data outweighs the risk of those users growing more annoyed and suspicious. Note that from a legal perspective, many red flags can be raised in this section about the legality of the processing purposes provided or how users are enticed into giving consent and prevented from updating their preferences (cf. [38]), but we will leave that topic to governing bodies to sort out.

Problem 6: Repeated Consents

So far, our discussion has been limited to a single interaction between a user and one particular website in a specific setting. The problems show that in each individual instance, users deal with cookie banners that they consider of low value to them; users are often unable to estimate whether they want to use functions for which particular cookies are necessary; design patterns favor the wishes of the data processor, and the activity of giving consent involves a high cognitive load that inhibits careful deliberation. The final problem is that the same user has to "make" their choices again when accessing the same website from a different browser on the same device or from a different end device. Considered rationally, it is unlikely that the user will really make a different decision when using different browsers or devices. For some reason, data processors have developed the persistent belief that by forcing users to set individual preferences per website, users will be more liberal in giving consent because they have to constantly make their choice anew. This is especially unlikely and might only hold true if the user is able to become more affectionate and trustful toward a website [9]. Just as with interpersonal relationships, self-disclosure only increases after that relationship has had time to evolve [13].

4.2 Solutions for Improved User Experience

The six problems with cookie banners presented in Sect. 4.1 essentially revolve around aspects of user experience and the time given to users to build trust. When juxtaposing the needs of data subjects and data processors (see also the needs analysis methodology described in Sect. 4 of the chapter "Achieving Usable Security and Privacy Through Human-Centered Design"), we can derive possible design solutions that take both the psychological processes of the users and the intentions of the website's operator into account. Here we propose four initial solution ideas, which naturally presume that the operator of the digital ecosystem is benevolent toward the users and has already made the decision to avoid questionable activities such as the use of dark patterns. A common thread shared by all four solutions is that they support a fair interplay (i.e., a healthy relationship) between the user and the website, where the user is given a feeling of empowerment as the decisions are made together with the PIMS that will guide them through a pleasant process. After the user has been allowed to explore the website at their own pace, it equips them to make a truly *informed* decision, which ultimately has a higher probability of being more in favor of the website if that impression is positive.

Solution 1: Make Cookies Something of Later Concern

A first solution could be to reverse the current approach. Just as there are taboo topics during a first date, cookies are not necessarily the best opening line for a website. Instead, users are likely to respond positively to a fairly non-intrusive message that tells them the website operator will start by not collecting cookies. They should be offered the possibility to already change their settings to provide consent to unlock certain features, for example, through *in-line cookie options* [19], but would be free to continue using the website without investing the mental workload. The website has just one chance to make a good first impression, and by enabling the user to explore some more, the user is likely to build up more trust toward the website. The option to consent to more cookies could be postponed until after a kind of *trial period*, during which the user can get an impression of the value the website can give to them. Now, they will be able to make a far more informed decision, and because of greater trust and a more positive attitude overall, they might give more lenient consents than they would have given otherwise. This might especially be effective if it is demonstrated clearly how granting consent makes certain kinds of processing and thus certain features possible.

Solution 2: Reject Until Further Notice

Many mailboxes in the Netherlands famously have a sticker through which residents opt out of unsolicited advertising material and papers (NO/NO) or either one of them (NO/YES or YES/NO) getting delivered. This reflects their generic consent to or rejection of receiving printed materials.[6] In some cases, an extra sticker provides an exception that the residents do want to receive separately delivered advertisements, for example, from their local supermarket. Similarly, users could have predefined YES/NO settings across their devices that reflect the basic attitude of a user toward privacy (see, e.g., the user group profiles discussed in Sect. 5 of the chapter "Achieving Usable Security and Privacy Through Human-Centered Design"). This should enforce general preferences to be adopted by an individual website's cookie preferences, and the user would only need to make adjustments for that particular website to add exceptions. This strategy could take the form of allowlists in a PIMS, as described in Sect. 3.3. However, this case would exceed the limits of a digital ecosystem, as it would essentially be a more ubiquitous implementation of cookie banner blockers like *Consent-O-Matic* [30] (see Sect. 5).

Solution 3: Provide Differentiated Decision Support

Another factor that could positively influence the users' perceptions is the decision support provided by the interface. By using short and to-the-point descriptions, it should be easier for the user to understand what the (positive or negative) consequences of giving consent for a particular purpose are. This goes farther than the often-seen and quite meaningless statement "We respect your privacy" and should also demonstrate how cookies have been used for their intended purpose. One might wonder what some websites that have been collecting data "for website optimization" for years actually did with all that information. Ultimately, these measures help to achieve a "collaborative mixing & matching" of cookies suitable for the user instead of the often-seen trickery (e.g., dark patterns, see also the chapter "The Hows and Whys of Dark Patterns: Categorizations and Privacy", or nudging, see also the chapter "Privacy Nudges and Informed Consent? Challenges for Privacy Nudge Design") employed to seduce users to give their consent. Useful inspiration may be drawn from research on *explainable artificial intelligence* (AI) [20] and prompt a form of *explainable security and privacy* [7, 14]. In the design of ethical AI systems, the quality of *explainability* ensures transparency about why and how an algorithm caused a particular system behavior or recommendation, so that a user can make a more informed decision based on the AI's output. Similarly, *explainable*

[6] The Dutch Wikipedia page gives a good overview of the context: https://nl.wikipedia.org/wiki/Brievenbussticker. This opt-out practice is several decades old and is increasingly being replaced with opt-in solutions, such as a YES/YES sticker and the possibility for residents to register themselves in an index.

persuasion can inform a user of influencing techniques used in persuasive interfaces (e.g., on gambling websites) [10]. The principles and concepts from explainability could be used in a similar manner to guide informed consent regarding privacy.

Solution 4: Encourage Decision Review

A final factor could be to stimulate users to revise their decisions; not by forcing them through another cookie banner, but by suggesting to them in a non-intrusive way that they should review their cookie settings, akin to the way some digital ecosystems suggest performing an occasional *privacy check-up*.[7] If trust is built, here, too, the user might be more motivated to grant more consents upon review. This is also fairer to users who, in retrospect, realize that a dark pattern led them to granting consents they did not mean to grant. Provided that the user has given consent for their activities to be tracked, a PIMS could employ usage mining to establish a user profile and make personal recommendations on what settings and generic consents would befit the user specifically, thereby helping them arrive at the configuration that is optimal for them personally.

5 Feasibility of Technical Implementation

We have seen by now what legal requirements exist and which psychological improvements can be made. Now we would like to give a short introduction to the technical aspects. To investigate the feasibility of implementing our proposed consent handling solution, in this section we explore existing work and its relation to our proposed ideas.

Because our aim is to reduce the users' mental load associated with granting consent in digital ecosystems, we propose that all consents—but not necessarily the personal data of the users—be managed by one central platform that stores consents and provides an interface through which these consents can be managed (e.g., granting, objecting to, revoking, or otherwise altering a consent). The central platform then communicates the type of data processing to which the user has consented to the digital ecosystem's services. This is somewhat similar to concepts such as browser plug-ins, e.g., *Consent-O-Matic* [30], which allow users to preset their preferences regarding their consents in cookie banners and then automatically fill out these banners. However, our approach describes a more general way of managing consents, which is not limited to cookie banners but can be applied to different services in digital ecosystems. Although some of these approaches were

[7] Examples include Google (https://myaccount.google.com/privacycheckup), Facebook (https://www.facebook.com/privacy/checkup/?source=settings_and_privacy), or the third-party privacy checker of Kaspersky (https://privacy.kaspersky.com/).

initially tailored to websites, we consider our solution to be suitable for the services of a digital ecosystems, which may be websites, but could also be other services.

5.1 Consent Representation Formats

A first step toward a general framework for consent management—regardless of its form—is to ensure that consents are represented in a way that allows the managing platform and the digital ecosystem services to unambiguously communicate the types of processing and to determine whether consent to them was granted or denied by the user. To this end, various approaches and ontologies have already been defined and put into practice. An early approach was *P3P* [46], as mentioned in Sect. 3. On the one hand, its intention was to enable websites to inform their users of their specific data collection intentions; on the other hand, it was to enable users to set preferences for automatically accepting or denying these requests without having to read all of the policies. Unfortunately, this recommendation was only implemented by a small number of websites and became obsolete in 2018. Another proposal suggested combining requests for consent with classical methods of access control, for which Appenzeller et al. [4] suggested using the *eXtensible Access Control Markup Language (XACML)* [31]. A wide range of different ontologies exist for requesting consent and representing processing with no set standard, which Rantos et al. [37] consolidated automatically using machine learning algorithms.

A wholly different approach was proposed in the form of the *Tracking Preference Expression* [47], better known as *Do Not Track (DNT)*. As an extension to the HTTP protocol, which is mainly used for web communication, an addition to the communicated data (more specifically, a flag in the HTTP header) allowed users to express their preferences regarding tracking and servers to inform about their tracking behavior. But just like P3P, this concept was not widely adopted and thus failed to reach its aim. Recently, a similar concept called *Global Privacy Control (GPC)* [48] has emerged, which follows the same principles. It remains to be seen whether it will be more successful than its predecessor. These two concepts, DNT and GPC, already implement the first of two possible ways of handling consents: by forwarding consents to services or by forwarding data to services. These will be discussed in the following two sections.

5.2 Consent Forwarding

Assuming our initial setup of a centralized platform for managing consents that gets reflected in multiple digital ecosystem services, the obvious approach here is for users to give their consents in one central place, from where a digital ecosystem service retrieves the given consents when users interact with that service. This centralizes all requests for consents in one place but changes little with respect to the current practice of websites and services. It is up to the digital ecosystem service itself to act in accordance with the consents it is presented with.

Especially for websites and their analytics data (e.g., where did the user click, which videos did they play, and how long did they remain on the website), this is the only reasonable solution because the digital platform that manages a digital ecosystem usually does not collect these kinds of personal data for every associated service or website. The costs and efforts involved in implementing such a system would simply be too high. Thus, such consents can only be forwarded to a specific digital ecosystem service according to an ontology or through a standard like DNT or GPC. Based on the consents, the digital ecosystem service in turn collects the analytics data itself, providing no credible proof of adhering to these consents. Pathmabandu et al. [33] sought to mitigate this disadvantage by scanning the data transmissions between users and the digital ecosystem service and trying to recognize the consented data patterns, thereby verifying whether these patterns match the data and processing categories to which the users consented. Since they applied their framework to smart buildings, it remains an open question whether their proposed idea can be transferred to website analytics data.

This concept in its basic form is currently the way data processing and consents are typically handled in contexts where several (digital ecosystem) services come into play. Users are asked for their consent when they first start using the service, but they have no way of checking whether the service truly complies with this processing. It should be assumed that most services do adhere to these given consents—especially because legal statutes require them to—but some insecurity remains for the user regarding whether their personal data is processed lawfully and without malicious intent and that no data other than what they consented to is being processed.

5.3 Data Forwarding

One possible solution for eliminating the insecurity among users about what happens with their personal data is for the centralized platform to only forward the data to the (digital ecosystem) service for whose use the user has granted consent. This could prevent the services from being able to collect data for which no consent was given. However, it cannot be ensured that this data will subsequently be processed only for the purposes to which the user has consented.

A rather technical solution to addressing this challenge was put forth by Agrawal et al., who suggested *Hippocratic databases* [1]. These databases are meant to include an access control mechanism that allows systems to apply the users' data sharing preferences at the database level. Additional tables in the database encapsulate the data and only grant access when data is demanded (a) by the specified recipients and (b) with the associated processing purpose. Such policies representing the users' preferences do not necessarily have to be integrated into the database; for example, Appenzeller et al. [4] used XACML policies at a higher abstraction level to represent the users' consent and regulate the data that is forwarded to the services.

Sticky Policies [35] are another approach aimed at ensuring that only the data the users have consented to get forwarded. This concept ensures the encryption of the users' data and *sticks* a policy to the encrypted data that describes under what conditions and by whom it may be used. A (digital ecosystem) service that intends to use that data must prove its compliance with the policy to a trusted authority before receiving the key for decrypting the data. Ulbricht et al. [44] extended this idea with a knowledge graph for federated data sources, of which a service might not yet know which data is available. The knowledge graph consists of short descriptions of what is contained in the encrypted data (e.g., address, gender, age, or more general classes like demographic data), based on which the service can determine whether the data is of interest.

Just as with P3P, the success of any of the approaches described in this section depends on their implementation by the services. But while P3P and DNT were meant to be used in the World Wide Web with its millions of very diverse services and websites, a digital ecosystem provides a small, finite set of services that is more manageable and needs a much smaller number of implementations for successful application. Thus, we believe that digital ecosystems are an environment that is well suited for the successful implementation of these approaches.

6 Discussion

In this work, we have suggested *generic consents* as a user-friendly way of giving tailored informed consent to data processing with reduced mental load, greater trust, and better informedness. From a legal perspective, we assert that our proposed approach of a PIMS implemented as a consent management system in a digital ecosystem can increase usability and privacy. Combined with a trial period (presented in the context of websites, but also applicable to ecosystem services)—a time in which users can gain trust in a service and better inform themselves—we claim that generic consents greatly foster self-determined and better deliberated decisions by users to consent to sharing their data. We also discussed existing ontologies and standards for representing (requests for) consent through which our vision can be realized. However, there are still some open questions regarding our idea, which we would like to discuss in this section.

These generic consents can be considered an extension to the consents demanded in Article 6 GDPR, which should, among other things, be specific and unambiguous. Generic consents are inherently not as specific as explicit consents; they are, in fact, intentionally unspecific to a degree. In this regard, they do not strictly comply with the regulations of the GDPR. But we argue that fine-grained specific consents are unmanageable for users. Having to handle as many consents as we have seen with website cookies, for example, makes it impossible for data subjects to make truly *informed* decisions [11, 45]. This does not lead to informed consents. Consequently, we believe that a allowlist containing generic consents introduces a necessary abstraction level to help users contain the amount of data processing they are asked to consent to. Given that there is currently no case law on this topic (see Sect. 3), the lawfulness of our concept is yet to be determined.

6.1 *Allowlists Created by NGOs (Solution 1)*

Although both suggested alternatives for the practical implementation of generic consents and (predefined) allowlists are theoretically feasible, Solution 1, in particular, has some drawbacks: if a allowlist is provided by NGOs (as proposed by Stiemerling et al. [42]), keeping it up to date in implemented applications may be a challenge. The allowlists would have to be made available in a machine-readable way so that applications can automatically query them and detect any changes. Another challenge is the workload involved in creating comprehensive allowlists of services and websites around the world. Ensuring that each entry receives a justified and fair evaluation is far beyond the capabilities of any organization—let alone maintaining these lists, given that asset providers might change the privacy practices in their service at any time.[8] Even if it were possible to create and maintain such allowlists, a reasonable expectation from a privacy-concerned point of view is that for few services on the allowlist, there will be a guarantee that they will not misuse the data or protect it insufficiently. To make such allowlists usable, the bar for approved services would need to be lowered considerably—which defeats the initial goal of increasing privacy. Another possibility would be to create different allowlists, each according to their own privacy level. This raises another open question: how does one rank services for such a allowlist? By how much data they collect or by the kind of data they collect? By the guarantees they provide for securing the data processing? Or by the reputability and trustworthiness of the asset provider? How to best measure and weight these aspects in order to provide a meaningful indicator of the "privacy" they ensure is yet to be determined. Thus, for Solution 1, many hurdles have yet to be overcome.

6.2 *Allowlists Created by the User (Solution 2)*

Solution 2 is not subject to the same problems as Solution 1 because it encompasses allowlists that users have tailored to their own privacy preferences. However, it is not

[8] One could suggest automating the evaluation by encouraging services to report their intended data processing in a standardized way to these NGOs—not considering how many services would actually follow this suggestion. This would allow NGOs to use simple automated checklists or machine learning to evaluate the data processing. However, it would raise the question of whether this procedure is as fair, justified, and reasonable as a manually performed evaluation.

clear regarding what aspect(s) users should best generalize consents while making sure these are still fairly informed and specific. Generalization is theoretically possible for all five aspects related to consent—Data, Asset Provider, Digital Ecosystem Service, Processing Type, and Purpose (see Sect. 2). From a functional point of view, one might keep the asset provider and perhaps also the digital ecosystem service generalized while specifying specific data categories, processing types, and purposes (e.g., "I always want navigation services to be able to access my location for the purpose of navigation."). Based on trust, one could also specify the asset provider while generalizing all other aspects (e.g., "I allow my navigation service to perform any processing type it requests."). It is also possible to specify the data category while generalizing all other aspects (e.g., "Any service may process my current location for any purpose."). Each generalization has its advantages and drawbacks. Some subsume a large number of requests for consent while others contain only a small number. For the user experience to get the maximum benefit, it is most advantageous to cover a large number of requests, while from a legal point of view, a smaller number is better. Which generalization would work best in practice remains to be seen.

6.3 Blocklists

As a complement to generic consents in a allowlist, a blocklist might be a suitable counterpart through which users could add exceptions to their preferences (e.g., "I do not allow asset provider ShadyProvider to process any of my data."). Blocklists and their consecutive exceptions to generic consents help foster the users' self-determination. However, they also increase system complexity. For example, when a consent (e.g., allowing navigation services to access location) is at odds with an exception (e.g., not allowing ShadyProvider to do any data processing), the system must determine which of these has the upper hand. Should it base this decision on a heuristic in which the blocklist always prevails over the alowlist? Or does the most recent consent or exception take precedence? Regardless of the heuristic, both the system and the user will have greater difficulty managing and understanding how the configuration of consents plays out. A possible way to simplify this is to introduce a blocklist that sets the user's bottom line of configurations to which an exception should always be made in all subsequent requests for consent. The exceptions from this list would then automatically get inserted into all generic consents to be created. In this way, a user only needs to specify their exceptions once and only has to confirm their choice without needing to make any manual adjustments (e.g., "I always want navigation services, *except those from ShadyProvider,* to be able to access my location for the purpose of navigation," with the highlighted part automatically created from the blocklist).

6.4 *Usability*

A final aspect to discuss concerns when and how the allowlists and blocklists should be created. Although it would be beneficial if this were done right after a user started using the digital ecosystem, this might not be the best time if we consider Sect. 4.2. However, a prompt asking users to specify their consents at the central managing platform could be provided early on (e.g., during the registration process or upon the first log-on) for users who really want to grant their consent. The system could then check if a user has already configured their allowlists and blocklists and ask them to perform a privacy review. Importantly, the user should be able to adjust their preferences at any time, especially because it is impossible to thoroughly consider all digital ecosystem services they will ever encounter when they initially create these lists. Rather, it is more likely that the user will eventually come across a digital ecosystem service for which some or all of the necessary consents still need to be configured. In that case, the user would receive a request for consent for which they can create a specific consent, but they would also be given the option to specify it more broadly as a generic consent. This exposes one of the main weaknesses of our proposed idea: users would still receive requests for missing consents that have not been explicitly denied in a blocklist if this is needed for the interaction with a particular digital ecosystem service. When this occurs during the trial period suggested in Sect. 4, where consent is not yet provided, the result is that a user will receive more requests for consent than with the current practice of obtaining consent to all processing right at the beginning. Consequently, a control mechanism should ensure that users are not overwhelmed by requests and provide even less well-informed decisions because what we are aiming to achieve is the exact opposite. Hence, the system should adequately assist the user in creating generic consents that fit their personal preferences in order to decrease the number of requests they receive during their normal interactions with the digital ecosystem services.

7 Conclusion

In this chapter, we gave a short introduction to the use of generic consents in digital ecosystems. The challenges we highlighted show that in order to achieve a successful solution, careful and user-oriented design is crucial, and several open questions still need to be answered. When designed properly our proposed concept of generic consents in combination with a trial period can foster users' self-determined and informed decision-making regarding consenting to the processing of their personal data. Further research is needed on how to help users create suitable generic consents, while case law must develop in which the judiciary explores to what degree generic consents still sufficiently comply with data protection regulations such as the GDPR. The concept proposed in this chapter is a step toward ensuring that users can make truly informed and self-determined decisions when faced with the vast amount of data processing in our time.

Acknowledgments This work is funded by the German Federal Ministry of Education and Research (BMBF) (grant numbers 16KIS1507 and 16KIS1510). We thank Sonnhild Namingha for proofreading this chapter and Jannis von Albedyll for providing insightful comments.

References

1. Agrawal, R., Kiernan, J., Srikant, R., & Xu, Y. (2002). Chapter 14—Hippocratic databases. In P. A. Bernstein, Y. E. Ioannidis, R. Ramakrishnan, & D. Papadias (Eds.), *VLDB '02: Proceedings of the 28th International Conference on Very Large Databases* (pp. 143–154). Morgan Kaufmann.
2. Albayrak, T., Karasakal, S., Kocabulut, O., & Dursun, A. (2020). Customer loyalty towards travel agency websites: The role of trust and hedonic value. *Journal of Quality Assurance in Hospitality & Tourism, 21*(1), 50–77.
3. Ali, A. S., Zaaba, Z. F., Singh, M. M., & Hussain, A. (2020). Readability of websites security privacy policies: A survey on text content and readers. *International Journal of Advanced Science and Technology, 29*(6s), 1661–1672.
4. Appenzeller, A., Rode, E., Krempel, E., & Beyerer, J. (2020). Enabling data sovereignty for patients through digital consent enforcement. In *Proceedings of the 13th ACM International Conference on PErvasive Technologies Related to Assistive Environments*, PETRA '20. Association for Computing Machinery.
5. Assion, S. (2021). Stellungnahme als Sachverständiger zum Entwurf eines Gesetzes zur Regelung des Datenschutzes und des Schutzes der Privatsphäre in der Telekommunikation und bei Telemedien (TTDSG), BT-Drucksache 19/27441
6. Botta, J. (2021). Delegierte Selbstbestimmung? PIMS als Chance und Risiko für einen effektiven Datenschutz. *Zeitschrift für IT-Recht und Recht der Digitalisierung, 24*(12), 946–951.
7. Brunotte, W., Chazette, L., Kohler, L., Klünder, J., & Schneider, K. (2022). What about my privacy? Helping users understand online privacy policies. In *Proceedings of the International Conference on Software and System Processes and International Conference on Global Software Engineering*, ICSSP'22 (pp. 56–65). ACM.
8. Bundesgerichtshof (2018). Keine aggressive geschäftliche Handlung durch Werbeblocker mit Whitelisting-Funktion - Werbeblocker II. *Gewerblicher Rechtsschutz und Urheberrecht, 12*, 1251–1258.
9. Campbell, D. E. (2019). A relational build-up model of consumer intention to self-disclose personal information in e-commerce B2C relationships. *AIS Transactions on Human-Computer Interaction, 11*(1), 33–53.
10. Cemiloglu, D., Catania, M., & Ali, R. (2021). Explainable persuasion in interactive design. In *2021 IEEE 29th International Requirements Engineering Conference Workshops (REW)* (pp. 377–382). IEEE.
11. Cranor, L. F. (2022). Cookie monster. *Communications of the ACM, 65*(7), 30–32.
12. Degeling, M., Utz, C., Lentzsch, C., Hosseini, H., Schaub, F., & Holz, T. (2019). We value your privacy ... now take some cookies: Measuring the GDPR's impact on web privacy. In *Proceedings of the 2019 Network and Distributed System Security Symposium* (arXiv:1808.05096). Internet Society.
13. Derlega, V. J., Winstead, B. A., Wong, P. T. P., & Greenspan, M. J. (1987). *Self-disclosure and relationship development: An attributional analysis* (pp. 172–187). Sage.
14. Droste, J. R. C. (2022). Development of a concept for privacy explanations and its prototypical evaluation. Master's Thesis, Leibniz University Hanover, Hanover.
15. European Union. (2016). General Data Protection Regulation. Regulation (EU) 2016/679.
16. Ghandour, A., Parackal, M., & Deans, K. R. (2021). Relationship development process in ecommerce websites. In *2021 Proceedings of the 22nd International Arab Conference on Information Technology (ACIT)* (pp. 1–9). IEEE.

17. Golland, A. (2021). Das Telekommunikation-Telemedien-Datenschutzgesetz - Cookies und PIMS als Herausforderungen fü Website-Betreiber. *Neue juristische Wochenschrift, 31*, 2238–2243.
18. Grimm, R., & Rossnagel, A. (2000). Can P3P help to protect privacy worldwide? In *Proceedings of the 2000 ACM Workshops on Multimedia*, MULTIMEDIA '00 (pp. 157–160). ACM.
19. Habib, H., Li, M., Young, E., & Cranor, L. (2022). "okay, whatever": An evaluation of cookie consent interfaces. In *Proceedings of the 2022 CHI Conference on Human Factors in Computing Systems*, CHI '22. ACM.
20. Hagras, H. (2018). Toward human-understandable, explainable AI. *Computer, 51*(9), 28–36.
21. Hansen, H., & Brechtel, S. (2020). Zu den Anforderungen an die Einwilligung für Cookies und Werbung. *Praxis im Immaterialgüter- und Wettbewerbsrecht, 16–17*, 385.
22. Hassenzahl, M. (2007). *The hedonic/pragmatic model of user experience* (pp. 10–14). COST294-MAUSE, Lancaster, UK.
23. Jandt, S., & Steidle, R. (2018). *Datenschutz im Internet - Rechtshandbuch zu DSGVO und BDSG* (Vol. 1). Nomos Verlag.
24. Joinson, A. N., Reips, U.-D., Buchanan, T., & Paine Schofield, C. (2010). Privacy, trust, and self-disclosure online. *Human–Computer Interaction, 25*(1), 1–24.
25. Koch, M., Krohmer, D., Naab, M., Rost, D., & Trapp, M. (2022). A matter of definition: Criteria for digital ecosystems. *Digital Business, 2*(2), 100027.
26. Loy, C., & Baumgartner, U. (2021). Consent-Banner und Nudging - Tracking-Mechanismen: Wie viel "Anstupsen" ist erlaubt? *Zeitschrift für Datenschutz, 8/2021*, 404–408.
27. Machuletz, D. and Böhme, R. (2020). Multiple purposes, multiple problems: A user study of consent dialogs after GDPR. In *Proceedings on Privacy Enhancing Technologies 2020* (pp. 481–498). De Gruyter.
28. Mathur, A., Acar, G., Friedman, M. J., Lucherini, E., Mayer, J., Chetty, M., & Narayanan, A. (2019). Dark patterns at scale. *Proceedings of the ACM on Human-Computer Interaction*, 3(CSCW):article 81.
29. München, O. (2017). Vertrieb eines Werbeblockers mit "Whitelisting"-Funktion - Whitelisting I. *Gewerblicher Rechtsschutz und Urheberrecht, 11*, 1147–1157.
30. Nouwens, M., Bagge, R., Kristensen, J. B., & Klokmose, C. N. (2022). Consent-o-Matic: Automatically answering consent pop-ups using adversarial interoperability. In *Extended Abstracts of the 2022 CHI Conference on Human Factors in Computing Systems*, CHI EA '22. ACM.
31. OASIS. (2022). OASIS eXtensible Access Control Markup Language (XACML) TC. https://www.oasis-open.org/committees/tc_home.php?wg_abbrev=xacml
32. Court of Justice of the European Union. (2021). CJEU, justice of 11.11.2020 – C-61/19 – Orange România/ANSPDCP. *Zeitschrift für Datenschutz, 2*, 89–91.
33. Pathmabandu, C., Grundy, J., Chhetri, M. B., & Baig, Z. (2020). *An informed consent model for managing the privacy paradox in smart buildings* (pp. 19–26). ACM.
34. Paulsen, N., & Weiß, R. (2022). *Cookie-Banner spalten Internet-Nutzer*. Bitkom e.V. https://www.bitkom.org/Presse/Presseinformation/Cookie-Banner-spalten-Internetnutzer
35. Pearson, S., & Casassa-Mont, M. (2011). Sticky policies: An approach for managing privacy across multiple parties. *Computer, 44*(9), 60–68.
36. Przybylski, A. K., Murayama, K., DeHaan, C. R., & Gladwell, V. (2013). Motivational, emotional, and behavioral correlates of fear of missing out. *Computers in Human Behavior, 29*(4), 1841–1848.
37. Rantos, K., Drosatos, G., Demertzis, K., Ilioudis, C., Papanikolaou, A., & Kritsas, A. (2019). Advocate: A consent management platform for personal data processing in the IoT using blockchain technology. In J.-L. Lanet, C. Toma (Eds.), *Innovative Security Solutions for Information Technology and Communications* (pp. 300–313). Springer.
38. Santos, C., Rossi, A., Sanchez Chamorro, L., Bongard-Blanchy, K., & Abu-Salma, R. (2021). Cookie banners, what's the purpose? Analyzing cookie banner text through a legal lens. In *Proceedings of the 20th Workshop on Workshop on Privacy in the Electronic Society*, WPES '21 (pp. 187–194). ACM.

39. Shankar, V., Urban, G. L., & Sultan, F. (2002). Online trust: A stakeholder perspective, concepts, implications, and future directions. *The Journal of Strategic Information Systems, 11*(3), 325–344.
40. Soe, T. H., Nordberg, O. E., Guribye, F., & Slavkovik, M. (2020). Circumvention by design—dark patterns in cookie consent for online news outlets. In *Proceedings of the 11th Nordic Conference on Human-Computer Interaction: Shaping Experiences, Shaping Society*. ACM.
41. Song, I., Larose, R., Eastin, M. S., & Lin, C. A. (2004). Internet gratifications and internet addiction: On the uses and abuses of new media. *Cyberpsychology & Behavior, 7*(4), 384–394.
42. Stiemerling, O., Weiß, S., & Wendehorst, C. (2021). Forschungsgutachten zum Einwilligungsmanagement nach §26 TTDSG, Studie im Auftrag des Bundesministeriums für Wirtschaft und Energie.
43. Taeger, J., & Gabel, D. (2022). *Kommentar DSGVO—BDSG—TTDSG* (Vol. 4). C.H. Beck Verlag.
44. Ulbricht, M.-R., & Pallas, F. (2016). CoMaFeDS: Consent management for federated data sources. In *Proceedings of the International Conference on Cloud Engineering Workshop (IC2EW)* (pp. 106–111). IEEE.
45. Utz, C., Degeling, M., Fahl, S., Schaub, F., & Holz, T. (2019). (Un)informed consent: Studying GDPR consent notices in the field. In *Proceedings of the 2019 ACM SIGSAC Conference on Computer and Communications Security*, CCS '19 (pp. 973–990). ACM.
46. The World Wide Web Consortium (W3C). (2018). The platform for privacy preferences 1.0 (P3P1.0) specification. https://www.w3.org/TR/P3P/
47. The World Wide Web Consortium (W3C). (2019). Tracking preference expression (DNT). https://www.w3.org/TR/tracking-dnt/
48. The World Wide Web Consortium (W3C). (2022). Global privacy control (GPC). https://globalprivacycontrol.github.io/gpc-spec/
49. Westin, F., & Chiasson, S. (2021). "It's so difficult to sever that connection": The role of FoMO in users' reluctant privacy behaviours. In *Proceedings of the 2021 CHI Conference on Human Factors in Computing Systems*, CHI '21. ACM.
50. Wu, K.-W., Huang, S. Y., Yen, D. C., & Popovad, I. (2012). The effect of online privacy policy on consumer privacy concern and trust. *Computers in Human Behavior, 28*(3), 889–897.
51. Zhang, B., Wu, M., Kang, H., Go, E., & Sundar, S. S. (2014). Effects of security warnings and instant gratification cues on attitudes toward mobile websites. In *Proceedings of the SIGCHI Conference on Human Factors in Computing Systems*, CHI '14 (pp. 111–114). ACM.

Human-Centered Design for Data-Sparse Tailored Privacy Information Provision

Mandy Goram, Tobias Dehling, Felix Morsbach, and Ali Sunyaev

1 Motivation

In the age of information with its diverse data-driven business models [9], consumers provide and share much information about themselves and others. To prevent abuse of consumer information, data protection laws have become more restrictive and require informed consent for many uses of consumer data [41]. Hence, it should be inevitable for consumers to cut their way through the privacy notices jungle to get information on privacy practices [17]. However, an uninhabitable jungle would have to be conquered; privacy notices are just confusing and impractical for consumers [28].

The intended purpose of privacy notices is to inform consumers by providing information about the privacy practices of companies and the information systems they provide [30]. Consumers face two problems: first, the sheer volume of privacy notices [22] that need to be provided for each visited website and every other type of online or offline information system, and second, the extensive texts, which are usually difficult to understand and often formulated in a complicated manner [39].

M. Goram (✉)
Karlsruhe Institute of Technology, Institute of Applied Informatics and Formal Description Methods, Karlsruhe, Germany
e-mail: mandy.goram@kit.edu

T. Dehling · A. Sunyaev
Karlsruhe Institute of Technology, Institute of Applied Informatics and Formal Description Methods, Karlsruhe, Germany

KASTEL Security Research Labs, Karlsruhe, Germany
e-mail: dehling@kit.edu; sunyaev@kit.edu

F. Morsbach
KASTEL Security Research Labs, Karlsruhe, Germany
e-mail: felix.morsbach@kit.edu

N. Gerber et al. (eds.), *Human Factors in Privacy Research*,
https://doi.org/10.1007/978-3-031-28643-8_14

This results in consumers not taking notice of privacy notices at all and often giving broad consent to data processing and sharing without knowing what they consent to [3]. This is often to the disadvantage of consumers—yet it happens with their consent [29].

Different approaches have been developed to support consumers. Privacy-enhancing technologies (PETs) support consumers, for instance, with privacy-preserving configurations of applications [18] or disguising their identity [14]. The more focused transparency-enhancing technologies (TETs), a subclass of PETs, provide information on consequences of data disclosure and information system use [26] through different forms of privacy information provision, such as, visualization concepts [36], just-in-time notifications [36], privacy seals [35], and text summaries of privacy notices [43]. Supporting consumers in making decisions on application use and data disclosure with TETs requires more than just a technique or visualization concept [36] because privacy decision-making is context-dependent: "The rules people follow for managing privacy vary by situation, are learned over time, and are based on cultural, motivational, and purely situational criteria" [2, p. 511]. Hence, privacy decisions made in one context may not be applicable in another. Privacy information provision requires knowledge about the context in which decisions are made by consumers to provide information about privacy practices that really matter to consumers in their specific situation [36]. People will, for example, have quite different privacy concerns when being asked to share health information while talking to a physician during treatment—or during a job interview.

How to account for context in privacy information provision is a pressing issue for supporting consumers with TETs. Personalization strategies are required to give consumers seamless access to context-specific information on privacy practices. This requires flexible information systems that can detect and adapt to consumer preferences, for instance, based on consumer behavior, system interactions, or previous decisions. The remainder of this chapter will shed light on how to accomplish this.

This chapter is structured as follows, we start with an overview of extant TETs, their functionalities, and potentials for tailoring. We go on with outlining a solution space for tailored privacy information provision while protecting sensitive privacy preference information. After that, we describe TET solution archetypes for tailored privacy information provision by explaining what tailoring approaches are suitable and how feasible local and remote processing is.

2 Overview of Extant Transparency-Enhancing Technologies

Various TETs have emerged in research and practice. These can be divided into six different types in terms of their functionality and purpose: privacy practice scoring, privacy practice description, privacy practice monitoring, privacy risk assessment, privacy practice history, and privacy practice comparison TETs. See Table 1 for an overview.

Table 1 Overview of the TET types, their functionalities, and examples

TET type	Abstract functionality	Examples
Practice scoring	Calculate a single score which represents how good/bad (appropriate) privacy practices are based on information from privacy notices, system functionality, or system behavior	*PrivacyMonitoring* [33]: creates a score for a website and explains how the score was calculated; *PrivacyScore* [24]: compares websites and allows consumers to rate websites on a range of security and privacy features; *Privacy Rating* [4]: based on predefined privacy aspects, the tool calculates an overall score of a website
Practice description	Describe privacy practices in an information system or of a provider and how consumer data might be used	*Layered privacy notices* [36]: present consumers with a brief notice with high-level information and allows consumers to expand each section to access more detailed information; *PrivacyCheck* [43]: text summarization tool that analyzes privacy notices through a browser plug-in; *Just-in-time notifications* [36]: appear when consumers have to make privacy decisions and present information that may be relevant for the decision
Practice monitoring	Monitor information use or other privacy practices of an information system and may alert consumers if actual divert from intended/expected practices	*Privacy Cleaner* [32]: scans, tracks, and controls access to information about a consumer, *Privacy Evaluation* [10]: evaluates popular educational applications based on a wide range of legal requirements and best practices for data protection
Risk assessment	Calculate a risk assessment for consumers based on system interactions, information shared, or privacy settings	*Cover your tracks* [11]: shows the unique and identifying features of a browser that trackers can use for identification; *Privacy Analyzer* [34]: allows consumers to see what data their browser exposes
Practice history	Lists changes in privacy notices or practices in a chronological order	*Change history summary* [8]: summarizes changes between different versions of privacy notices; *Privacy notice differences* [13, 40]: displays all changes between a document and its previous version
Practice comparison	Compares privacy practices and other characteristics between information systems	*Privacy Matters* [37]: compares popular messenger apps; *Browser Comparison Tool* [6]: compares web browsers; *Privacy Risk Index* [7]: compares mHealth apps and its privacy practices

As illustrated by the overview in Table 1, TETs come in many flavors. Yet, an all-to-common denominator is the provision of standardized information. Adaptivity to consumers' context-specific privacy preferences is a facet of TETs that offers much room for improvement. In the following sections, we will explore this untapped potential of TET with respect to stronger adaptivity to consumers' privacy preferences while protecting the confidentiality of sensitive preference information.

2.1 Tailoring Potential of Transparency-Enhancing Technologies

The TET types included in Table 2 yield different rooms for improvement by tailoring privacy information provision. Some could, for instance, be more interactive to better adapt to context-specific consumer preferences. Others overload consumers with too much information and require a more focused design. Overall, there is a lack of tailored, privacy need-based information provision. Instead of offering standardized sets of information, tailored TETs can take consumers' individual privacy preferences into account. For the tailoring, it is necessary to have information about the consumer to tailor TETs accordingly. This information can be provided by the consumer or detected automatically. Potential for tailoring information on privacy practices depends on the TET type. An overview of tailoring potentials of the different TET types is presented in Table 2.

Table 2 shows that the TET types yield room for improvement by tailoring privacy information provision to consumer privacy preferences. However, this requires access to preference information and other consumer information (Fig. 1), which poses privacy risks that should be addressed. Figure 1 shows categories of necessary consumer data for tailored information provision on privacy practices. The specific data required for tailoring depends on the TET, for example, the data required for tailored privacy practice descriptions could be a consumer's interest on data sharing practices. The privacy risks can be addressed by protecting the confidentiality of the additional information required for tailoring. To do so, technical privacy-preserving mechanisms can be used. Once information about the context-dependent preferences of consumers is available and confidentiality of that information is protected through technical privacy-preserving mechanisms, tailored privacy information provision becomes possible without introducing additional privacy risks.

3 Solution Space for Tailoring Challenges

The solution space for tailored privacy information provision requires access to privacy preferences and confidentiality protection of preference information so that tailored TETs can be made available to consumers.

Table 2 Overview of what can be tailored in the TET types

TET type	Practice scoring	Practice description	Practice monitoring	Risk assessment	Practice history	Practice comparison
Analysis features	Privacy practices considered in score	N/A	Privacy practices to be monitored	Risks to be assessed; consumer practices and device characteristics to be analyzed	Privacy practices of interest	Systems to be compared; privacy practices of interest
Analysis weights	Weights for privacy practice relevance	N/A	Relevance of monitored privacy practices	Relevance of privacy risks	N/A	N/A
Displayed information	N/A	Privacy practices on which information is displayed	N/A	N/A	N/A	Representation of monitored privacy practices
Aggregation level	Granularity of explanations for calculated score	Granularity of information presented on privacy practices	N/A	Granularity of explanations for determined risks	Level of detail on which changes should be tracked	Level of detail on which differences should be reported
Notification	N/A	N/A	Means to inform the consumer about deviations from intended privacy practices	N/A	N/A	N/A
Visualization	score (e.g., categorical)	privacy practices (e.g., icons)	monitored privacy practices	assessment outcomes	changes over time	differences between privacy practices

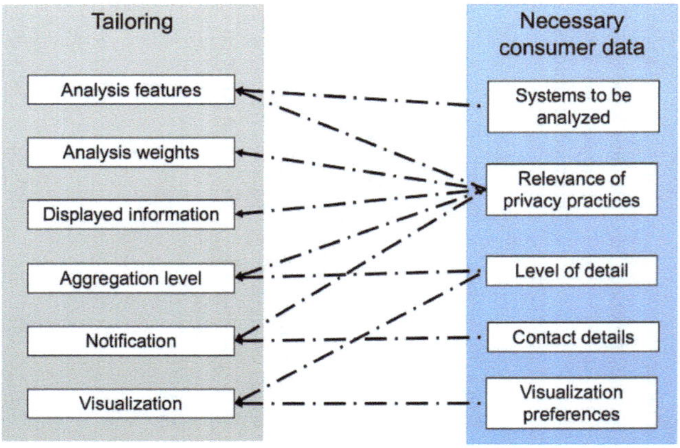

Fig. 1 Mapping of necessary consumer data for tailoring approaches

3.1 Privacy Preferences

For tailored privacy information provision, information about privacy preferences is required to tailor information provision to consumers' privacy needs. Consumer preferences can be elicited via three general approaches: (1) standardized preferences, (2) consumer-specified preferences, and (3) automatic detection of preferences.

Standardized preferences are specified by software designers or the software developer when the consumer interface is designed and cannot be changed by consumers. Preferences are represented statically in the design, for instance, what information is made available [22], how a privacy score is calculated [4], or what consumer archetypes are predefined [12]. Provision of information on privacy practices may be based in part on consumer studies investigating which privacy practices are important and should be considered when providing and preparing information on privacy practices [17]. However, the results of consumer studies do not capture the diverse situation-specific factors and circumstances that will be present when a consumer is actually using the TET [27]. Hence, standardized preferences are likely to not match the actual privacy preferences in real use contexts [25].

Consumer-specified preferences are more likely to match actual privacy preferences in real use contexts. Preference information is stored in consumer profiles [38] or collected as part of a session (e.g., through search queries and filters). Preference profiles can be created with various explicit preference elicitation approaches and require consumers to manually set their preferences [23], for example, through situation-specific questionnaires, preference menus, search queries, search filters, or ratings. Explicit preference elicitation approaches burden consumers with additional effort to decrease the gap between captured and real preferences [20]. Thus, they bear the risk to overwhelm consumers by being overly complex. On the other hand, simple explicit approaches may not offer sufficient degrees of freedom to close the gap far enough, so that outcomes may not improve much over preference standardization [20].

Automatic detection of preferences avoids this trade-off. Here, preferences are derived from consumer interactions with the system to better match real use contexts without requiring additional consumer effort [38]. Automatic detection approaches detect various consumer characteristics through observation of system interactions (e.g., mouse movements, content clicked, reading time, or location) or may leverage data collected beyond the application boundary (e.g., physical reactions, facial expressions, or eye movements) [44]. Based on the collected data, consumer preferences can be derived in a more context-specific manner. However, this requires complex technical procedures and extensive data collection [38]. In addition, consumers may find the subliminal data collection inappropriate and there is always the risk of false classification. We will discuss PETs suitable for tailored TETs in the next section.

3.2 Technical Privacy-Preserving Mechanisms

In a classic information system architecture, the necessary data required for tailored information provision is collected on consumer devices (e.g., a mobile phone or laptop) and transmitted to a server operated by the information system provider. The provider processes the data to generate tailored privacy information and sends this back to the consumer device. In this architecture, the information system provider has full access to the necessary data required for tailored privacy information provision, information that is in itself sensitive [38]. This poses a privacy risk for consumers. While the information system provider may have limited data use to prespecified purposes and may have consent according to the General Data Protection Regulation (GDPR) [15], recent data breaches and scandals have shown that these practices do not guarantee the protection of consumer data against misuse [19].

An alternative approach is to not collect or process the raw data on a central server in the first place. The possibility of misuse is significantly reduced when data is not collected by a third party. The raw data stays with the consumer and the tailoring of information provision can happen in such a way that the system provider has no direct access to consumer data. The privacy-enhancing technologies community [31] developed multiple approaches and techniques that can be used to mitigate these privacy risks by protecting the confidentiality of privacy preferences. In the following paragraphs, we will briefly describe and outline the potential use of local computation, homomorphic encryption (HE), and secure multiparty computation (SMPC)—three common options for privacy-preserving computing [21].

Local computing restricts processing of consumer preferences and tailoring of privacy information to the consumer device itself. For some use cases and applications, it is not necessary to process the data on a third-party server. For example, the weighting of privacy practice score criteria is not a complicated or resource-intensive operation and can easily be done on a mobile device. Therefore, whenever possible, the tailoring of privacy information should happen only on the consumer device. However, some data processing may either require too much data to be done on mobile devices, for example, data about privacy practices of multiple information systems, or require access to some central component that cannot be stored on consumer devices, for example, to protect intellectual property. In this case, there are technical methods for privacy-preserving computation that allow for processing of data while protecting confidentiality of consumer data.

Homomorphic encryption (HE) [1] allows to perform calculations on encrypted data. The input data is encrypted and the operation is executed on the cipher text. The result of this blindfolded operation will be decrypted and will then match the output of the operation as if it had been performed directly on plain data. HE can be used in TETs to compute operations on confidential data. A consumer can, for example, encrypt their private data locally on their device using local encryption keys and send the encrypted data to an information system provider. The provider computes the desired operation, here, the tailored privacy information provision, on the encrypted data and sends the constantly encrypted results back to the consumer device. The consumer can then decrypt and use them using the local keys. In this way, the information system provider has no access to sensitive consumer data (neither input nor output of the tailoring operation) but can still perform its job, even if application of proprietary code is required. While HE allows for private computation on sensitive data, it comes with a high computation overhead on the information system providers' side, especially, with respect to memory consumption. This makes homomorphic-encrypted calculations very expensive and limits its attractiveness for ubiquitous application.

Secure multi-party computation (SMPC) is a collection of methods and algorithms in which a group of consumers wants to compute a joint function on their private data without revealing their private inputs. For example, in the millionaire's problem, two persons want to determine who of them is richer without revealing their own wealth to the other [42]. This setting is also a special case and called secure two-party computation (S2PC). S2PC is especially interesting due to its high relevance in many real-world scenarios such as private database queries. S2PC can be used to protect consumer preference data by computing an SMPC function, while consumer preferences are stored on the consumer devices and serve only as input to the shared SMPC function. This way, the inputs remain hidden from the information system provider. By also encrypting the result with a secret key only known to the consumer using the SMPC function, the tailored output would also remain hidden from the information system provider. While generally any function can be implemented in an SMPC fashion [16] and general-purpose compilers for doing so exist, the applicability of SMPC is often severely limited by its high communication bandwidth requirements. There exist multiple approaches to realize

SMPC, but approaches based on Yao's garbled circuits [5] are said to be the most widely applicable ones, in which the function to be evaluated is transformed into a Boolean circuit. In this approach the execution cost scales linearly with the size of the circuit. This makes SMPC often less suitable for scenarios with resource-constraint devices, such as mobile phones.

The exact overhead and resources requirements of HE and SMPC highly depend on the concrete implementation and the computations required. HE is generally said to be cheap for client devices but computationally expensive for the server side, especially in memory consumption. SMPC, however, is generally said to be computationally cheap but requires a high communication bandwidth between the participating parties.

4 Solution Archetypes for Tailored Privacy Information Provision

4.1 Suitability of Tailoring Approaches

To provide consumers with easy and quick access to privacy information, it is important to take their individual information needs into account. However, it is not always appropriate to apply consumer-specified or detected tailoring to all TET types. Table 3 shows which tailoring approaches are suitable for which TETs.

Privacy practice scoring TETs provide an overview and summary of privacy practices of an information system. A standardized privacy practice scoring TET uses evaluation criteria specified by TET providers. Standardization of privacy practice scoring is appropriate when consumers want to get a general or first impression of a system or its provider without having to make elaborate settings on their own (Type Practice Scoring TETs:standardized). If consumers want to include specific aspects in the app score or set their own weights, scoring TETs must allow for customizability, as is possible with consumer-specified approaches (Type Practice Scoring TETs:consumer-specified). Preference detection is not recommended

Table 3 Comparison of the usefulness and applicability of standardized approaches, consumer-specified approaches, and detection approaches for TET type tailoring. Legend: −− very unsuitable, − unsuitable, + suitable, ++ very suitable, *N/A* not applicable

	Standardized	Consumer-specified	Detected
Practice scoring	+	++	−
Practice description	−−	+	++
Practice monitoring	−−	++	+
Risk assessment	++	+	N/A
Practice history	−−	+	++
Practice comparison	+	++	+

because it is not transparent to the consumer how the rating was calculated and what preferences are included (Type Practice Scoring TETs:detected).

Privacy practice description TETs inform consumers about privacy practices. Standardization means that all information about privacy practices considered relevant by the standardization body is provided (Type Practice Description TETs:standardized), which may lead to mismatches between communicated information and consumers information needs [39]. To provide consumers with quick and easy access to relevant privacy practice information, it makes sense to tailor privacy information to consumers information needs. Consumer-specified preference information can be used to filter for relevant privacy information (Type Practice Description TETs:consumer-specified). But consumers may not know what to look for when they are faced with filters, key words, or other kinds of proxies because most of the consumers are not privacy experts. Preference detection is a better way to provide relevant privacy information. Consumers must not know specific search terms or filter criteria because preference detection makes the connection between their privacy preferences and the underlying privacy information without any explicit user engagement (Type Practice Description TETs:detected).

Privacy practice monitoring TETs provide consumers with an overview of activities of an information system. A standardized monitoring includes information defined by TET providers. Consumers get only information others find relevant but cannot tailor monitoring to their own information needs, which is why standardized privacy practice monitoring is not consumer-friendly (Type Practice Monitoring TETs:standardized). A consumer-specified view of the processed data helps consumers to find the relevant information faster and tailor the monitoring to their own needs (Type Practice Monitoring TETs:consumer-specified). Preference detection is suitable too and offers faster access to relevant information because no input is required from consumers. However, proper working privacy practice monitoring based on preference detection requires suitable data to infer privacy preferences, which is hard to come by for monitoring (Type Practice Monitoring TETs:detected).

Privacy risk assessment TETs aim to make consumers aware of privacy risks. Standardization of the information provided is therefore appropriate, as risks unknown to consumers are also considered (Type Risk Assessment TET:standardized). Consumer-specified preference information can, however, be used to focus the assessment (Type Risk Assessment TETs:consumer-specified). Instead of providing access to all browser, app, or device content, it should be possible to make a dedicated decision about access and the scope of the evaluation. Preference detection (Type Risk Assessment TET:detected) is far too complicated for such a specific TET, as it is far too indeterminate to infer preferences for risk assessment from interaction data.

Privacy practice history TETs indicate changes in privacy practices through brief summaries or a comparison between old and new privacy practices. Standardization of privacy practice histories cannot account for individual consumer preferences. Therefore, a standardized privacy practice history does not add value to privacy information provision (Type Practice History TETs:standardized). Consumers should be able to choose how and about what they are informed, which is possible through consumer-specified approaches (Type Practice History TETs:consumer-specified). Even better would be to communicate also information on novel privacy practices, which would be possible via preference detection without need for manual effort and additional knowledge by the consumer (Type Practice History TETs:detected).

Privacy practice comparison TETs allow consumers to compare privacy practices between different information systems. Standardization of comparison features supports consumers in getting an overview over privacy practices (Type Practice Comparison TETs:standardized). But consumers should at least choose by themselves which information systems to compare against each other. The consumer-specified approach has an advantage, since a targeted selection of criteria gives consumers quicker access to information that is of interest to them (Type Practice Comparison TETs:consumer-specified). Preference detection is suitable too because of the quicker facilitation of access to relevant information. However, preferences detection makes it harder for consumers to keep track of changes in comparison criteria (Type Practice Comparison TETs:detected).

4.2 Feasibility of Local and Remote Processing

After having had a look on what types of TET tailoring approaches are a suitable solution for better provision of privacy information, we now move on to possible implementation approaches that can be deployed either locally or remotely, with different confidentiality-protecting mechanisms. Table 4 shows an overview of possible implementation approaches and their applicability for tailored TETs.

For privacy practice scoring TETs, which provide an overview and summary of privacy practices of an information system, and privacy practice description TETs, which inform consumers about privacy practices, the standardized approach is best realized with remote processing, as no adjustments based on user data are made. Consumer-specified and detected preferences can be processed locally, as the necessary calculations are not too computationally intensive. Hence, remote processing using HE is preferable if remote processing is necessary.

Privacy practice monitoring TETs provide consumers with an overview of activities of an information system. They can use local and remote processing for the standardized approach. It is important to keep in mind that in a local setting, only the locally available data and activities are available for monitoring; the same applies to remote approaches, which can only monitor provider activities. For consumer-specified tailoring, HE is preferable to SMPC as the preferences will likely only change very infrequently and the encrypted preferences can be reused. With a detection approach, changes will be more frequent and diminish this advantage, resulting in more overhead. Thus, SMPC should be a more suitable choice.

Table 4 Comparison of the applicability of standardized approaches, consumer-specified approaches, and detection approaches for TET type tailoring in local and remote environments. Legend: −− very unsuitable, − unsuitable, 0 not useful, + suitable, ++ very suitable, *N/A* not applicable

		Local	Remote	Remote with HE	Remote with SMPC
Practice scoring	Standardized	−−	++	N/A	N/A
	Consumer-specified	+	−−	++	+
	Detected	+	−−	++	+
Practice description	Standardized	−−	++	N/A	N/A
	Consumer-specified	+	−−	++	+
	Detected	+	−−	++	+
Practice monitoring	Standardized	+	+	N/A	N/A
	Consumer-specified	+	−−	++	+
	Detected	++	−−	+	++
Risk assessment	Standardized	++	−−	0	0
	Consumer-specified	++	−−	0	0
	Detected	++	−−	0	0
Practice history	Standardized	−−	++	N/A	N/A
	Consumer-specified	++	−	+	+
	Detected	++	−−	+	++
Practice comparison	Standardized	−−	++	N/A	N/A
	Consumer-specified	+	−−	++	+
	Detected	+	−−	+	++

Privacy risk assessment TETs use consumer data to calculate an individual risk score. Tailoring can be used to specify the analysis activity more precisely. For the standardized, consumer-specified, and detected approach, the necessary analyses can take place locally on the consumer device. The use of remote approaches is therefore not justified. HE and SMPC could be applied but without any benefits and would, therefore, constitute a waste of resources.

Tailoring privacy practice history TETs, which indicate changes in privacy practices through brief summaries or a comparison of past and current privacy practices, is best realized remotely when using standardized preferences, as there is no need for every device to calculate the same tailoring. Tailoring using consumer-specified or detected preferences can be done best locally. If the processing has to be done by the TET provider, the data should be protected: HE should be used when using consumer-specified preferences, as they are unlikely to change often and SMPC is more appropriate to handle the frequent changes when detecting preferences.

Privacy practice comparison TETs require lots of data about different providers in order to allow consumers to compare privacy practices between different information systems. This makes local processing for the standardized approach difficult; instead, remote processing is the most suitable choice, as no data needs to be collected from the consumer. In case of consumer-specified and detected tailoring, processing can be done locally, but it needs access to many data sources, which provide content for the tailoring that must be stored locally. Hence, encrypted remote processing makes sense to avoid storing multiple redundant copies of the same data. In the case of consumer-specified tailoring, HE should be used because consumers are unlikely to change their preferences once specified for the comparison to be made. In the case of detection, preferences are adapted more frequently, so SMPC is most likely a better choice.

5 Conclusions

In the beginning of this chapter, we set out to find a way through the privacy notice jungle. The good news is that there is a way. Even if revelation of privacy preferences is a “No-Go!” for consumers, we can realize confidentiality of privacy preferences through information systems design and offer tailored privacy information provision with confidentiality of privacy preferences. However, depending on the concrete use case and implementation, there might be a significant computational overhead compared to designs that do not provably protect the confidentiality of privacy preferences. A long road lies ahead; it should be kept in mind that there are no out-of-the-box solutions for tailored privacy information provision, nor do all approaches work equally well. Implicit detection approaches need very comprehensive data to perform reliable preference detection, which is not always technically feasible (e.g., tracking diverse sensor data in every situation) or practical (e.g., collecting a high amount of data for simple tailoring approaches like applying a filter criteria). Explicit consumer-specified preferences also have a drawback. Consumers have to think about and decide for themselves which settings they want in which situations. This may lead to frustration and rejection among consumers when privacy settings have to be repeatedly configured. Therefore, a sophisticated approach for using privacy preferences across a variety of information systems and a mix of implicit and explicit approaches is needed to provide consumers with real value and a path through the privacy notice jungle. On a more abstract level, the key takeaway of this chapter is that we should put more thought into what we are building and using our systems for to allow for privacy through human-centered design instead of static, predefined solutions which do not meet consumer needs. Since consumer privacy preferences are context-dependent [36], TETs need to be context-sensitive. Making this possible requires, however, even more consumer data more consumer data, which may cue additional privacy concerns. Yet, this is not as bad as it seems. In this chapter, we have outlined the parameters that can be adjusted for TETs and how privacy-preserving approaches can be implemented. The new and further development of TETs is in the hands of privacy researchers and privacy practitioners.

Acknowledgments This work was partially funded by a research grant for the project PANDIA: platform for the analysis of privacy notices of interactive assistance systems in the health care domain—consumer-centered privacy communication (German Federal Ministry of Education and Research (BMBF), funding reference number: 16SV8398). This work was also supported by funding from the topic Engineering Secure Systems of the Helmholtz Association (HGF) and by KASTEL Security Research Labs.

References

1. Acar, A., Hidayet Aksu, A. U., & Conti, M. (2018). A survey on homomorphic encryption schemes: Theory and implementation. *ACM Computing Surveys, 51*(4), 1–79.
2. Acquisti, A., Brandimarte, L., & Loewenstein, G. (2015). Privacy and human behavior in the age of information. *Science, 347*(6221), 509–514.
3. Acquisti, A., Brandimarte, L., & Loewenstein, G. (2020). Secrets and likes: The drive for privacy and the difficulty of achieving it in the digital age. *Journal of Consumer Psychology, 30*(4), 736–758.
4. Barth, S., Ionita, D., Jong, M. D., Hartel, P. H., & Junger, M. (2021). Privacy rating: A user-centered approach for visualizing data handling practices of online services. *IEEE Transactions on Professional Communication, 64*(4), 354–373.
5. Beaver, D., Micali, S., & Rogaway, P. (1990). The round complexity of secure protocols. In *Proceedings of the twenty-second annual ACM symposium on Theory of computing - STOC '90* (pp. 503–513). ACM Press.
6. Browser comparison tool | avoidthehack! https://avoidthehack.com/util/browser-comparison
7. Brüggemann, T., Hansen, J., Dehling, T., & Sunyaev, A. (2016). An information privacy risk index for mhealth apps. In *Proceedings of the 4th Annual Privacy Forum* (pp. 190–201). Springer.
8. Change history for Microsoft privacy statement–Microsoft privacy. https://privacy.microsoft.com/en-us/updates
9. Clemons, E. K. (2019). *New patterns of power and profit: A strategist's guide to competitive advantage in the age of digital transformation* (1st ed.). Palgrave Macmillan.
10. Common sense privacy evaluations. https://privacy.commonsense.org/evaluations/1
11. Cover your tracks. https://coveryourtracks.eff.org/
12. Dehling, T., Schmidt-Kraepelin, M., Demircan, M., Szefer, J., & Sunyaev, A. (2016). User archetypes for effective information privacy communication. In *Proceedings of the Pre-ICIS Workshop on Information Security and Privacy, AIS*.
13. Difference check tool. https://www.man7.org/linux/man-pages/man1/diff.1.html
14. Dingledine, R., Mathewson, N., & Syverson, P. (2004). *Tor: The second-generation onion router*. Naval Research Lab Washington DC.
15. General Data Protection Regulation (GDPR). (2016). https://gdprinfo.eu/
16. Goldreich, O., Micali, S., & Wigderson, A. (1987). How to play any mental game. In *Proceedings of the Nineteenth ACM Symp. on Theory of Computing, STOC* (pp. 218–229).
17. Habib, H., Zou, Y., Yao, Y., Acquisti, A., Cranor, L., Reidenberg, J., Sadeh, N., & Schaub, F. (2021). Toggles, dollar signs, and triangles: How to (in)effectively convey privacy choices with icons and link texts. In *Proceedings of the 2021 CHI Conference on Human Factors in Computing Systems* (pp. 1–25). ACM.
18. Heurix, J., Zimmermann, P., Neubauer, T., & Fenz, S. (2015). A taxonomy for privacy enhancing technologies. *Computers & Security, 53*, 1–17.
19. Hu, M. (2020). Cambridge analytica's black box. *Big Data & Society, 7*(2), 205395172093809.
20. Jawaheer, G., Weller, P., & Kostkova, P. (2014). Modeling user preferences in recommender systems: A classification framework for explicit and implicit user feedback. *ACM Transactions on Interactive Intelligent Systems (TiiS), 4*, 2:1–26.

21. Kaaniche, N., Laurent, M., & Belguith, S. (2020). Privacy enhancing technologies for solving the privacy-personalization paradox: Taxonomy and survey. *Journal of Network and Computer Applications, 171*, 102807.
22. Kelley, P. G., Cesca, L., Bresee, J., & Cranor, L. F. (2010). Standardizing privacy notices: An online study of the nutrition label approach. In *Proceedings of the SIGCHI Conference on Human Factors in Computing Systems*. Association for Computing Machinery.
23. Loepp, B., Hussein, T., & Ziegler, J. (2014). Choice-based preference elicitation for collaborative filtering recommender systems. In *Proceedings of the SIGCHI Conference on Human Factors in Computing Systems (CHI '14), Association for Computing Machinery* (pp. 3085–3094).
24. Maass, M., Wichmann, P., Pridöhl, H., & Herrmann, D. (2017). Privacyscore: Improving privacy and security via crowd-sourced benchmarks of websites. arXiv:1705.05139 [cs].
25. Milne, G. R., Pettinico, G., Hajjat, F. M., & Markos, E. (2017). Information sensitivity typology: Mapping the degree and type of risk consumers perceive in personal data sharing. *Journal of Consumer Affairs, 51*(1), 133–161.
26. Murmann, P., & Fischer-Hubner, S. (2017). Tools for achieving usable ex post transparency: A survey. *IEEE Access, 5*, 22965–22991.
27. Nissenbaum, H. F. (2010). *Privacy in context: Technology, policy, and the integrity of social life*. Stanford Law Books.
28. Obar, J. A., & Oeldorf-Hirsch, A. (2020). The biggest lie on the Internet: Ignoring the privacy policies and terms of service policies of social networking services. *Information, Communication & Society, 23*(1), 128–147.
29. Peppet, S. R. (2011). Unraveling privacy: The personal prospectus and the threat of a full-disclosure future. *Northwestern University Law Review, 105*(3), 1153–1204.
30. Pollach, I. (2006). Privacy statements as a means of uncertainty reduction in www interactions. *Journal of Organizational and End User Computing, 18*(1), 23–49.
31. PoPETs/P.E.T.S. (2022). https://petsymposium.org/
32. Privacy cleaner. https://chrome.google.com/webstore/detail/privacy-cleaner/liiikhhbkpmpomjmdofandjmdgapiahi
33. Privacy Score Guide. Privacy monitor. https://www.privacymonitor.com/score/
34. Privacy test & analyzer: See what information websites know about you. https://privacy.net/analyzer/
35. Rodrigues, R., Wright, D., & Wadhwa, K. (2013). Developing a privacy seal scheme (that works). *International Data Privacy Law, 3*(2), 100–116.
36. Schaub, F., Balebako, R., Durity, A., & Cranor, L. (2015). A design space for effective privacy notices. In *Eleventh Symposium On Usable Privacy and Security (SOUPS 2015)* (pp. 1–17). USENIX Association.
37. Secure messaging apps comparison | privacy matters. https://www.securemessagingapps.com/
38. Shanmugarasa, Y., Paik, H.-y., Kanhere, S. S., & Zhu, L. (2022). Automated privacy preferences for smart home data sharing using personal data stores. *IEEE Security Privacy, 20*(1), 12–22.
39. Sunyaev, A., Dehling, T., Taylor, P. L., & Mandl, K. D. (2014). Availability and quality of mobile health app privacy policies. *Journal of the American Medical Informatics Association, 22*(e1), 28–33.
40. Updates: Privacy policy—privacy & terms–Google. https://policies.google.com/privacy/archive?hl=en-US
41. Woods, D. W., & Böhme, R. (2022). The commodification of consent. *Computers & Security, 115*, 102605.
42. Yao, A. C. (1982). Protocols for secure computations. In *23rd Annual Symposium on Foundations of Computer Science* (pp. 160–164).
43. Zaeem, R. N., German, R. L., & Barber, K. (2018). Privacycheck: Automatic summarization of privacy policies using data mining. *ACM Transactions on Internet Technology, 18*(4), 1–18.
44. Zhang, S., Feng, Y., Bauer, L., Cranor, L. F., Das, A., & Sadeh, N. (2021). "Did you know this camera tracks your mood?" Understanding privacy expectations and preferences in the age of video analytics. In *Proceedings on Privacy Enhancing Technologies 2021* (Vol. 2, pp. 282–304).

Acceptance Factors of Privacy-Enhancing Technologies on the Basis of Tor and JonDonym

Sebastian Pape and David Harborth

1 Introduction and Background

Bruce Schneier states [49]: "Surveillance is the business model of the internet. Everyone is under constant surveillance by many companies, ranging from social networks like Facebook to cellphone providers." One of the reasons for the surveillance of users is a rising economic interest in the Internet [3]. However, users are not helpless and can make use of privacy-enhancing technologies (PETs) to protect them. Examples of PETs include services that allow anonymous communication, such as Tor [68] or JonDonym [40].

Tor and JonDonym are low-latency anonymity services that redirect packets in a certain way to hide metadata (the sender's and optionally—in case of a hidden service—the receiver's Internet protocol (ip) address) from passive network observers. While Tor and JonDonym differ technically, they are highly comparable with respect to the general technical structure and the use cases. Tor offers an adapted browser including the Tor client for using the Tor network, the "Tor Browser." Similarly, the "JonDoBrowser" includes the JonDo client for using the JonDonym network.

However, the entities who operate the PETs are different. Tor is operated by a non-profit organization with thousands of voluntarily operated servers (relays) and an estimated 2 million daily users by the Tor Project [68] and an estimated 8 million daily users by Mani et al. [46]. Tor is free to use with the option that users can donate to the Tor project. JonDonym is run by a commercial company with servers (mix cascades) operated by independent and non-interrelated organizations or private individuals who all publish their identity. A limited service is available

S. Pape (✉) · D. Harborth
Chair of Mobile Business and Multilateral Security, Goethe University Frankfurt, Frankfurt, Germany
e-mail: sebastian.pape@m-chair.de; david.harborth@m-chair.de

N. Gerber et al. (eds.), *Human Factors in Privacy Research*,
https://doi.org/10.1007/978-3-031-28643-8_15

for free, and different premium rates allow to overcome the limitations. The actual number of users is not known since the service does not keep track of this. While the number of users of anonymization services is large enough to conduct studies and evaluate the running systems, it is quite low compared to the number of Internet users in total, which was estimated to 4.13 billion in 2019 [7]. Far less than 1% of the users use anonymization networks.

In order to investigate why there is not a broader adoption of anonymization services, some user research seems to be necessary: Investigating users' privacy concerns and their technology acceptance to find factors promoting the use of PETs. Since Tor is one of the most prominent PETs, the hope is that the insights can also be transferred to other PETs.

Besides the users' perspective, it is also important to investigate the economic side: Are users willing to pay for PETs and which incentives and hindrances exist for companies to implement PETs?

For PETs such as anonymization networks such as Tor [68] or JonDonym [40] that allow anonymous communication, there has been a lot of research [50, 64], but the large majority of it is of technical nature and does not consider the users and their perceptions. However, the number of users is essential for anonymization networks since an increasing number of (active) users also increases the anonymity set. The anonymity set is the set of all possible subjects who might be related to an action [58], and thus, a larger anonymity set may make it more difficult for an attacker to identify the sender or receiver of a message. Therefore, it is crucial to understand the reasons for the users' intention to use a PET or obstacles preventing it [1].

However, for the propagation of a PET, it is not only important to understand the users' intentions to use the PET, but also the users' willingness to pay for the service, which would allow companies to build a business model upon the provision of the service. The main challenge in motivating the user to pay for PET, i. e., an anonymization service, is that the user can barely notice a working PET directly. Noticing an anonymization network is in most cases the result of a limitation of throughput, performance, or response time. Indirect effects such as fewer profiling are also hard to detect, but even harder to connect to the PET in place. This makes it hard for a company as well as the user to sell or, respectively, understand the advantages for these types of PETs. As a consequence, it is hard for a company to come up with a business model, and thus the further distribution of PETs is prevented [52].

Therefore, besides investigating the users' intention to use a PET on the basis of Tor in Sect. 3.1 and JonDonym in Sect. 3.2, we also investigate in Sect. 3.4 the economic sides from the perspective of the users' willingness to pay for Tor or JonDonym and in Sect. 3.5 from the perspective of a business owner to provide a PET in general as service.

2 Methodology

In this section, we first describe how the questionnaire was built and how the data were collected and evaluated (cf. Sects. 2.1–2.3). In the second part, we briefly sketch how we conducted and evaluated experts' interviews (cf. Sects. 2.4 and 2.5).

2.1 Questionnaire Composition

To investigate the users intention to use Tor or JonDonym, we made use of two different popular structural equation [19] models:

Internet Users' Information Privacy Concerns (IUIPC) is a construct by Malhotra et al. [45] for measuring and explaining privacy concerns of online users that is embedded in a larger nomological net with other privacy-related variables. IUIPC is operationalized as a second-order construct[1] of the sub-constructs collection, awareness, and control (please refer also to the chapter "Toward Valid and Reliable Privacy Concern Scales: The Example of IUIPC-8" for a detailed discussion of the IUIPC). That means the user's concerns are determined by concerns about data on the user in relation to the value or received benefits, by concerns about the control users have over their own data, and by concerns about his or her awareness regarding organizational privacy practices. The privacy concerns then influence trusting beliefs and risk beliefs that in turn influence the user's behavior. The use behavior was the release of personal information to a marketing service provider in the original research. The trusting and risk beliefs refer to the users' perceptions about the behavior of online firms (in general) to protect or lose the users' personal information.
The IUIPC construct has been used in various contexts, such as Internet of Things [51], Internet transactions [39], and mobile apps [59]. Furthermore, it has recently been re-evaluated in several studies [54, 55]. But so far it had not been applied to a PET such as an anonymization service. There is a major difference between PETs and other services, i. e., apps [30, 35, 53] or games [24, 33] regarding the application of the IUIPC instrument. The other services had a certain use for their customer (primary use), and the users' privacy concerns were investigated for the use of the service. The concepts of trusting and risk beliefs matched that in a way that they were referring to "general companies" that may provide a service to the user based on data they receive. However, for anonymization services, providing privacy is the primary purpose. Therefore, it is necessary to distinguish between trusting and risk beliefs with respect to technologies that aim to protect personal data (PETs) and regular Internet

[1] For an extensive discussion on second-order constructs, see Steward [66].

services. As a consequence, the trust model within IUIPC's causal model was extended by trusting beliefs in Tor/JonDonym.

Technology Acceptance Model (TAM) was developed by Davis [9, 10] based on the the theory of reasoned action (TRA) by Fishbein and Ajzen [12] and the theory of planned behavior (TPB) by Ajzen [2] (see also the chapter "From the Privacy Calculus to Crossing the Rubicon: An Introduction to Theoretical Models of User Privacy Behavior"). According to the TRA, a person's behavioral intention determines that person's behavior. The behavioral intention itself is influenced by the person's subjective norms and attitude toward the behavior. The subjective norms refer to a person's normative beliefs and normative pressure to perform or not perform the behavior. The attitude relies on the person's beliefs about the behavior and its consequences. TPB is an extension of the TRA with the same overall structural process: the behavioral intention is influenced by several components and influences the behavior. However, the TPB adds perceived behavioral control that refers to a person's perception regarding the ease or difficulty of performing a given behavior in a given situation.

2.2 Questionnaire Data Collection

We conducted a *survey among users of the anonymization services JonDonym and Tor*. For both surveys, we conducted the study with German- and English-speaking users. Thus, we administered two questionnaires for each service. All items for the German questionnaire had to be translated into German since all of the constructs are adapted from the English literature [26, 27]. To ensure content validity of the translation, we followed a rigorous translation process [23, 24]. First, we translated the English questionnaire into German with the help of a certified translator (translators are standardized following the DIN EN 15038 norm). The German version of the questionnaire was then translated back to English by a second independent certified translator. This step was done to ensure the equivalence of the translation. Third, a group of five academic colleagues checked the two English versions with regard to this equivalence. All items were found to be equivalent.

Since we investigate the effects of privacy concerns, trust and risk beliefs on the use of JonDonym and Tor, we collected data of actual users of the PET. We installed the surveys on a university server. For JonDonym, the links to the surveys were distributed with the beta version of the JonDonym browser and published on the official JonDonym homepage. For Tor, the links to the English and German version were distributed over multiple channels on the Internet (cf. [29, Appendix A]). Surprisingly, although there are approximately two million active Tor users, it was more difficult to gather the necessary number of complete answers for a valid and reliable quantitative analysis for Tor users. After deleting all incomplete sets and sets from participants who answered a test question in the middle of the survey incorrectly, 124 usable data sets remained for Tor [29] and 141 usable data sets remained for JonDonym [28] for our analysis. The questionnaires and the answers to Likert scale questions are available online [31, 32].

For both services, the demographic questions were not mandatory. This was done on purpose since we assumed that most of the participants are highly sensitive with respect to their personal data. Therefore, we had to resign from a discussion of the demographics in our research context. This decision is backed up by Singh and Hill, who found no statistically significant differences across gender, income groups, educational levels, or political affiliation in the desire to protect one's privacy [65]. However, other studies also showed that technological knowledge is not equally distributed in different age groups [17, 53], and users with a better education are more likely to use PETs [60]. In the end, our decision is a trade-off between the ability to take demographic effects in consideration and the chance to have highly privacy-aware participants who might have aborted answering the questionnaire (or lied) if demographic questions had been mandatory.

2.3 Questionnaire Evaluation

We made use of a mixed method approach consisting of quantitative and qualitative methods. We start by describing the quantitative methods and then describe the qualitative part.

Quantitative Methods

We applied a standard statistical analysis approach called *structural equation modeling* (SEM) to assess our research model and the corresponding hypotheses regarding the cause–effect relationships among these constructs. SEM can reveal how much of the variance in the dependent variables (effects) can be explained by the independent variables (causes). There are two main approaches for SEM, namely covariance-based SEM (CB-SEM) and partial least squares SEM (PLS-SEM). Since our research goal is to predict the dependent variables (effects) *behavioral intention* and *actual use behavior* of PETs and maximize the explained variance for these dependent variables, we use PLS-SEM [19] for our analysis (Hair et al. extensively discuss on the use of PLS-SEM [18]). For that purpose, we first built our models for IUIPC-10 [28, 29, 34] and TAM [25, 37, 38] based on the existing literature. We then tested our model using SmartPLS [63]. To assess the quality of all different models, we investigated the structural model (e.g., possible collinearity problems) and the measurement model (internal consistency reliability, convergent validity, and discriminant validity). For all of the models, the structural model and the measurement model were consistent and checks were fine for reliability and validity on both data sets. For details, we refer to the respective papers [25, 28, 29, 34, 37, 38].

Since JonDonym and Tor are different with respect to the pricing schemes and the organizational structure of the providers, we are interested whether there are significant differences in the hypothesized relationships between the variables. To compare JonDonym and Tor users in the TAM, we split the data set into two parts and analyzed the results for Tor and JonDonym separately. For that, we conducted a *multigroup analysis* in SmartPLS and tested whether there are statistically significant differences for each of the hypotheses.

As a last step, we conducted a *logistic regression* [21] to find out which factors influence users' willingness to pay for privacy (in our case willingness to pay for JonDonym and willingness to donate to Tor). We used the logistics regression to build the model because our dependent variable is a binary variable. A linear regression is not an appropriate model here due to the violation of the assumption that the dependent variable (WTP) is continuous, with errors that are normally distributed [48]. Willingness to pay for JonDonym is defined as the binary classification of JonDonym users' actual behavior. The regression was conducted with the open-source statistic software R.

We use a less conservative level of statistical significance of 10% here since the p value is sensitive to the relatively small sample sizes when comparing results for Tor and JonDonym. Thus, we provide this level of statistical significance in this analysis to indicate potential statistically significant differences between the effects for Tor and JonDonym. In addition, the oftentimes referenced statistical significance level of 5% only indicates a "convenient" threshold for judging statistical significance [13] and can be considered a rule of thumb.

Qualitative Methods

The questionnaire contained four open questions from which we aimed to get deeper insights into certain aspects of the quantitative analysis described above. We asked if users have any concerns, which additional features they would like, and why they would (not) recommend JonDonym or Tor. JonDonym users were additionally asked under which circumstances they would choose one of the premium tariffs. Two researchers analyzed the statements independently from each other and abstracted the individual answers to codes. Codes summarize the data and present different dimensions of a concept. For example, we find that *usability* is an important concept for both technologies. However, the results indicate that the code *usability* can be found with a negative as well with a positive characteristic depending on the user and the respective context (e. g., users praising or complaining about the usability of the PETs depending on what they intend to achieve).

Altogether 626 statements were collected. The coding was done in two stages, following a method from sociology [6, 16], which comprises two or three coding phases, namely initial coding, axial coding, and focused coding. We only used initial and focused coding since this level of structuring is sufficient for our data [6]. First, we initially coded each of the statements. These initial codes in itself provide a sorting and structuring for the data. Initial codes represent topics that occur frequently in the data, i. e., topics often mentioned by participants. In our case, we decided to name these codes "Subconcepts" in our results since they already provide one level of abstraction. After the initial coding phase, we compared the different codings of the researchers and discussed the individual codes. Thereby, we agreed upon certain subconcepts that were similar or the same but expressed differently by the coders. In a next step, we calculated the intercoder reliability. We did not use a common codebook or a predefined set of codes to do the initial coding. Therefore, the known reliability measures such as Cohen's Kappa [8] are not usable for our case since these measures are relying on predefined categories. Consequently, we used a very simple calculation in order to provide a reliability measure dividing the number of equally coded statements by the total number of statements to be coded. We had 226 matches for Tor and 242 matches for JonDonym, which yield intercoder reliabilities of 68.69% and 81.48%, respectively, for the total number of statements for each PET. Thus, the intercoder reliability is equal to 74.76% for both PETs. These numbers are relatively large considering that we coded independently from each other without agreeing to fixed subconcepts beforehand. We also counted the incidents in which one of the coders had at least one more code assigned to a statement than the other coder in order to provide more transparency of our coding process. This happened 52 times (coder 1 had 29 times more codes, coder 2 had 23 times more codes) for Tor and 44 times for JonDonym (coder 1 had 27 times more codes, coder 2 had 17 times more codes). These instances are counted toward the mismatches in the intercoder reliability measures. In the second step, we structured the most occurring themes in these initial codes and came up with the focused codes. We name these codes "Concepts" and find that users primarily make statements about either technical issues, their beliefs and perceptions, or economic issues.

2.4 *Interview Data Collection*

For the *interviews of privacy experts*, we designed a semi-structured interview guide that we used to conduct the interviews. Semi-structured in this context means that the interview is significantly influenced by the respondent's interaction and answers. The questionnaire only records particularly relevant questions that definitely need to be addressed from the researcher's point of view. This has the advantage of being able to obtain the deepest possible insights and most detailed answers from the participant. The questionnaire can be divided into three main topics. First, general questions about the person and the company are asked. This is followed by questions about privacy and PETs. The second part covers technical questions about the status quo and possible future developments. The third part covers economic and societal issues. We interviewed experts and professionals who are involved with privacy-enhancing technologies (PETs) in their companies or in whose products or services privacy plays a special role. The experts are from companies that directly offer PETs or in which privacy plays an important role in the value proposition. Examples include the telecommunications sector, payment providers, or eCommerce solution providers. We conducted and analyzed ten interviews, varying in duration from 44 to 180 min. The demographic information can be found in our respective article [20].

2.5 Interview Evaluation

The *expert interviews* were all recorded and then transcribed word for word. The transcripts were then analyzed using what is known as open coding and selective coding [6, 16, 67]. Open coding is the first step of data analysis and is closely oriented to the data (the transcripts). In the next step, codes are summarized and abstracted (selective coding). These steps are performed separately for each interview and then between interviews. This so-called comparative method [6, 16, 67] is an elementary component of the qualitative research methodology. By constantly comparing across interviews, we derived abstract categories from the data that provide a diverse picture of incentives and disincentives. These coding steps were performed by two authors to identify and resolve any discrepancies in the analysis of the data.

3 Results

We first present the results for the two different structural equation models based on IUIPC (cf. Sect. 3.1) and TAM (cf. Sect. 3.2). Then, we briefly discuss the evaluation of the open questions (cf. Sect. 3.3). Besides users' concerns and factors influencing their technology use acceptance, it is also important to consider factors for a successful business model built on a PET. For that purpose, we additionally investigated the users' willingness to pay or donate for a PET (cf. Sect. 3.4) and also considered the perspective of companies by investigating their incentives and hindrances to implement PETs (cf. Sect. 3.5).

3.1 Internet Users Information Privacy Concerns

The basic idea of investigating users' privacy concerns was to learn how they influence users' behavioral intention to use the service. Figure 1 shows the SEM for JonDonym users and Fig. 2 for Tor users. The models for JonDonym and Tor users turned out to be very similar. Most of the relations were as expected, somewhat surprising was the result that general trusting and risk belief had no significant effect on the use behavior. However, for the rather small effect sizes, it might be that the sample size was simply not large enough to show a significant relationship. In any case, the trust in JonDonym or Tor had by far a larger influence on the use behavior, respectively, the behavioral intention. The result shows that the reputation of being

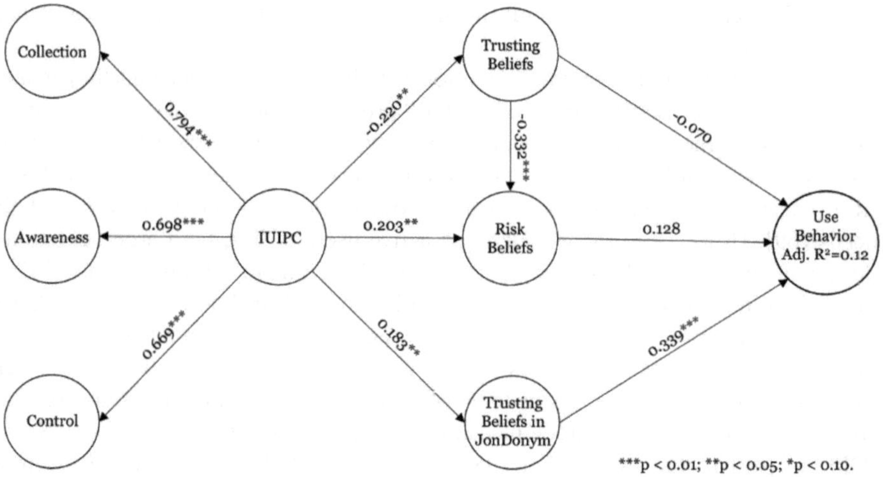

Fig. 1 JonDonym users, IUIPC, path estimates, and adjusted R^2 values of the structural model [28]

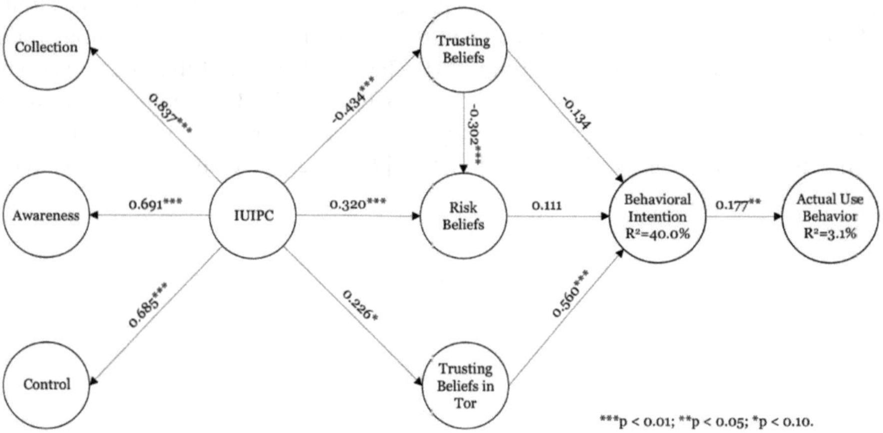

Fig. 2 Tor users, IUIPC, path estimates, and adjusted R^2 values of the structural model, figure taken from Harborth and Pape [29] licensed under CC BY-NC-ND 4.0

a trustworthy provider, respectively, service, is crucial for an anonymization service provider. The results also show that users with a higher level of privacy concerns rather tend to trust their anonymization service provider, which might be affected by the fact that we only asked users of the respective PET.

In general, if there is a reliable measure of the use behavior, it is a better indicator than the users' behavioral intention to use a service. Since we questioned actual users, we could use their use frequency of the services. However, the results indicate that the influence of the behavioral intention on the actual use behavior was rather small for Tor users.

Users' attitudes and behavioral intention can differ from the decisions they make. This phenomenon is often denoted as the "privacy paradox" [15]. Two possible explanations come to mind to explain the privacy paradox: (i) users balance between potential risks and benefits they gain from the service (privacy calculus) [11] and (ii) users are concerned but lack knowledge to react in a way that would reflect their needs [69]. However, since we surveyed active users of Tor, both argumentations do not fit. Regarding the privacy paradox, we have already discussed how PETs differ from regular Internet services. Regarding the lack of knowledge, users have already installed the PET and use it. However, it is still important to investigate the users' capabilities since users need a certain amount of knowledge in order to adequately evaluate the given level of privacy [57, 69]. For that purpose, we added the users' privacy literacy measured with the *Online Privacy Literacy Scale* (OPLIS) [47] to the model. For that purpose, we slightly adapted the original questionnaire since it aimed at the German population and contains questions about German and European data protection laws. With our sample of Tor users possibly spread from all over the world, it does not make sense to ask them for German or even European privacy laws. As a consequence, we omitted the respective questions about national laws, and we extrapolated our results from 15 to 20 questions for a comparison with the reference group [34]. The results showed that users' privacy literacy positively influences trusting beliefs in Tor (cf. Fig. 3). Therefore, educating users and increasing their privacy literacy should add to the behavioral intention of using Tor. Built on our work, Lux and Platzer [44] investigated the relation between

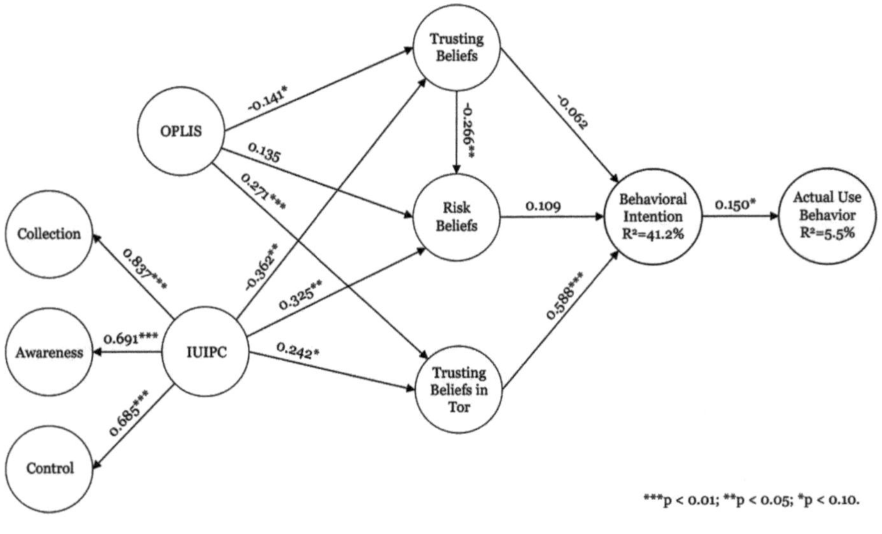

Fig. 3 Tor users, IUIPC and OPLIS, path estimates, and adjusted R^2 values of the structural model [34]

online privacy literacy and the usage of Tor in more detail following our approach to use only 15 items and to extrapolate the result. We will further investigate the influence of the behavioral intention on the actual use behavior by making use of the TAM model in the next subsection.

3.2 Technology Acceptance Model

Within the same survey, we also asked the participants about certain constructs we could use in a TAM model [27]: How they perceived the usefulness, the ease of use, and the anonymity of the PET. Since we had already identified trust in the PET as a major driver for the behavioral intention, we included it too. The resulting model is shown in Fig. 4 including JonDonym and Tor users [37].

The model shows significant relationships for all paths as already known from the TAM model with three noteworthy observations:

- There are three main drivers of the PETs' perceived usefulness: perceived anonymity, trust, and perceived ease of use that explain almost two-thirds of its variance. This demonstrates that for PETs the two newly added variables perceived anonymity and trust in the PETs can be important antecedents in technology acceptance models for PETs.
- Similar than in the IUIPC model, trust in the PET is the most important factor for behavioral intention. This underlines the importance of trust in the PETs as

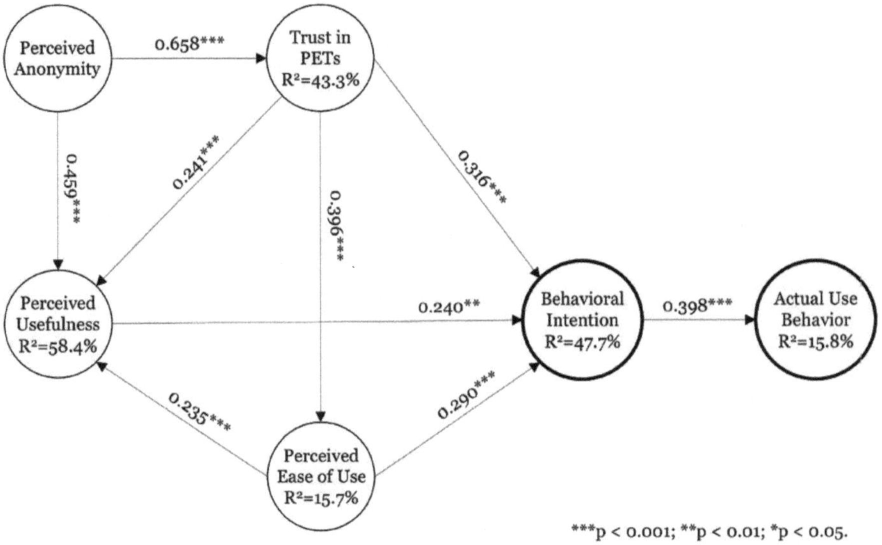

Fig. 4 TAM-based research model with path estimates and R^2 values of the structural model for PETs, figure taken from Harborth et al. [37] licensed under CC BY-NC-ND 3.0

a highly relevant concept when determining the drivers of users' use behavior of PETs.

- Since the effects of perceived anonymity and trust in the PETs on behavioral intention and actual use behavior were partially indirect, we calculated the total effects. All of the effects were highly statistically significant (p value <0.001), and the total effects on behavioral intention are relatively large (PA $\rightarrow$ BI: 0.446; Trust_{PETs} $\rightarrow$ BI: 0.511), while the effects on the actual use are as expected smaller (PA $\rightarrow$ USE: 0.177; Trust_{PETs} $\rightarrow$ USE: 0.203).

To investigate the differences between JonDonym and Tor and also to further investigate the small effect of behavioral intention on actual use behavior, we conducted a multigroup analysis to test whether there are statistically significant differences between JonDonym and Tor users as shown in Table 1. The table also shows the path coefficients for both PETs individually.

These results indicate that the most significant difference between JonDonym and Tor users was the effect size between behavioral intention and actual use, which is 0.679 for JonDonym and 0.179 for Tor. Less significant observations were that the effects of trust on behavioral intention and perceived anonymity on perceived usefulness were slightly larger for JonDonym users. A possible explanation could be the structure of the two services, as JonDonym is a profit-oriented company that charges for the unlimited use of the PET [40], while Tor is a community-driven project based on donations.

Table 1 Results of the MGA analysis (gray background indicates statistical significance at least at the 10% level) [37]

	Original path coefficient		*P* values		Path coefficient difference	*P* value
Relationships	JonDonym	Tor	JonDonym	Tor	JonDonym vs Tor	
PA $\rightarrow$ Trust_{PETs}	0.597	0.709	<0.001	<0.001	0.112	0.865
PA $\rightarrow$ PU	0.543	0.369	<0.001	<0.001	0.174	0.088
Trust_{PETs} $\rightarrow$ BI	0.416	0.232	<0.001	0.010	0.184	0.064
Trust_{PETs} $\rightarrow$ PU	0.173	0.304	0.035	0.008	0.131	0.823
Trust_{PETs} $\rightarrow$ PEOU	0.378	0.431	<0.001	<0.001	0.053	0.657
PU $\rightarrow$ BI	0.183	0.300	0.046	0.002	0.117	0.805
PEOU $\rightarrow$ BI	0.206	0.371	0.011	<0.001	0.165	0.929
PEOU $\rightarrow$ PU	0.182	0.300	0.039	<0.001	0.118	0.830
BI $\rightarrow$ USE	0.679	0.179	<0.001	0.029	0.500	<0.001

BI behavioral Intention, *PEOU* perceived ease of use, *PA* perceived anonymity, *USE* actual use frequency, *PU* perceived usefulness

3.3 *Evaluation of Open Questions*

To gather some reasons for the observed differences and possibly identify other differences of the services from a user perspective, we included five open questions in the survey. The results of their coding are shown in Table 2. In the left column, we have the three concepts technical issues, beliefs and perceptions, and economical issues. Each of them includes several subconcepts. The results were then clustered into statements common to both PETs, such as feature requests (**Tor.1**, **Jon.1**), statements only referring to Tor, such as statements about malicious exit nodes (**Tor.2**), and statements only referring to JonDonym, such as concerns about the location of mix cascades (**Jon.2**). For each statement, we selected at least one quote shown at the bottom of the table.

The result for user perceptions shows that both services differ not that much with respect to technical issues but in the users' beliefs. Unsurprisingly, economical issues were only concerning JonDonym. Three main differences might be able to explain the observed different effect sizes in the structural equation model. As already discussed, trust models between the services were different in the way that for JonDonym, users have to trust a company (**Jon.13**), while Tor users have to trust their community (**Tor.12**). While the concept for both technologies is that the users' anonymity does not rely on a single malicious server, there is still trust necessary since only a minority of the users will inspect the programs they are running. For JonDonym users, the size of the user base was also an issue (**Jon.11**). However, the most interesting observation also in terms of explaining the weak effect of behavioral intention on actual use behavior for Tor users was that many Tor users were concerned about looking like a criminal (**Tor.13**, **Tor.14**).

3.4 *Customers' Willingness to Pay or Donate*

Within the same survey as already described in the previous subsection, we also asked JonDonym users about their recent tariff and Tor users if they ever have donated to Tor [21]. It showed that the majority of users was not willing to pay or donate for the services: 85 out of 141 users (60%) used JonDonym's free tariff and 93 out of 124 (75%) Tor users have never donated to Tor.

For JonDonym, we also compared the users' preferences for certain tariff structures depending on factors such as data volume, pricing, and contract duration. We were comparing users' preferences toward existing tariffs: a high-data-volume tariff, a low-price tariff, and a low-anonymity tariff and two newly created tariffs adding a lower data volume than the low-price tariff and a higher volume than the high-data-volume tariff. Free users were neutral to all tariffs but showed a slight preference to the newly created low-traffic tariff. Already paying users preferred the existing and newly created high-data-volume tariffs over the others. This indicates that free users would prefer the cheapest tariff if they decide to pay at all. This

Table 2 Results of the coding for the open questions including quotes [37]

Concepts	Subconcepts	Common to both PETs	Specific Subconcepts for Tor	Specific Subconcepts for JonDonym
Technical Issues	PET design	Feature Requests (**Tor.1, Jon.1**)	Malicious exit nodes (**Tor.2**)	Location of mix cascades (**Jon.2**)
	Compatibility	Accessibility of websites (**Tor.3, Jon.3**)		
	Usability	Documentation (**Tor.4, Jon.4**) Ease of use (**Tor.5, Jon.5**) Missing knowledge to use it correctly (**Tor.6,Jon.6**)		
	Performance	Latency (**Tor.7, Jon.7, Jon.8**)		
Beliefs and Perceptions	Anonymity	Concerns about deanonymization (**Tor.8, Jon.9**) Reason of use (**Tor.9, Jon.10**)		Size of the user base (**Jon.11**)
	Consequences	Fear of investigations (**Tor.10, Tor.11, Jon.12**)	Beliefs about social effects (**Tor.13, Tor.14**)	
	Trust		Trust in the community (**Tor.12**)	Trust in technology (**Jon.13**)
	Substitute technologies	Best available tool (**Tor.15, Jon.14**)		Tor as reference technology (**Jon.3, Jon.8, Jon.11**)
Economical Issues	Costs			Lower costs, other pricing schemes (**Jon.15**)
	Payment methods			Easy, anonymous payment options (**Jon.15**)
	Use cases		Circumvent Censorship (**Tor.16**)	Willingness to pay in certain scenarios (**Jon.16, Jon.17**)

Tor.1 *TCP support for name resolution via Tor's DNSPort [...]*

Tor.2 *Many exit nodes are run by governmental intelligence organizations. Exit notes can collect unencrypted data.*

Tor.3 *It can't be used on all websites; therefore it is of limited use to me*

Tor.4 *Easy to understand instructions for users with different levels of knowledge.*

Tor.5 *Tor protects privacy while on the web and is easy to use.*

Tor.6 *An unexperienced user may not understand the technical limitations of Tor and end up losing [...] privacy.*

Tor.7 *Increased latency makes the experience painful at times*

Tor.8 *It may fail to provide the expected level of anonymity because of attacks which may not even be known at the time they are performed (or commonplace).*

Tor.9 *It is a key component to maintaining one's privacy when browsing on the Internet.*

Tor.10 *Tor usage "Stands out"*

Tor.11 *[...] having a cop boot at my door because of Tor.*

Tor.12 *An end user needs to trust the network, the persons running Tor nodes and correct implementations [...]*

Tor.13 *Only social backlash from people thinking that Tor is mostly used for illegal activities.*

Tor.14 *For the same reason I don't hang out in brothels, using Tor makes you look like a criminal*

Tor.15 *While not perfect, Tor is the best option for reliable low-latency anonymization*

Tor.16 *It can be used as a proxy / VPN to get past censorship*

Jon.1 *Larger number of Mix Cascades, more recent software, i.e. pre-configured browser, faster security updates*

Jon.2 *First and last server of the mix cascade should not be located in the same country*

Jon.3 *Unlike Tor, JonDonym is not blocked by some websites. (Google for example among others)*

Jon.4 *Clearer explanations and instructions for JonDoFox*

Jon.5 *Easy to use, outside the mainstream like i.e. Tor*

Jon.6 *Privacy is less than expected because of wrong configuration settings.*

Jon.7 *[...] Even if it is quite slow without a premium tariff*

Jon.8 *[...] sometimes it's a little bit to slow, but compared with Tor...*

Jon.9 *Defeat of your systems by government agencies.*

Jon.10 *It provides a minimum level of personal data protection and online safety.*

Jon.11 *Tor is better due to having a much larger user base. More users results in greater anonymity*

Jon.12 *By using the service, am I automatically marked by intelligence authorities as a potential terrorist, supporter of terrorist organizations, user [...] for illegal things?*

Jon.13 *How can I trust JonDonym? How can JonDonym proof that servers are trustworthy?*

Jon.14 *It appeared to be the least worst option for anonymization when I researched anonymization services*

Jon.15 *Fair pricing, pre-paid is an easy payment option.*

Jon.16 *For use it in a country where it's difficult surf the net*

Jon.17 *If I would use the computer for work-related tasks*

suggests that providers of PETs should offer tariffs with a low monetary barrier to convert free users into paying users. However, even with a low monetary barrier, there would still be the need to resolve the payment barrier, which regularly shows in e-commerce when customers are abandoning their shopping cart before the payment process [61].

We also built a regression model to identify significant factors contributing to the willingness to pay. For that purpose, we defined a binary classifier for the willingness to pay (JonDonym), being 0 if the respondent was using a free tariff and being 1 if the respondent was using a premium tariff. Analogous, we defined the willingness to donate (Tor), being 0 if the respondent has never donated and being 1 if the respondent has donated at least once. As independent variables, we considered risk propensity (RP), frequency of improper invasion of privacy (VIC), trusting beliefs in online companies (TRUST), trusting beliefs in JonDonym (TRUST_{PET}), and knowing of Tor / JonDonym (TOR/JD) and derived the following research model:

$$WTP/WTD_i = \beta_0 + \beta_1 \cdot RP_i + \beta_2 \cdot VIC_i + \beta_3 \cdot TRUST_i + \beta_4 \cdot TRUST_{PET,i} + \beta_5 \cdot TOR/JD_i + \epsilon_i.$$

The results are shown in Table 3, and one more time indicates that trust in the PET is the prevalent factor. On a highly significant level, the regression model suggests that a one unit increase in trust results in a roughly 12% higher likelihood that users choose a premium tariff (JonDonym) or donate (Tor). Besides that, the only significant variables were risk propensity for JonDonym and past privacy victim experiences for Tor. Surprisingly, risk propensity had a negative coefficient, indicating that more risk-averse users are less likely to choose a premium tariff for JonDonym. This contradicts previous findings [14] that risk aversion can act as a driver to protect an individual's privacy. For Tor, bad experiences with privacy breaches lead to a higher probability of donating money, even though on a more marginal level of roughly 5% per unit.

Table 3 Results of the logistic regression model for users' willingness to pay/donate [21]

	WTP for JonDonym		WTD for Tor		
	Coeff	Avg. marg. effects	Coeff	Avg. marg. effects	Difference
(Intercept)	−0.0376	−0.0081	6.1455***	−0.9768	0.9687
RP	−0.4967**	−0.1067	−0.1492	−0.0237	−0.083
VIC	−0.0397	−0.0085	0.3352**	0.0533	−0.0618
TRUST	−0.0868	−0.0187	−0.1222	−0.0194	0.0007
TRUST_{PET}	0.5661***	0.1217	0.7835***	0.1245	−0.0028
TOR/JD	−0.5792	−0.1245	0.488	0.0776	−0.2021

$^*p < 0.1$, $^{**}p < 0.01$, $^{***}p < 0.001$

3.5 Companies' Incentives and Hindrances to Implement PETs

Equally important to the user perspective for the broad distribution of PETs is the perspective of the companies since users can only order services if they are offered. Therefore, we investigated the incentives and hindrances of companies to implement PETs either in their existing products or as a stand-alone product.

For that purpose, we conducted semi-structured interviews with 12 experts and managers from companies dealing with privacy and PETs in their daily business [20]. Our interview guide consisted of three relevant parts about general questions on the interviewees and their companies, technical questions on the status quo, and questions on economic and societal issues. The interviews were recorded, transcribed, openly coded, and in a second round selectively coded. The selective coding was done first separately and then among all interviews to consolidate the developed codings [6, 16]. We identified the following categories:

Technical Optimization: PETs help to optimize the company within an organization and technical dimension and can get the company a technological lead. For that purpose, the *integration into the business process* was named as a necessary condition, and it was criticized that it is in general hard to get information about the practical use of PETs. PETs were also seen as a tool for *data management and avoidance* to improve business processes.

Business model: The category considering business models was by far the largest. Here, the interviewees saw the largest incentives but also the largest hindrances. With the implementation of PETs, companies intend to *further develop their services*. How and if that works depends on the customers' requirements, on the level of convenience for the existing service (if it depends on customer data) as well as on the PET's handling. Customers' awareness of privacy was also seen as an important factor. However, the interviewees were discordant if raising it should be the task of the company. PETs were also seen as a chance to *enlarge the company's clientele* by addressing "nerds." The mass market was seen from the viewpoint that most customers do not request PETs but would accept them and that there is a chance to implement PETs in existing products that are already widespread. Interviewees also did not agree on the *development of new business models* in terms of offering privacy as a premium feature. While some considered it as naturally to ask for a fee for the additional effort on the company's side, others questioned that approach by referring to the perception of the "non-premium" customers that they do not have sufficient security and privacy levels when using the company's service. As a last incentive, a better *positioning for the future* was named, which could gain the company an advantage over its competitors.

Corporate perception: The particular technology was considered to be less important, but a positive perception by business partners was considered to be highly useful to gain *trust*. Using PETs to have a communicable unique selling point enables the company to *profile itself through PETs*. *Business ethics* was considered from multiple viewpoints. Based on the assumption that anonymity and the use of PETs are independent of moral value positions, the question was raised if informative awareness campaigns are morally defensible or a way of using the customer's fear to sell them PETs. On the other hand, it was advocated for integrating PETs independently of the economic value but rather because it seems to be the right thing to do.

Our results do not draw a clear picture in some areas since the perceptions differ a lot, i. e., on the question if privacy can be sold to the customers as a premium service. This shows that more research is necessary to determine underlying factors and elaborate precise recommendations to companies on how they can integrate PETs in their products while having a proper business model in mind.

4 Discussion and Conclusion

Our results indicate that for models based on IUIPC the traditional influence of trusting and risk beliefs is overruled by trust in the respective PET. With the newly introduced constructs perceived anonymity and trust in the PET, technology acceptance models are applicable for PETs also. Most of the existing variables in the TAM were also found in the participants' statements (e. g., usability, performance, anonymity, and trust). Trust in the PET also plays a major role when it comes to paying for or donating to the service. For companies, the introduction of PETs offers a huge chance but also rises challenges, in particular about a profitable business model. However, our results can only be a first insight into issues of hindering a broader adoption of PETs, where more details have to be brought to light in future work.

Future work could also investigate PETs that are integrated into regular services, e. g., the use of machine learning to help users with the privacy preferences [42], integration of PETs into physical services such as payment and shipment for e-commerce [56], or the integration of PETs into the Internet infrastructure eliminating the users' effort to set up PETs themselves [22]. However, this would raise additional challenges as it needs to be clearly investigated if users refer to the PET part of the service or the traditional part. Moreover, as already discussed in the introduction, an ideal PET would be barely noticeable, which would raise questions regarding suitable business models and the opportunity to "sell" privacy as a feature. It has also been shown that if users are aware that a tool should protect their privacy, they are getting biased and tend toward being more concerned about potential privacy issues of the tool than for non-privacy tools [4, 5]. Further problems of integrating PETs into existing services are that, on the one hand, it is hard to decide which of the many PETs is the best choice [43, 62] and that, on the other hand, it is hardly possible to ask the users about their preferences since in most cases the users do not notice the main achievement of the PET to protect their privacy, but rather things such as increased latency, more complex processes, or similar side effects.

While the adding of online privacy literacy did not improve the explanatory power of the model a lot, research in other areas such as the Corona Warning App [36, 53] (please refer to the chapter "Privacy Research on the Pulse of Time: COVID-19 Contact-Tracing Apps" for an overview of research in this area) or inferences of voice recordings [41] suggests that knowledge and awareness play a fundamental role in the users' perception. Thus, in this case, the used OPLIS construct might not have been specific enough to relate the users' knowledge with their concerns and behavior.

Summing up, while there has been lots of progress on the cryptographic side and the technical implementation of PETs, there is still a gap concerning the understanding of factors influencing users to use PETs. From a company perspective, it is equally important to address the question on how to embed which PET in a service and which business model supports a monetization strategy of this privacy feature.

Acknowledgments This work was supported by the European Union's Horizon 2020 research and innovation program from the project CyberSec4Europe (grant agreement number 830929).

References

1. Abu-Salma, R., Sasse, M. A., Bonneau, J., Danilova, A., Naiakshina, A., & Smith, M. (2017). Obstacles to the adoption of secure communication tools. In *IEEE security & privacy* (pp. 137–153).
2. Ajzen, I. (1991). The theory of planned behavior. *Organizational Behavior and Human Decision Processes, 50*(2), 179–211.
3. Bédard, M. (2016). The underestimated economic benefits of the Internet. Regulation series, The Montreal Economic Institute. Economic Notes.
4. Bracamonte, V., Pape, S., & Kiyomoto, S. (2021). Investigating user intention to use a privacy sensitive information detection tool. In *Symposium on Cryptography and Information Security (SCIS)*.
5. Bracamonte, V., Pape, S., & Löbner, S. (2022). "All apps do this": Comparing privacy concerns towards privacy tools and non-privacy tools for social media content. *Proceedings on Privacy Enhancing Technologies (PoPETs), 2022*(3), 57–78.
6. Charmaz, K. (2014). *Constructing grounded theory* (2nd ed.) Sage Publications.
7. Clement, J. (2020). Number of Internet users worldwide 2005–2019. https://www.statista.com/statistics/273018/number-of-internet-users-worldwide/
8. Cohen, J. (1968). Weighted kappa: Nominal scale agreement provision for scaled disagreement or partial credit.
9. Davis, F. D. (1985). *A technology acceptance model for empirically testing new end-user information systems: Theory and results*. Massachusetts Institute of Technology.
10. Davis, F. D. (1989). Perceived usefulness, perceived ease of use, and user acceptance of information technology. *MIS Quarterly, 13*(3), 319–340.
11. Dinev, T., & Hart, P. (2006). An extended privacy calculus model for e-commerce transactions. *Information Systems Research, 17*(1), 61–80.
12. Fishbein, M., & Ajzen, I. (1975). *Belief, attitude, intention and behavior: An introduction to theory and research*. Addison-Wesley.
13. Fisher, R. A. (1970). *Statistical methods for research workers* (14th ed.). Oliver & Boyd.

14. Frik, A., & Gaudeul, A. (2016). The relation between privacy protection and risk attitudes, with a new experimental method to elicit the implicit monetary value of privacy. *CEGE Discussion Papers, Number.*
15. Gerber, N., Gerber, P., & Volkamer, M. (2018). Explaining the privacy paradox: A systematic review of literature investigating privacy attitude and behavior. *Computers & Security, 77*, 226–261.
16. Glaser, B. G., & Strauss, A. L. (1967). *The discovery of grounded theory*. Aldine Publishing.
17. Graeff, T. R., & Harmon, S. (2002). Collecting and using personal data: Consumers' awareness and concerns. *Journal of Consumer Marketing, 19*(4), 302–318.
18. Hair, J., Hult, G. T. M., Ringle, C. M., & Sarstedt, M. (2017). *A primer on partial least squares structural equation modeling (PLS-SEM)*. SAGE Publications.
19. Hair, J., Ringle, C. M., & Sarstedt, M. (2011). PLS-SEM: Indeed a silver bullet. *Journal of Marketing Theory and Practice, 19*(2), 139–152.
20. Harborth, D., Braun, M., Grosz, A., Pape, S., & Rannenberg, K. (2018). Anreize und Hemmnisse für die Implementierung von Privacy-Enhancing Technologies im Unternehmenskontext. In *Sicherheit 2018: Sicherheit, Schutz und Zuverlässigkeit, Beiträge der 9. Jahrestagung des Fachbereichs Sicherheit der Gesellschaft für Informatik e.V. (GI), 25.-27. April 2018, Konstanz* (pp. 29–41).
21. Harborth, D., Cai, X., & Pape, S. (2019). Why do people pay for privacy-enhancing technologies? The case of Tor and JonDonym? In *ICT Systems Security and Privacy Protection—34th IFIP TC 11 International Conference, SEC 2019, Lisbon, Portugal, June 25–27, 2019, Proceedings* (pp. 253–267).
22. Harborth, D., Herrmann, D., Köpsell, S., Pape, S., Roth, C., Federrath, H., Kesdogan, D., & Rannenberg, K. (2017). Integrating privacy-enhancing technologies into the Internet infrastructure. https://arxiv.org/abs/1711.07220. Also available via https://epub.uni-regensburg.de/36346/
23. Harborth, D., & Pape, S. (2017). Exploring the hype: Investigating technology acceptance factors of Pokémon GO. In W. Broll, H. Regenbrecht, & J. E. Swan II (Eds.), *2017 IEEE International Symposium on Mixed and Augmented Reality, ISMAR 2017, Nantes, France, October 9–13, 2017* (pp. 155–168).
24. Harborth, D., & Pape, S. (2017). Privacy concerns and behavior of Pokémon GO players in Germany. In M. Hansen, E. Kosta, I. Nai-Fovino, & S. Fischer-Hübner (Eds.), *Privacy and Identity Management. The Smart Revolution—12th IFIP WG 9.2, 9.5, 9.6/11.7, 11.6/SIG 9.2.2 International Summer School, Ispra, Italy, September 4–8, 2017, Revised Selected Papers*, volume 526 of *IFIP Advances in Information and Communication Technology* (pp. 314–329). Springer.
25. Harborth, D., & Pape, S. (2018). Examining technology use factors of privacy-enhancing technologies: The role of perceived anonymity and trust. In *24th Americas Conference on Information Systems, AMCIS 2018, New Orleans, LA, USA, August 16–18, 2018*. Association for Information Systems.
26. Harborth, D., & Pape, S. (2018). German translation of the concerns for information privacy (CFIP) construct. Technical report, SSRN.
27. Harborth, D., & Pape, S. (2018). German translation of the unified theory of acceptance and use of technology 2 (UTAUT2) questionnaire. Technical report, SSRN.
28. Harborth, D., & Pape, S. (2018). JonDonym users' information privacy concerns. In *ICT Systems Security and Privacy Protection—33rd IFIP TC 11 International Conference, SEC 2018, Held at the 24th IFIP World Computer Congress, WCC 2018, Poznan, Poland, September 18–20, 2018, Proceedings* (pp. 170–184).
29. Harborth, D., & Pape, S. (2019). How privacy concerns and trust and risk beliefs influence users' intentions to use privacy-enhancing technologies—the case of Tor. In *52nd Hawaii International Conference on System Sciences (HICSS) 2019* (pp. 4851–4860).
30. Harborth, D., & Pape, S. (2019). Investigating privacy concerns related to mobile augmented reality applications. In H. Krcmar, J. Fedorowicz, W. F. Boh, J. M. Leimeister, & S. Wattal (Eds.), *Proceedings of the 40th International Conference on Information Systems ICIS 2019, Munich, Germany, December 13–15, 2019.*

31. Harborth, D., & Pape, S. (2020). Dataset on actual users of the privacy-enhancing technology JonDonym.
32. Harborth, D., & Pape, S. (2020). Dataset on actual users of the privacy-enhancing technology Tor.
33. Harborth, D., & Pape, S. (2020). Empirically investigating extraneous influences on the "APCO" model—childhood brand nostalgia and the positivity bias. *Future Internet, 12*(12), 220.
34. Harborth, D., & Pape, S. (2020). How privacy concerns, trust and risk beliefs and privacy literacy influence users' intentions to use privacy-enhancing technologies—the case of Tor. *ACM SIGMIS Database: the DATABASE for Advances in Information Systems, 51*(1), 51–69.
35. Harborth, D., & Pape, S. (2021). Investigating privacy concerns related to mobile augmented reality apps—a vignette based online experiment. *Computers in Human Behavior, 122*, 106833.
36. Harborth, D., & Pape, S. (2022). A privacy calculus model for contact tracing apps: Analyzing the German corona-Warn-App. In *ICT Systems Security and Privacy Protection—37th IFIP TC 11 International Conference, SEC 2022*, volume 648 of *IFIP Advances in Information and Communication Technology* (pp. 3–19).
37. Harborth, D., Pape, S., & Rannenberg, K. (2020). Explaining the technology use behavior of privacy-enhancing technologies: The case of Tor and JonDonym. *Proceedings on Privacy Enhancing Technologies (PoPETs), 2020*(2), 111–128.
38. Harborth, D., Pape, S., & Rannenberg, K. (2021). Explaining the technology use behavior of privacy-enhancing technologies: The case of Tor and JonDonym (poster). In *17th Symposium on Usable Privacy and Security (SOUPS 2021).*
39. Heales, J., Cockcroft, S., & Trieu, V.-H. (2017). The influence of privacy, trust, and national culture on Internet transactions. In G. Meiselwitz (Ed.), *Social computing and social media. Human behavior* (pp. 159–176). Springer.
40. JonDos Gmbh. (2018). Official Homepage of JonDonym. https://www.anonym-surfen.de
41. Kröger, J. L., Gellrich, L., Pape, S., Brause, S. R., & Ullrich, S. (2022). Personal information inference from voice recordings: User awareness and privacy concerns. *Proceedings on Privacy Enhancing Technologies (PoPETs), 2022*(1), 6–27.
42. Löbner, S., Tesfay, W. B., Nakamura, T., & Pape, S. (2021). Explainable machine learning for default privacy setting prediction. *IEEE Access, 9*, 63700–63717.
43. Löbner, S., Tronnier, F., Pape, S., & Rannenberg, K. (2021). Comparison of de-identification techniques for privacy preserving data analysis in vehicular data sharing. In B. Brücher, C. Krauß, M. Fritz, H. Hof, & O. Wasenmüller (Eds.), *CSCS '21: ACM Computer Science in Cars Symposium, Ingolstadt, Germany, November 30th, 2021* (pp. 7:1–7:11). ACM.
44. Lux, A., & Platzer, F. (2022). Online-Privatheitskompetenz und Möglichkeiten der technischen Umsetzung mit dem Anonymisierungsnetzwerk Tor. In *Selbstbestimmung, Privatheit und Datenschutz* (pp. 129–149). Springer Vieweg.
45. Malhotra, N. K., Kim, S. S., & Agarwal, J. (2004). Internet users' information privacy concerns (IUIPC): The construct, the scale, and a causal model. *Information Systems Research, 15*(4), 336–355.
46. Mani, A., Wilson-Brown, T., Jansen, R., Johnson, A., & Sherr, M. (2018). Understanding Tor usage with privacy-preserving measurement. In *2018 Internet Measurement Conference (IMC'18)* (pp. 1–13).
47. Masur, P. K., Teutsch, D., & Trepte, S. (2017). Entwicklung und Validierung der Online-Privatheitskompetenzskala (OPLIS). *Diagnostica.*
48. McKelvey, R. D., & Zavoina, W. (1975). A statistical model for the analysis of ordinal level dependent variables. *Journal of Mathematical Sociology, 4*(1), 103–120.
49. Mineo, L. (2017). On Internet privacy, be very afraid (Interview with Bruce Schneier). https://news.harvard.edu/gazette/story/2017/08/when-it-comes-to-internet-privacy-be-very-afraid-analyst-suggests/
50. Montieri, A., Ciuonzo, D., Aceto, G., & Pescapé, A. (2017). Anonymity services Tor, I2P, JonDonym: Classifying in the dark. In *Teletraffic Congress (ITC 29), 2017 29th International* (Vol. 1, pp. 81–89). IEEE.

51. Naeini, P. E., Bhagavatula, S., Habib, H., Degeling, M., Bauer, L., Cranor, L., & Sadeh, N. (2017). Privacy expectations and preferences in an IoT world. In *Symposium on Usable Privacy and Security (SOUPS)*.
52. Pape, S. (2020). Requirements engineering and tool-support for security and privacy. Habilitation thesis, submitted to the Faculty of Computer Science and Mathematics of the Johann Wolfgang Goethe University, Frankfurt am Main, Germany in September 2020.
53. Pape, S., Harborth, D., & Kröger, J. L. (2021). Privacy concerns go hand in hand with lack of knowledge: The case of the German Corona-Warn-App. In A. Josang, L. Futcher, & J. Hagen (Eds.), *ICT Systems Security and Privacy Protection—36th IFIP TC 11 International Conference, SEC 2021*, volume 625 of *IFIP Advances in Information and Communication Technology* (pp. 256–269). Springer.
54. Pape, S., Ivan, A., Harborth, D., Nakamura, T., Kiyomoto, S., Takasaki, H., & Rannenberg, K. (2020). Open materials discourse: Re-evaluating Internet users' information privacy concerns: The case in Japan. *AIS Transactions on Replication Research, 6*(22), 1–7.
55. Pape, S., Ivan, A., Harborth, D., Nakamura, T., Kiyomoto, S., Takasaki, H., & Rannenberg, K. (2020). Re-evaluating Internet users' information privacy concerns: The case in Japan. *AIS Transactions on Replication Research, 6*(18), 1–18.
56. Pape, S., Tasche, D., Bastys, I., Grosz, A., Laessig, J., & Rannenberg, K. (2018). Towards an architecture for pseudonymous e-commerce—applying privacy by design to online shopping. In *Sicherheit 2018: Sicherheit, Schutz und Zuverlässigkeit, Beiträge der 9. Jahrestagung des Fachbereichs Sicherheit der Gesellschaft für Informatik e.V. (GI), 25.-27. April 2018, Konstanz* (pp. 17–28).
57. Park, Y. J. (2013). Digital literacy and privacy behavior online. *Communication Research, 40*(2), 215–236.
58. Pfitzmann, A., & Hansen, M. (2010). A terminology for talking about privacy by data minimization: Anonymity, unlinkability, undetectability, unobservability, pseudonymity, and identity management.
59. Raber, F., & Krueger, A. (2017). Towards understanding the influence of personality on mobile app permission settings. In *IFIP Conference on Human-Computer Interaction* (pp. 62–82). Springer.
60. Rainie, L., Kiesler, S., Kang, R., Madden, M., Duggan, M., Brown, S., & Dabbish, L. (2013). Anonymity, privacy, and security online. *Pew Research Center, 5*.
61. Rajamma, R. K., Paswan, A. K., & Hossain, M. M. (2009). Why do shoppers abandon shopping cart? Perceived waiting time, risk, and transaction inconvenience. *Journal of Product & Brand Management, 18*(3), 188–197.
62. Rannenberg, K., Pape, S., Tronnier, F., & Löbner, S. (2021). Study on the technical evaluation of de-identification procedures for personal data in the automotive sector. Technical report, Goethe University Frankfurt.
63. Ringle, C. M., Wende, S., & Becker, J. M. (2015). SmartPLS 3. www.smartpls.com
64. Saleh, S., Qadir, J., & Ilyas, M. U. (2018). Shedding light on the dark corners of the Internet: A survey of Tor research. *Journal of Network and Computer Applications, 114*, 1 – 28.
65. Singh, T., & Hill, M. E. (2003). Consumer privacy and the Internet in Europe: A view from Germany. *Journal of Consumer Marketing, 20*(7), 634–651.
66. Stewart, K. A., & Segars, A. H. (2002). An empirical examination of the concern for information privacy instrument. *Information Systems Research, 13*(1), 36–49.
67. Strübing, J. (2013). Zum Verhältnis von Theorien und Methoden. *Qualitative Sozialforschung. Eine Einführung* (pp. 27–52).
68. The Tor Project. (2018). https://www.torproject.org
69. Trepte, S., Teutsch, D., Masur, P. K., Eicher, C., Fischer, M., Hennhöfer, A., & Lind, F. (2015). Do people know about privacy and data protection strategies? Towards the online privacy literacy scale (OPLIS). In *Reforming European Data Protection Law* (pp. 333–365). Springer.

Increasing Users' Privacy Awareness in the Internet of Things: Design Space and Sample Scenarios

Sarah Prange and Florian Alt

1 Introduction

In the era of ubiquitous computing [57], data collection and, as such, potential privacy intrusions are omnipresent. Computing devices do not only inflate users' everyday lives at home, but also in semi-public to public spaces. Examples include, but are not limited to, vacuum cleaning robots collecting floor maps of our homes, smart door locks providing access to our workspaces, digital ordering stations in restaurants, and security cameras in highly frequented places. In addition, the variety of devices and functionality, along with the concrete privacy implications, is huge. For instance, a particular smart TV might only provide access to online streaming services, while other smart TVs might additionally allow for voice interaction using built-in microphones.

As a result, it becomes increasingly challenging for users to stay aware of where their personal data are collected, and with whom it is shared. Moreover, not only device owners are affected, but also incidental users, even without explicit interaction [9].

In this chapter, we shed light on these challenges and illustrate current privacy awareness mechanisms (Sect. 2). However, existing mechanisms, such as, e.g., device indicators, tend to be overlooked [7, 44]. Other mechanisms, such as, e.g., labels on devices' packaging [17, 20, 30], mainly target those who purchase and

S. Prange (✉)
University of the Bundeswehr, Munich, Germany

LMU Munich, Munich, Germany
e-mail: sarah.prange@unibw.de

F. Alt
University of the Bundeswehr, Munich, Germany
e-mail: florian.alt@unibw.de

N. Gerber et al. (eds.), *Human Factors in Privacy Research*,
https://doi.org/10.1007/978-3-031-28643-8_16

set up the devices but are rarely available to other target groups such as visitors of the environment [39] or passers-by. At the same time, privacy awareness is a prerequisite for users to be able and act upon their privacy needs [9, 40, 41]. As such, increasing privacy awareness is a necessary first step.

To address this, we set out with a design space on how and in which contexts privacy-relevant information could be brought to users (Sect. 3). We illustrate three sample scenarios in which privacy-relevant information should be easily accessible for users, along with sample applications from our prior work (Sect. 4): providing privacy-relevant information on computing devices during purchase decisions, providing privacy-relevant information on demand, and providing privacy-relevant information within the environment. Note that the scenarios cover device purchase decisions as well as devices that are already installed and in use. The chapter is complemented with directions for future research (Sect. 5) and a summary (Sect. 6).

2 Background and Related Work

An increasing number of everyday objects are equipped with computing power and interconnected, commonly being referred to as the *Internet of Things (IoT)* [2, 4]. Think about, e.g., smart home appliances, but also smart cars, or surveillance systems in public spaces. While providing great benefits and features, these devices pose new threats to users' privacy [62].

In the following, we discuss the privacy challenges that arise from an IoT-infused world (Sect. 2.1) and current mechanisms aiming at increasing users' privacy awareness (Sect. 2.2).

2.1 Privacy Challenges

Privacy, which is individual control over when, where, and how personal data are being collected and shared [13], becomes increasingly challenging as sensing and computing technologies are seamlessly integrated into our daily lives [57]. The number of devices capable of collecting personal data is steadily rising, and sensing technology is placed in both private and public places.

The variety of devices is huge. For instance, smart vacuum cleaning robots scan floor maps of our homes to operate;[1] smart fridges reorder groceries; smart electricity meters monitor energy consumption and can thus infer users' activities [48]; smart voice assistants listen to our conversations [35]; cameras record and analyze

[1] https://www.technologyreview.com/2017/07/25/150346/your-roomba-is-also-gathering-data-about-the-layout-of-your-home/, last accessed August 31, 2022.

semi-public and public spaces for security purposes; smart door locks provide access to homes or offices via biometric features [42].

Also, devices come with various functionality and data collection capabilities, with different impacts on users' privacy. For instance, conversations—as potentially captured by a smart speaker—might be, from a privacy perspective, of different values as compared to grocery orders by a smart fridge. As a consequence, it is hard for users to correctly assess the privacy implications of specific devices, even if they have a general understanding of the technology [39].

Moreover, IoT devices are usually shared among multiple users, and the ecosystem of stakeholders is complex [23, 27, 61]. It not only includes device owners as those who set up and primarily use devices, but also secondary users such as, e.g., co-inhabitants of a smart home [9, 23, 24, 34], guests in a rental apartment [9, 38, 40], or passers-by in semi-public and public spaces [9, 46]. Manufacturers of devices, as well as providers of single services, are also relevant parties. This makes it unclear as to who is responsible for even providing privacy-relevant information and to whom.

Lastly, it is unclear what information is relevant to users in which context, for them to be able to make informed privacy decisions.

2.2 Privacy Awareness Mechanisms

An increasing number of devices in our environments are capable of collecting personal data about us with built-in sensors. This may happen inconspicuously and without direct interaction [9]. Even worse, users are oftentimes unaware of this, let alone the privacy implications of this data collection [3, 9, 34, 62].

Users, however, want to be informed about data being collected about them and shared with device providers [18, 28, 43, 52]. Moreover, awareness of privacy implications is a prerequisite for users to be able and preserve their privacy, and to decide with whom they are willing to share their personal data [9, 40, 41]. As such, there is a need to design suitable mechanisms that help increase privacy awareness [52, 58] among all affected individuals [9, 60].

Prior work suggested mechanisms that provide *general* privacy information (to, e.g., support purchase decisions) and information on *installed devices* (i.e., that are already in use and collecting data).

General Privacy Information

Prior to data collection, providers of devices and services must provide privacy-relevant information. The default approach to this is privacy notices [11, 21], a textual description of which data are collected and how it is processed. These policies, however, tend to be long, are hard to understand for users, and thus are oftentimes not read thoroughly [56].

Research tried to address this challenge and make privacy-relevant information more accessible to users, to ultimately increase their awareness. Ebert et al. found that more concise and salient privacy notices can successfully increase users' privacy awareness [15]. Others suggested ways to make privacy policies more appealing and understandable. *Polisis* is a framework for automated analysis of privacy policies, to, e.g., assign icons [25]. Building upon this framework, the *PriBot* is a chat agent that provides privacy-relevant information and can answer users' questions [26]. Kitkowska et al. suggested visual and appealing designs for privacy policies and showed that these can successfully spark users' curiosity and ultimately create an understanding of privacy policies [31]. Another opportunity is the use of icons based on a risk assessment [16]. Mozilla's "Privacy not included guide" provides an emoji-based scale, assessing the privacy implications of computing devices ranging from "not creepy" to "super creepy," based on crowd-sourced data.[2]

Privacy Labels To particularly target purchase decisions of computing devices, Kelley et al. introduced the "privacy label", which acts similar to nutrition labels for groceries but includes information on data collection and sharing of a device. They found this representation to be easier and more comprehensible than privacy policies based on natural language [30]. Such privacy labels also make privacy information more accessible and can thus inform purchase decisions, avoiding concerns rising later on [20]. Moreover, Emami-Naeini et al. showed that critical information should be included in a primary layer (e.g., directly on a device's packaging), while details can be moved to additional sources (such as, e.g., a website) and linked on the label [17]. These types of labels became obligatory for IoT devices in several countries (e.g., UK,[3] Singapore[4]), and for applications on Apple's iOS.[5]

Privacy Information on Installed Devices

Many devices that collect data communicate their status through *device indicators* while being in use. For example, webcams indicate via small LEDs whether they are currently on. Amazon's Alexa provides feedback on its recording status via a light ring (e.g., red refers to "muted") [8, 35]. Research also suggested alternatives such as, e.g., physical webcam indicators in the form of a flower [33] or an eye that mimics gaze (i.e., recording) direction [53].

---

[2] https://foundation.mozilla.org/en/privacynotincluded/, last accessed September 1, 2020.

[3] https://www.gov.uk/government/consultations/consultation-on-regulatory-proposals-on-consumer-iot-security/consultation-on-the-governments-regulatory-proposals-regarding-consumer-internet-of-things-iot-security#designing-a-security-label, last accessed September 1, 2020.

[4] https://www.csa.gov.sg/Programmes/certification-and-labelling-schemes/cybersecurity-labelling-scheme/about-cls, last accessed June 17, 2022.

[5] https://mashable.com/article/apple-privacy-nutrition-labels-ios14/?europe=true, last accessed September 1, 2020.

To help users *detect* devices in their environment, Song et al. suggested attaching visual or auditory cues to devices [51]. *Lumos* is an augmented reality interface that can be employed on users' personal devices and help them detect hidden IoT devices in their environment [50]. Sami et al. used smartphones emitting laser signals to detect hidden cameras via the reflection of the laser [49]. Funk et al. guided users to smart objects using smart glasses [22]. Thakkar et al. suggested four different privacy awareness mechanisms for the smart home context: a physical data dashboard, a mobile application, ambient colored light, and voice messages on privacy via a smart speaker. These mechanisms aim at targeting device owners, but also potential bystanders, with detailed information being preferred by both target groups [54].

2.3 Summary and Limitations

In times where data collection is ubiquitously present, it becomes increasingly hard for users to even be aware of potential privacy intrusions and ultimately be able to protect their privacy. Research tried to tackle these challenges by designing mechanisms that target users' *privacy awareness*. However, current privacy awareness mechanisms are only effective to a limited extent. Users might overlook or not realize or understand the meaning of privacy indicators [7, 44]. Moreover, information on devices is oftentimes only available for those who purchase and configure devices, but not for potential *bystanders* (e.g., guests in a smart environment), who might likewise be affected. As a result, especially bystanders are uncertain about device states [1].

In addition, the exact device position and/or area of operation is oftentimes unclear, let alone the concrete privacy implications of certain devices and data being collected. The increasing number of devices being installed further exacerbates this issue. This calls for further research on *privacy awareness mechanisms* that target *device owners* and *bystanders* alike.

3 Design Space

Users' privacy perceptions are influenced by many factors, including, e.g., the environment in which data are collected in and type of data that is collected. We argue that this information is privacy-relevant and should be made available to users, to increase privacy awareness. Based on these factors, we derive a design space for privacy awareness mechanisms for the IoT. In the following, we discuss contextual factors that impact users' privacy perceptions, as well as types of information that are ultimately privacy-relevant and how this information could be provided.

3.1 Contextual Factors

Individual privacy perceptions and (dis)comfort with personal data being recorded are highly impacted by contextual factors, as highlighted in our previous work [46]:

Social Aspects and Trust: Users consider trust and relationships when deciding with whom to share their personal data [19, 36, 59, 60]. For instance, users rely on friends' opinions regarding data sharing [19] and consider *who* is collecting their data [36] as well as who is the owner of a particular device [40].

Environment: Also, users' current environment impacts their concerns. As such, data collection in *private* spaces (e.g., the home) is less acceptable as compared to data collection in other spaces, such as restaurants (*semi-public*) or *public* spaces [18, 37]. It is also important to users whether they are familiar with the environment [46]. In unfamiliar settings, users are particularly concerned about (hidden) data collection, especially when they consider the space private at the same time, as is the case for, e.g., rental apartments [38, 46, 51].

3.2 Privacy-Relevant Information

Privacy-relevant information can comprise various content and be made available to users in various ways.

Content

Depending on users' current context, various information could become relevant for users to decide whether or not they are willing to share their personal data:

Type of Sensor(s): The type of sensors—and, respectively, the type of data being collected—impacts users' privacy perceptions. For instance, cameras and microphones (i.e., video and audio recordings) are usually considered particularly sensitive [32].

Tracking Space: The area of data collection can further help users assess privacy intrusions, particularly bystanders who are not familiar with the space devices are in [9].

Device Owner: The relationship to the device owner crucially impacts users' willingness to be recorded by devices [9, 19, 36, 41, 59, 60]. For instance, users are more comfortable with devices being placed in trusted environments (e.g., in friends' homes) [39, 40, 46] as compared to devices being installed by (unknown) hosts of rental apartments [9, 38].

Purpose: Users are more likely to accept data collection if it suits their own needs and purpose. For instance, for health-related purposes, even long-term data tracking is acceptable [5]. This particularly holds true for personal physiological data [45]. In contrast, video and audio recordings are less acceptable, regardless of the purpose [37].

Availability and Output

The privacy-relevant information could be made available to users in various ways. For instance, information could be provided in relation to the environment, e.g., on a personal device such as a smartphone or tablet [50], or using contextual images [51]. Another opportunity is to provide information only on specific devices similar to, e.g., the privacy labels [17, 30].

Accordingly, privacy-relevant information is available at different times. For instance, information that is bound to the device's packaging [17, 30] is available to support purchase decisions. Hence, users would need to *actively search* for and inform themselves about devices to receive this information. Information that is provided independently on a personal device, however, would be *always available* to users as they are moving around. Lastly, privacy mechanisms can act in various degrees of proactivity (e.g., low, medium, high in the context of smart homes [29]). Privacy-relevant information could thus be provided *actively*, e.g., through *push notifications* on personal devices, e.g., when entering an unfamiliar area with data collection being in place.

4 Sample Scenarios

To further emphasize the relevance of increasing privacy awareness in the IoT, we illustrate three concrete scenarios in the following, along with sample applications. In particular, privacy awareness can and should be increased, in various ways, in the following cases: (1) supporting decisions for purchasing IoT devices with privacy-relevant information (*PriCheck* [55]); (2) allowing users to consult privacy-relevant information on demand (e.g., using their mobile phones, *PriView (mobile)* [46]); (3) providing privacy-relevant information and guidance within the environment (e.g., by means of augmented reality, *PriView (HMD)* [46]). For an overview of relevant design space dimensions per scenario, refer to Table 1.

4.1 Privacy-Relevant Information for Purchase Decisions

Prior work already identified device purchases as a relevant starting point and suggested means to support users' decision-making with privacy-relevant information, e.g., by labels on devices' packaging [17, 20, 30]. However, devices are also oftentimes purchased online, where users are not in the hands of the actual device

Table 1 Scenarios vs. Design Space: We see several scenarios in which privacy-relevant information is needed (left, Sect. 4), and how the design space dimensions would come into play in each scenario (right, Sect. 3)

Scenario	Context	Privacy-Relevant Information	
		Content (Visualization)	Availability and Output
Purchase Decisions	active search for (new) devices	device, sensors, data policies, security standards	on-demand, browser extension
On-Demand Information	active search for installed devices in arbitrary environments	device position (all); sensors, tracking space, recording state, device owner (some)	on-demand or push, mobile application
In Situ Information	information in arbitrary environments	device position (all); sensors, tracking space, recording state, device owner (some); or simple general warning	always-on or push, head-mounted display

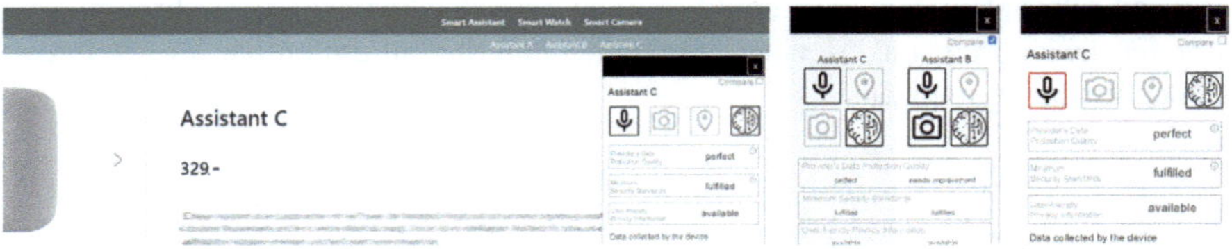

Fig. 1 *PriCheck* is a browser extension supporting purchase decisions with privacy-relevant information on smart devices. Figure from [55]

packaging. Users who *actively* search for devices should have access to privacy-relevant information during purchase decisions. As such, a promising approach is to provide privacy-relevant information in the form of a *browser extension*, to be easily accessible for users when forming a decision. A sample browser extension with privacy-relevant information is the *Privacy Bird* that notifies users if a website's privacy policy violates their preferences [12]. This could be similarly applied to online purchase decisions as well.

PriCheck as suggested by Volk et al. [55] provides privacy-relevant information, comparable to the privacy labels [17, 20, 30], in the form of a browser extension in an online shop (see Fig. 1). In particular, it shows the name of the device along with built-in sensors and functionality visualized as icons (black refers to "included"), data protection quality, security standards, and availability of data protection information. The extension also allows to compare two devices (see Fig. 1, center) and to highlight mismatches with pre-configured privacy preferences (see Fig. 1, right). In an exploratory study ($N = 11$), participants comparing devices in a mock online shop using *PriCheck* appreciated the usability of the extension as well as the information provided and agreed that it helped them considering privacy-relevant information for their decisions [55].

Supporting Purchase Decisions with *PriCheck*
To summarize, *PriCheck* [55] supports users as follows:

Context	online purchase of smart devices, *active search*
Device(s)	search for *one (new) device* at a time, and *comparison* between two devices
User(s)	one user who is about to become the *owner*
Content	built-in sensors and functionality, data protection quality and security standards, availability of data protection information
Availability	*on-demand*, but within the situation (online shop)

4.2 Carrying and Consulting Privacy-Relevant Information on Demand

Users might also want to *actively search* for devices that are already installed and in use. Indicators in the form of, e.g., LEDs or beep sounds [51], can help users discover devices, yet yield little additional information. Other mechanisms, such as the IoT assistant,[6] list devices in users' vicinity and allow to communicate privacy choices but do not cover other information such as the exact device position in users' environment.

PriView, employed as a mobile application using a thermal camera dongle [46], allows users to actively scan the environment for devices (see Fig. 2 left, top). In several visualizations, it shows: device position (red frame), textual information, tracking space (bubble), device state (segmentation via the thermal camera), or built-in sensors (Fig. 2 left, bottom). This can particularly help users in unfamiliar environments that are considered private (e.g., a rental apartment), to detect devices they are uncomfortable with. Participants in an exploratory user study ($N = 21$) appreciated the innovative and easy-to-use mobile application. They also liked *PriView* being available on their personal mobile devices, while also having the possibility to put it away anytime [46].

[6] https://play.google.com/store/apps/details?id=io.iotprivacy.iotassistant&hl=de&gl=US, last accessed May 26, 2022.

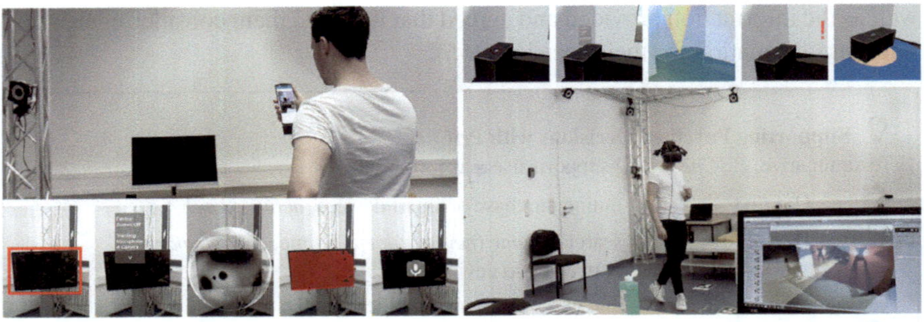

Fig. 2 *PriView* is a concept for privacy visualizations meant to increase users' awareness. *PriView* can, e.g., be employed as a mobile application for scanning the environment on demand (left) or in a head-mounted display (HMD), enabling to provide privacy-relevant information in the environment (right). Figure from [46]

Privacy-Relevant Information on Demand with *PriView* (mobile)

To summarize, *PriView (mobile)* [46] supports users as follows:

Context	*active* device search, scanning the (unfamiliar/untrusted) environment
Device(s)	potentially multiple devices that are already *installed* and in use
User(s)	primary users as well as bystanders, potentially unknown device owners
Content	device position (all visualizations); built-in sensors, textual information (including device owner), tracking space, recording state
Availability	on-demand, push notifications possible

4.3 Providing Privacy-Relevant Information and Guidance In Situ

To provide users with privacy-relevant information in arbitrary environments, augmented reality (AR) can serve as a means for in situ information and guidance. For instance, *PriView* employed in a head-mounted display (HMD) provides users with visualizations of potential privacy intrusions within the environment [46]. Similar to the mobile application, it shows: device position (red frame), textual information, tracking space in 3D, a general warning icon, and tracking space on the floor (Fig. 2 right). This can particularly help users in arbitrary environments to increase privacy awareness, particularly when they are new to a place. Participants of our study ($N = 21$) liked the visualizations being available in situ using the HMD. They wished for more details in spaces they considered private (e.g., a rental apartment), while simpler indications were sufficient in places where data collection is obvious (e.g., security cameras at a train station) [46].

In Situ Privacy-Relevant Information with *PriView* (HMD)

To summarize, *PriView (HMD)* [46] supports users as follows:

Context	information within the (unfamiliar/untrusted) environment
Device(s)	potentially multiple devices that are already *installed* and in use
User(s)	primary users as well as bystanders, potentially unknown device owners
Content	device position (all visualizations); built-in sensors, textual information (including device owner), tracking space, recording state
Availability	always-on, push notifications possible

5 Directions for Future Research

In the following, we illustrate and discuss interesting directions for future research that arise from privacy awareness challenges and mechanisms within the IoT.

5.1 Amount of Information

An interesting question for future research is *how much information* on IoT devices users will need to make informed privacy decisions. Is a simple device indicator enough to increase awareness, or would users prefer a deeper understanding of data collection and policies?

Moreover, the preferred amount of information varies depending on the environment [46]. For instance, in environments with multiple devices, including such that are firmly installed as well as such carried by passers-by, there is a high potential for an awareness mechanism to cause visual overload. As such, the amount of information should most likely be reduced, with the opportunity to still receive details on demand.

5.2 Contextualize and Adapt

As a next step, privacy awareness mechanisms could automatically *adapt* to the context and/or their users. For instance, different scenarios (cf. Sect. 4) might require different support for users' privacy awareness. For purchasing a new device to install it within their own environment, users might need awareness as to how it can be configured in a privacy-preserving way. Being in unfamiliar environments with installed devices, however, rather calls for information on spaces being covered by data collection, for users to be able to avoid these as they wish. Also, for scenarios that users encounter more often (e.g., visiting a certain place), awareness cannot be assumed at first but might increase over time as a mechanism is being used in this scenario. Moreover, an awareness mechanism could also adapt to users' prior knowledge (e.g., reduce information that users already received earlier) or technical expertise (e.g., use simpler versions for lay users, while providing more details for advanced users).

5.3 Enabling Control

While awareness is a prerequisite for users to be able to make informed privacy decisions [9, 40, 41], it is only a first step. In particular, users need to be given means to execute (or: enforce) these decisions on nearby devices. For instance, *PARA* is an augmented reality interface that provides privacy controls and allows users to filter data being collected about them [6]. Mobile applications, such as, e.g., the *IoT assistant,*[6] likewise allow users to control nearby devices but require to do so for each and every device or sensor separately, increasing complexity as the number of devices rises. The *PriKey* tries to tackle this challenge by summarizing privacy decisions per sensor in a tangible device [47]. *Personalized privacy assistants* [10, 14] can recommend privacy settings or even act autonomously based on users' privacy preferences or desired standards. This approach, however, needs to find a balance between awareness and control [10]. Future research should further look into how to build upon users' awareness and enable privacy *control*, particularly for those who do not have access to a device's interface.

6 Summary and Conclusion

In this chapter, we highlight the need for *increasing users' privacy awareness* within the *Internet of Things (IoT)*. In particular, the increasing number of devices with increasing functionality and sensors makes it challenging for users to stay aware of their personal data being collected. We shed light on design opportunities for bringing privacy-relevant information to users, as well as sample scenarios and applications: supporting purchase decision with *PriCheck* [55], consulting privacy-relevant on demand using *PriView (mobile)* [46], and providing in situ information and guidance using *PriView (HMD)* [46]. Promising directions for future research include investigating the necessary amount of information, adapting privacy awareness mechanisms to context, and enabling privacy control as a necessary next step.

References

1. Ahmad, I., Farzan, R., Kapadia, A., & Lee, A. J. (2020). Tangible privacy: Towards user-centric sensor designs for bystander privacy. *Proceedings of the ACM on Human-Computer Interaction, 4*(CSCW2), 1–28.
2. Alaa, M., Zaidan, A., Zaidan, B., Talal, M., & Kiah, M. (2017). A review of smart home applications based on Internet of Things. *Journal of Network and Computer Applications, 97*, 48–65.
3. Ali, B., & Awad, A. I. (2018). Cyber and physical security vulnerability assessment for IoT-based smart homes. *Sensors*, 18(3), 817.
4. Atzori, L., Iera, A., & Morabito, G. (2017). Understanding the Internet of Things: Definition, potentials, and societal role of a fast evolving paradigm. *Ad Hoc Networks, 56*, 122–140.
5. Barua, D., Kay, J., & Paris, C. (2013). Viewing and controlling personal sensor data: What do users want? In S. Berkovsky & J. Freyne (Eds.), *Persuasive technology* (pp. 15–26). Springer.
6. Bermejo Fernandez, C., Lee, L. H., Nurmi, P., & Hui, P. (2021). *PARA: Privacy management and control in emerging IoT ecosystems using augmented reality* (pp. 478–486). Association for Computing Machinery.
7. Chow, R., Egelman, S., Kannavara, R., Lee, H., Misra, S., & Wang, E. (2015). HCI in business: A collaboration with academia in IoT privacy. In F. Fui-Hoon Nah & C.-H. Tan (Eds.), *HCI in business* (pp. 679–687). Springer.
8. Chung, H., Iorga, M., Voas, J., & Lee, S. (2017). Alexa, can I trust you? *Computer, 50*(9), 100–104.
9. Cobb, C., Bhagavatula, S., Garrett, K. A., Hoffman, A., Rao, V., & Bauer, L. (2021). "I would have to evaluate their objections": Privacy tensions between smart home device owners and incidental users. *Proceedings on Privacy Enhancing Technologies, 4*, 54–75.
10. Colnago, J., Feng, Y., Palanivel, T., Pearman, S., Ung, M., Acquisti, A., Cranor, L. F., & Sadeh, N. (2020). Informing the design of a personalized privacy assistant for the Internet of Things. In *Proceedings of the 2020 CHI Conference on Human Factors in Computing Systems*, CHI '20 (pp. 1–13). Association for Computing Machinery.
11. Cranor, L. F. (2012). Necessary but not sufficient: Standardized mechanisms for privacy notice and choice. *Journal on Telecommunications and High Technology Law, 10*, 273.
12. Cranor, L. F., Guduru, P., & Arjula, M. (2006). User interfaces for privacy agents. *ACM Transactions on Computer-Human Interaction, 13*(2), 135–178.
13. Culnan, M. J., & Armstrong, P. K. (1999). Information privacy concerns, procedural fairness, and impersonal trust: An empirical investigation. *Organization Science, 10*(1), 104–115.
14. Das, A., Degeling, M., Smullen, D., & Sadeh, N. (2018). Personalized privacy assistants for the Internet of Things: Providing users with notice and choice. *IEEE Pervasive Computing, 17*(3), 35–46.
15. Ebert, N., Alexander Ackermann, K., & Scheppler, B. (2021). Bolder is better: Raising user awareness through salient and concise privacy notices. In *Proceedings of the 2021 CHI Conference on Human Factors in Computing Systems*, CHI '21. Association for Computing Machinery.
16. Efroni, Z., Metzger, J., Mischau, L., & Schirmbeck, M. (2019). Privacy icons. *European Data Protection Law Review, 5*(3), 352–366.
17. Emami-Naeini, P., Agarwal, Y., Cranor, L. F., & Hibshi, H. (2020). Ask the experts: What should be on an IoT privacy and security label? In *2020 IEEE Symposium on Security and Privacy (SP)* (pp. 447–464). IEEE.
18. Emami-Naeini, P., Bhagavatula, S., Habib, H., Degeling, M., Bauer, L., Cranor, L., & Sadeh, N. (2017). Privacy expectations and preferences in an IoT world. In *Proceedings of the Symposium on Usable Privacy and Security*, SOUPS '17 (pp. 399–412). USENIX Association.

19. Emami Naeini, P., Degeling, M., Bauer, L., Chow, R., Cranor, L. F., Haghighat, M. R., & Patterson, H. (2018). The influence of friends and experts on privacy decision making in IoT scenarios. *Proceedings of the ACM on Human-Computer Interaction, 2*(CSCW), 1–26.
20. Emami-Naeini, P., Dixon, H., Agarwal, Y., & Cranor, L. F. (2019). Exploring how privacy and security factor into IoT device purchase behavior. In *Proceedings of the CHI Conference on Human Factors in Computing Systems*, CHI '19 (pp. 534:1–534:12). ACM.
21. Feng, Y., Yao, Y., & Sadeh, N. (2021). A design space for privacy choices: Towards meaningful privacy control in the Internet of Things. In *Proceedings of the 2021 CHI Conference on Human Factors in Computing Systems*, CHI '21. Association for Computing Machinery.
22. Funk, M., Boldt, R., Pfleging, B., Pfeiffer, M., Henze, N., & Schmidt, A. (2014). Representing indoor location of objects on wearable computers with head-mounted displays. In *Proceedings of the 5th Augmented Human International Conference*, AH '14. Association for Computing Machinery.
23. Garg, R., & Moreno, C. (2019). Understanding motivators, constraints, and practices of sharing Internet of Things. *Proceedings of the ACM on Interactive, Mobile, Wearable and Ubiquitous Technologies, 3*(2), 1–21.
24. Geeng, C., & Roesner, F. (2019). Who's in control? Interactions in multi-user smart homes. In *Proceedings of the 2019 CHI Conference on Human Factors in Computing Systems*, CHI '19 (pp. 1–13). ACM.
25. Harkous, H., Fawaz, K., Lebret, R., Schaub, F., Shin, K. G., & Aberer, K. (2018). Polisis: Automated analysis and presentation of privacy policies using deep learning. In *27th USENIX Security Symposium (USENIX Security 18)* (pp. 531–548). USENIX Association.
26. Harkous, H., Fawaz, K., Shin, K. G., & Aberer, K. (2016). PriBots: Conversational privacy with chatbots. In *Twelfth Symposium on Usable Privacy and Security (SOUPS 2016)*. USENIX Association.
27. He, W., Golla, M., Padhi, R., Ofek, J., Dürmuth, M., Fernandes, E., & Ur, B. (2018). Rethinking access control and authentication for the home Internet of Things (IoT). In *27th USENIX Security Symposium (USENIX Security 18)* (pp. 255–272). USENIX Association.
28. Jakobi, T., Ogonowski, C., Castelli, N., Stevens, G., & Wulf, V. (2017). The catch(es) with smart home: Experiences of a living lab field study. In *Proceedings of the CHI Conference on Human Factors in Computing Systems*, CHI '17 (pp. 1620–1633). ACM.
29. Jin, H., Guo, B., Roychoudhury, R., Yao, Y., Kumar, S., Agarwal, Y., & Hong, J. I. (2022). Exploring the needs of users for supporting privacy-protective behaviors in smart homes. In *CHI Conference on Human Factors in Computing Systems*, CHI '22. Association for Computing Machinery.
30. Kelley, P. G., Bresee, J., Cranor, L. F., & Reeder, R. W. (2009). A "nutrition label" for privacy. In *Proceedings of the 5th Symposium on Usable Privacy and Security*, SOUPS '09. Association for Computing Machinery.
31. Kitkowska, A., Warner, M., Shulman, Y., Wästlund, E., & Martucci, L. A. (2020). Enhancing privacy through the visual design of privacy notices: Exploring the interplay of curiosity, control and affect. In *Sixteenth Symposium on Usable Privacy and Security (SOUPS 2020)* (pp. 437–456). USENIX Association.
32. Klasnja, P., Consolvo, S., Choudhury, T., Beckwith, R., & Hightower, J. (2009). Exploring privacy concerns about personal sensing. In H. Tokuda, M. Beigl, A. Friday, A. J. B. Brush, & Y. Tobe (Eds.), *Pervasive computing* (pp. 176–183). Springer.
33. Koelle, M., Wolf, K., & Boll, S. (2018). Beyond led status lights—design requirements of privacy notices for body-worn cameras. In *Proceedings of the Twelfth International Conference on Tangible, Embedded, and Embodied Interaction*, TEI '18 (pp. 177–187). Association for Computing Machinery.
34. Koshy, V., Park, J. S. S., Cheng, T.-C., & Karahalios, K. (2021). "We just use what they give us": Understanding passenger user perspectives in smart homes. In *Proceedings of the 2021 CHI Conference on Human Factors in Computing Systems*, CHI '21. Association for Computing Machinery.
35. Lau, J., Zimmerman, B., & Schaub, F. (2018). Alexa, are you listening? Privacy perceptions, concerns and privacy-seeking behaviors with smart speakers. *Proceedings of the ACM Conference on Human-Computer Interaction, 2*(CSCW), 102.

36. Lederer, S., Mankoff, J., & Dey, A. K. (2003). Who wants to know what when? Privacy preference determinants in ubiquitous computing. In *CHI '03 Extended Abstracts on Human Factors in Computing Systems*, CHI EA '03 (pp. 724–725). Association for Computing Machinery.
37. Lee, H., & Kobsa, A. (2016). Understanding user privacy in Internet of Things environments. In *2016 IEEE 3rd World Forum on Internet of Things (WF-IoT)* (pp. 407–412). IEEE.
38. Mare, S., Roesner, F., & Kohno, T. (2020). Smart devices in Airbnbs: Considering privacy and security for both guests and hosts. *Proceedings on Privacy Enhancing Technologies, 2020*(2), 436–458.
39. Marky, K., Prange, S., & Alt, F. (2021). Roles matter! understanding differences in the privacy mental models of smart home visitors and inhabitants. In *Proceedings of the 20th International Conference on Mobile and Ubiquitous Multimedia*, MUM'21. ACM.
40. Marky, K., Prange, S., Krell, F., Mühlhäuser, M., & Alt, F. (2020). "You just can't know about everything": Privacy perceptions of smart home visitors. In *19th International Conference on Mobile and Ubiquitous Multimedia* (pp. 83–95). Association for Computing Machinery.
41. Marky, K., Voit, A., Stöver, A., Kunze, K., Schröder, S., & Mühlhäuser, M. (2020). "I don't know how to protect myself": Understanding privacy perceptions resulting from the presence of bystanders in smart environments. In *Proceedings of the 11th Nordic Conference on Human-Computer Interaction: Shaping Experiences, Shaping Society*, NordiCHI '20. Association for Computing Machinery.
42. Mecke, L., Pfeuffer, K., Prange, S., & Alt, F. (2018). Open sesame! user perception of physical, biometric, and behavioural authentication concepts to open doors. In *Proceedings of the 17th International Conference on Mobile and Ubiquitous Multimedia*, MUM 2018 (pp. 153–159). Association for Computing Machinery.
43. Mikusz, M., Houben, S., Davies, N., Moessner, K., & Langheinrich, M. (2018). Raising awareness of IoT sensor deployments. In *Proceedings of the Living in the Internet of Things: Cybersecurity of the IoT*. IET.
44. Portnoff, R. S., Lee, L. N., Egelman, S., Mishra, P., Leung, D., & Wagner, D. (2015). Somebody's watching me? Assessing the effectiveness of webcam indicator lights. In *Proceedings of the 33rd Annual ACM Conference on Human Factors in Computing Systems*, CHI '15 (pp. 1649–1658). Association for Computing Machinery.
45. Prange, S., Mayer, S., Bittl, M.-L., Hassib, M., & Alt, F. (2021). Investigating user perceptions towards wearable mobile electromyography. In *Proceedings of the 18th IFIP TC 13 International Conference on Human-Computer Interaction*, INTERACT '21. Springer.
46. Prange, S., Shams, A., Piening, R., Abdelrahman, Y., & Alt, F. (2021). PriView—exploring visualisations to support users' privacy awareness. In *Proceedings of the 2021 CHI Conference on Human Factors in Computing Systems*. Association for Computing Machinery.
47. Rodriguez, S. D., Prange, S., Ossenberg, C. V., Henkel, M., Alt, F., & Marky, K. (2022). PriKey—investigating tangible privacy control for smart home inhabitants and visitors. In *Proceedings of the 12th Nordic Conference on Human-Computer Interaction*, NordiCHI '22. Association for Computing Machinery.
48. Saha, M., Thakur, S., Singh, A., & Agarwal, Y. (2014). EnergyLens: Combining smartphones with electricity meter for accurate activity detection and user annotation. In *Proceedings of the 5th International Conference on Future Energy Systems*, e-Energy '14 (pp. 289–300). Association for Computing Machinery.
49. Sami, S., Tan, S. R. X., Sun, B., & Han, J. (2021). LAPD: Hidden spy camera detection using smartphone time-of-flight sensors. In *Proceedings of the 19th ACM Conference on Embedded Networked Sensor Systems*, SenSys '21 (pp. 288–301). Association for Computing Machinery.
50. Sharma, R. A., Soltanaghaei, E., Rowe, A., & Sekar, V. (2022). Lumos: Identifying and localizing diverse hidden IoT devices in an unfamiliar environment. In *31st USENIX Security Symposium (USENIX Security 22)*. USENIX Association.
51. Song, Y., Huang, Y., Cai, Z., & Hong, J. I. (2020). I'm all eyes and ears: Exploring effective locators for privacy awareness in IoT scenarios. In *Proceedings of the 2020 CHI Conference on Human Factors in Computing Systems*, CHI '20 (pp. 1–13). Association for Computing Machinery.

52. Tabassum, M., Kosiński, T., & Lipford, H. R. (2019). “I don’t own the data”: End user perceptions of smart home device data practices and risks. In *Proceedings of the Fifteenth USENIX Conference on Usable Privacy and Security*, SOUPS’19 (pp. 435–450). USENIX Association.
53. Teyssier, M., Koelle, M., Strohmeier, P., Fruchard, B., & Steimle, J. (2021). Eyecam: Revealing relations between humans and sensing devices through an anthropomorphic webcam. In *Proceedings of the 2021 CHI Conference on Human Factors in Computing Systems*, CHI ’21. Association for Computing Machinery.
54. Thakkar, P. K., He, S., Xu, S., Huang, D. Y., & Yao, Y. (2022). “It would probably turn into a social faux-pas”: Users’ and bystanders’ preferences of privacy awareness mechanisms in smart homes. In *CHI Conference on Human Factors in Computing Systems*, CHI ’22. Association for Computing Machinery.
55. Volk, V., Prange, S., & Alt, F. (2022). PriCheck—an online privacy assistant for smart device purchases. In *Extended Abstracts of the 2022 CHI Conference on Human Factors in Computing Systems*, CHI EA ’22. Association for Computing Machinery.
56. Waddell, T. F., Auriemma, J. R., & Sundar, S. S. (2016). Make it simple, or force users to read? Paraphrased design improves comprehension of end user license agreements. In *Proceedings of the 2016 CHI Conference on Human Factors in Computing Systems*, CHI ’16 (pp. 5252–5256). Association for Computing Machinery.
57. Weiser, M., Gold, R., & Brown, J. S. (1999). The origins of ubiquitous computing research at PARC in the late 1980s. *IBM Systems Journal, 38*(4), 693–696.
58. Yao, Y. (2019). Designing for better privacy awareness in smart homes. In *Conference Companion Publication of the 2019 on Computer Supported Cooperative Work and Social Computing*, CSCW ’19 (pp. 98–101). Association for Computing Machinery.
59. Yao, Y., Basdeo, J. R., Kaushik, S., & Wang, Y. (2019). Defending my castle: A co-design study of privacy mechanisms for smart homes. In *Proceedings of the 2019 CHI Conference on Human Factors in Computing Systems*, CHI ’19 (pp. 1–12). ACM.
60. Yao, Y., Basdeo, J. R., Mcdonough, O. R., & Wang, Y. (2019). Privacy perceptions and designs of bystanders in smart homes. *Proceedings of the ACM on Human-Computer Interaction, 3*(CSCW), 1–24.
61. Zeng, E., & Roesner, F. (2019). Understanding and improving security and privacy in multi-user smart homes: A design exploration and in-home user study. In *28th USENIX Security Symposium (USENIX Security 19)* (pp. 159–176). USENIX Association.
62. Zhou, W., Jia, Y., Peng, A., Zhang, Y., & Liu, P. (2019). The effect of IoT new features on security and privacy: New threats, existing solutions, and challenges yet to be solved. *IEEE Internet of Things Journal, 6*(2), 1606–1616.

Challenges, Conflicts, and Solution Strategies for the Introduction of Corporate Data Protection Measures

Christian K. Bosse, Denis Feth, and Hartmut Schmitt

1 Introduction

Safeguarding and exercising data subjects' rights by implementing technical and organizational measures are highly important. Accordingly, data protection laws such as the General Data Protection Regulation [24] and the California Privacy Rights Act [18] address these measures. However, it must be considered that privacy and data protection are not only about technical and organizational aspects. There is also a third sphere that has to be considered: the social sphere. Within and between these three spheres—technical, organizational, and social—a variety of conflicts can arise, e.g., due to different interests of various stakeholders [7]. In particular, one must be aware that any data protection measure can also have undesired side effects. For example, backups can negatively influence data minimization or deletion processes in a company. Of course, this does not mean that backups are to be avoided. However, if such dependencies and conflicts are not explicitly considered when designing data protection measures, this can lead to a complete rejection by employees in the worst case [8]. In this chapter, we discuss these challenges and offer appropriate solutions. We focus on the business context, in particular the relationship between employees and employers, and illustrate our discussion with a specific example [47].

C. K. Bosse (✉)
Institut für Technologie und Arbeit e.V., Kaiserslautern, Germany
e-mail: christian.bosse@ita-kl.de

D. Feth
Fraunhofer IESE, Kaiserslautern, Germany
e-mail: denis.feth@iese.fraunhofer.de

H. Schmitt
HK Business Solutions GmbH, Friedrichsthal, Germany
e-mail: hartmut.schmitt@hk-bs.de

N. Gerber et al. (eds.), *Human Factors in Privacy Research*,
https://doi.org/10.1007/978-3-031-28643-8_17

Chapter Overview First, Sect. 2 creates an overview of related research in the two relevant topic areas of socio-technological adoption of new technologies and the usability of security and data protection measures. Then, in Sect. 3, we argue why digital transformation needs to be viewed holistically and present our sphere model that shows the multiple interactions between the three spheres. In the following Sect. 4, we address challenges that may arise, whether due to a lack of consideration of the interactions between these spheres, deliberate manipulation of individuals' behavior, or privacy-intrusive data protection measures. In Sect. 5, we use an example to describe the operationalization of our models before drawing a final conclusion in Sect. 6.

2 Related Work

Our work is primarily related to research from two areas: socio-technical aspects of the introduction of new technologies and the usability of security and data protection measures. In the following, we will distinguish ourselves from these works or put them in context.

2.1 Technology Introduction and Acceptance

The adoption of new technologies is not a new field of research in science, although initially, the framework conditions were still different: As early as the 1950s, studies were conducted on the adoption of new technologies in agriculture and their diffusion processes [5]. The diffusion theory resulting from this work describes, among other things, the social system as a relevant factor for the diffusion of an innovation, consisting of its norms, organizational rules, structures, as well as opinion leaders [41]. After this, the effects of various factors on users' attitude regarding the new technology and their interaction became the subject of research. A basic technology acceptance model [19, 20], which has been further developed and supplemented over the years [45, 53], analyzes and describes these. This includes initial approaches to structuring the introduction process as well as controlling interventions by the organization [52].

Due to the dynamics associated with rapid technological progress and the modern megatrend of digitalization, this work is gaining relevance once again. Influencing factors that can increase the success of implementation processes can be derived from this work. These factors include, for example: active involvement of users of the new technology in the introduction process [3], support from managers [57], well-designed training courses [48], or the involvement of internal and/or external experts [30] who actively accompany and help shape the change. These factors must be seen in the context of the current change of work and the digital transformation that goes hand in hand with it [27]. The focus is increasingly shifting toward employees, who are recognized as a central factor that acts in a self-determined and self-organized manner. In addition to corporate goals regarding costs and quality, work design increasingly addresses employee-related goals such as personality development, even if these goals are sometimes in conflict with corporate goals [43]. The increase in self-determination and privacy regarding data in the workplace, which can be enabled by the use of a privacy dashboard, should also be seen in this context [50]. However, previous work has primarily focused on the use of privacy-enhancing technologies (see the chapter "Acceptance Factors of Privacy-Enhancing Technologies on the Basis of Tor and JonDonym") at the interface between companies and end users, mostly with a focus on the latter [6]. The design of a fair exchange of information between companies and employees supported by a technological solution has not been comprehensively researched yet.

2.2 Usable Security and Usable Privacy

Existing literature on usable security shows that the user is an important part of modern security chains. The strongest technical security measure is not effective if attackers can circumvent it by means of social engineering, for example. Well-known case studies have analyzed the usability of email encryption with PGP [56], of file sharing with Kazaa [26], and of authentication mechanisms and password policies [16, 28]. However, such case studies are specific to one technology or application and do not consider conflicts arising from the technologies. Design principles for usable, yet secure systems [23, 33] focus on the development of usable security systems by supporting developers and emphasizing the importance of considering the user. However, these principles ignore the area of technology introduction.

In the area of data protection measures, the so-called privacy dashboards are becoming increasingly important, also in the enterprise context [22, 40]. In general, various projects evaluate the applicability and usability of privacy dashboards. In the myneData project [34], for example, a user-controlled data market for personal data was created. A decentralized solution is offered by the MyData project [38], where a cockpit is only used for transparency and control, but the data remain with the services and can be exchanged via (existing) channels after user consent. In the SPECIAL project [32], a holistic approach was developed where data from various sources are aggregated and harmonized based on machine learning and semantic technologies. Even though usability is an important aspect of these projects, challenges and conflicts were not explicitly considered. For a more detailed summary of research on usable privacy, please refer to the chapter "Empirical Research Methods in Usable Privacy and Security".

3 Digital Transformation as a Holistic Challenge

Companies in all sectors and industries are affected by digital transformation [58], and so are the working environments of their employees. Driven by the rapid progress in technology, traditional jobs are changing, business processes are being re-oriented, and innovative digital business models are emerging. In industrial production, for example, digital innovations often lead to radical change, which is also called digital disruption [7]. The analysis of data, including a lot of personal data, offers the possibility to optimize existing processes and workflows. To successfully master the key challenge of digital transformation, all three of the spheres mentioned in Sect. 1 must be considered as shown in Fig. 1 [10].

The *organizational sphere* roughly comprises everything that has to do with regulations and processes within a company, such as works council agreements, data protection regulations, incentive systems, standards, and laws. This sphere is so relevant because it defines how a company works. Problems within the organizational sphere therefore usually have a direct impact on the effectiveness and/or efficiency of an organization.

The *technical sphere* deals with the tools for implementing organizational regulations. A high level of usability of the tools used according to ISO 9241-11:2018 [29] is essential. This is shown, for example, by a study conducted in Germany among 1000 employees [36], according to which 55% of the participants bypassed their company's security measures at least once a week and 17% even did so daily. The reason: the use of IT security systems is perceived as too complicated and time-consuming. Accordingly, aspects such as ergonomics, interface design, and interaction design of security and data protection tools—summarized under the term "usable security and privacy"—must be taken seriously. Problems with the use of technical tools have a direct impact on their acceptance or hinder employees in the

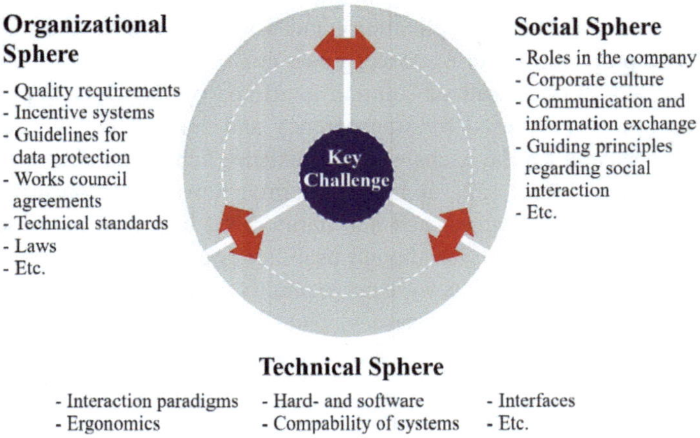

Fig. 1 Interaction in our three-sphere model

performance of their tasks. This can even go as far as employees actively exploring and establishing ways to perform their tasks without the use of the new technology, even if this behavior can be harmful for the company [7, 21].

In the *social sphere*, primarily interpersonal aspects come into play. The attitude of employees toward digital transformation in general and the introduction of new processes or technologies have a significant influence on the success of the implementation. Corporate culture and good communication play a major role here. Problems in this sphere can lead to mistrust and a lack of acceptance and condemn a digitalization project to failure from the outset. Similarly, power struggles or rivalries between individuals or groups in the social sphere of the company, for example, can lead to the success of a technology introduction being jeopardized.

4 Challenges in the Operational Introduction of Data Protection Measures in Companies

In this section, we use three examples to illustrate the challenges that can arise in the context of introducing data protection measures in companies. In doing so, we draw on the 3-sphere model already presented, in which the challenges can be located. We also highlight apparent contradictions that can arise in this context. The three challenges presented are underpinned with the help of fictitious examples based on practical experience, so that the relevance for practice becomes more apparent.

4.1 Lack of Considering the Interactions of the Spheres

It is easy to understand that each of the three spheres is relevant individually, and however, strong interactions between the spheres exist. If only one sphere, e.g., the technical one, is considered when introducing data protection measures, gaps and backdoors can arise due to the close links with the other spheres. The interaction of the spheres offers a wide range of opportunities to obtain sensitive employee data or personal information even without direct technical access [46]. For this reason, when implementing a new technology, various domain- and company-specific regulations, standards, and legal requirements must be considered. It may even become necessary to adjust internal regulations or processes to support the new technologies [10, 54].

Also, all relevant employees must be involved as early as possible. Involvement in the selection and design of technology is just as important as training on their application. Without employee participation and process adaptation, the monitoring that employees may perceive can have a variety of unintended effects. For example, employees may feel that they are under constant scrutiny and may adapt their actions or behaviors in ways that may be detrimental to organizational processes and workflows. Under certain circumstances, this can even pose a risk to the company if, for example, the protection of employee data is not ensured as a result [9, 10]. How quickly is sensitive employee data printed out shortly before the weekend and taken home instead of being retrieved from home via a protected connection to the company's IT system, whose use is both cumbersome and logged?

The emotional impact of new technologies should also not be underestimated. While some employees welcome them in principle, others reject them completely or even fear for their jobs. Such fears must be addressed openly, taken seriously, and resolved. Otherwise, fronts can quickly form that can only be overcome with great difficulty. In practice, however, the social impact is often neglected or considered much too late, possibly resulting in user requirements not being met, users being overwhelmed, or the works council intervening [9].

Example

To illustrate the extent to which the technical, organizational, and social spheres of a company interlock and influence each other, one might consider the example of the necessarily hasty establishment of remote work during the Covid-19 pandemic. If employees are expected to work from home, the company must provide the necessary technical equipment and make sure that it is usable, privacy-friendly, and secure. Furthermore, it has to ensure compliance with legal regulations, such as the Working Hours Act or occupational safety, as well as data protection [54]. In addition, works council agreements and, if necessary, further company standards and processes must be adapted accordingly and complied with [1]. Employees must also be trained on how to access company data securely from home and how to handle internal data in a private or publicly viewable environment—for example, when working with mobile devices in the home office or on business trips [13, 31]. Furthermore, effects on cooperation among colleagues as well as on the corporate culture are to be expected, necessitating guiding intervention by the management level. Managing at a distance, as is needed in decentralized and digitally working teams, presents a new challenge for managers. Strict guidelines and control no longer represent the contemporary style of leadership. A manager must be a supporter of the team and is responsible for promoting the ability to work [26, 37].

4.2 Exploiting the Gray Areas of Data Protection

New possibilities for data collection and processing in connection with employees' personal data are arousing new desires, not least on the employers' side [7]. For example, changed models of work like the home office boom triggered by the Covid-19 pandemic are fueling the desire of many employers to monitor those employees who are no longer working on the company's premises [35]. In order to obtain the desired data, employers often use practices that are not prohibited but are nonetheless ethically questionable because they violate the basic principles of self-determination and privacy protection. This can be achieved by exploiting basic psychological principles, exploiting the so-called privacy paradox or by a deceptive design of the user interface (see also the chapter "The Hows and Whys of Dark Patterns: Categorizations and Privacy"). In the following subsections, we describe these gray areas in more detail.

The practices described are comparable to practices that are referred to as social engineering in IT security. Social engineering refers to methods of behavioral manipulation in which human characteristics such as helpfulness, trust, or respect for authority are exploited to gain unauthorized access to information or IT systems [2]. However, the target of attacks is usually not employees' personal data, but other companies' data of high value. In most cases, the attackers are also external to the company, such as industrial spies, blackmailers, competitors, or disruptors.

Recognizing that gray areas are being entered can lead employees to reject newly introduced technologies, and the damage done may be greater than the benefits hoped for. In addition, there are also several examples where actual data protection violations became known and were also fined. For example, H&M was fined 35 million euros for illegal surveillance of its employees [44].

Exploitation of Basic Psychological Principles Possible points of attack that employers can exploit in a rather subtle way to obtain their employees' data are certain psychological principles, which the social psychologist Cialdini called "weapons of influence" [17]:

- *Reciprocity:* When someone does us a favor or gives us a gift, we feel obligated to return the favor and often give back even more than we initially received.
- *Scarcity:* We consider things that are only available in limited quantities or only for a certain time to be particularly valuable.
- *Authority:* We are more likely to agree with people we consider authorities because they are assumed to have more knowledge, experience, or expertise than we do.
- *Consistency:* Once we have made a decision or taken a position on something, we tend to stick to it.
- *Liking:* We are more likely to help other persons out if we like them. Similarity, compliments, and physical attractiveness contribute to liking.
- *Social Proof:* When we are uncertain, we often look at how others behave. The more people behave in a certain way, the higher the chance we consider this behavior appropriate. In other words, humans adapt to the (supposed) social norm.

In addition, there are several similar factors [15] that employers can use, such as:

- Appealing to values such as helpfulness and loyalty
- Exploiting personal or professional trust

- Short reflection time for requests, so that the individual cannot think about possible consequences of their action
- Greek gifts (example: permission for private use of company cell phones, which are then used to spy on employees)

The following fictitious example shows how these principles and factors can be used to compromise employee data protection.

Example

Christine E. Owens presides over approximately 70 employees as the chief executive officer of a start-up company. She would like to make company processes more efficient using data analysis. Her data protection officer, who has since been dismissed, said that because of the personal reference to employees, she may only use certain data with their consent. Christine is confident that all her employees will consent. She writes the following email to her employees:

"Most start-ups evaluate process data. A random survey in our company showed that 89% of the respondents think it would be good if we also evaluated process data. By giving your consent, you help to save costs, which contributes to the success of the company. The success of our company is very important to all of us. Please give me your consent for the collection and analysis of the data by 3 p.m. today. Tomorrow morning, I will approach everyone whose consent I have not received until then to find out more about the reasons for this. As your CEO, I am counting on you! Yours, Christine E. Owens"

There are several forms of influence in this fictitious example:

- *Authority: Christine emphasizes her position as CEO to gain the consent of the employees and builds up a threatening gesture ("I will approach everyone").*
- *Social proof: Using phrases such as "most startups" and "89% of respondents," Christine points to the social norm.*
- *Short reflection time: Instructing people to respond on the same day builds up time pressure.*
- *Appealing to loyalty: Christine points out the common vision ("success of the company") and that everyone's consent is expected ("I am counting on you!").*

Exploiting the Privacy Paradox The privacy paradox [4] describes a discrepancy between what users want and what users do regarding their privacy. Several studies [42] confirm that users do care about their privacy but do not act accordingly. There are several reasons for this: For example, security and data protection measures typically require a certain level of knowledge and certain skills, which some users do not possess [25]. Solutions for resolving the privacy paradox are still being heavily researched.

Deceptive Design of User Interfaces Further opportunities for behavioral manipulation to lower the level of employee privacy are opened up by digital *nudging* [55] (see also the chapter "Privacy Nudges and Informed Consent? Challenges for Privacy Nudge Design") and the use of dark patterns [11] (see also the chapter "The Hows and Whys of Dark Patterns: Categorizations and Privacy"). These phenomena can be exploited to weaken employee data protection already in the design of internally used IT systems. The aim of nudging is to (subtly) give an impetus to certain socially desirable behavior, i.e., to bring about "better" decisions [49]. This is done without coercion or financial incentives. One of the most effective digital nudges is the setting of default rules and preferences, such as the privacy-friendly defaults required in Art. 25 (2) GDPR. However, the same techniques can also be used to make users act contrary to their actual intentions, such as agreeing to permissive privacy settings. *Dark patterns* are patterns that are used in the design of user interfaces to mislead or entice the user to perform unwanted actions. These are actually anti-patterns—examples of how things should not be done—but are deliberately used in an unethical or deceptive manner. The systematic use of such dark patterns is described by Bösch et al. [14] as Dark Strategies.

4.3 Data Protection Measures Counteracting Privacy

It may seem counter-intuitive, but it is actually a real risk: Data protection measures can counteract privacy. We give three examples originally presented in [39]:

Transparency vs. Surveillance Data subjects may have the desire to know who is processing their personal data. Providing this information to the data subject can affect the privacy of data users (e.g., employees in customer service). For example, if the exact time and person of a data use is revealed, the data subject can draw conclusions about the data user's work behavior. Anonymization can at least partially resolve this conflict.

Trust vs. Mistrust Technical and organizational measures normally increase trust in an information system and its provider. However, information meant to increase transparency could cause resentment as data subjects become aware of the use of their personal data. Also, the sudden introduction of privacy-enhancing technology could arouse mistrust. Data subjects may wonder whether there has been a privacy incident that led to this rollout. Therefore, the objectives of the introduction of privacy-enhancing technology should be made clear.

Self-determination vs. Social Pressure Data subjects have the right of self-determination. For example, they could specify that their usage data must not be analyzed for the purpose of system optimization, or they may object to the publication of a picture on social media, which the marketing department would love to share. If data subject and data user know each other—for instance, if they are colleagues or have a business relationship—the data subject may experience social pressure to provide these data. This can be especially critical if the data user is an authority. A respectful work culture or respectful business relationship could resolve such a conflict.

These examples illustrate that an "ideal" solution does not or cannot exist. Even if a security or data protection measure initially appears ideal from the users' point of view and the users also employ it to implement their data protection, the introduction of such a tool alone may lead to new problems.

5 Operationalization in Practice

A research project [51] examined the challenges described above and developed application-oriented solutions, with the overarching goal of balancing the interests of employees and their employers and helping to strengthen a culture of trust in companies by improving employee privacy. Through the interaction of the various spheres, data protection is to be ensured in the long term not only through fair reconciliation of interests, but also with the help of extensive user awareness. The following example will further illustrate this.

Example

In the development project for a business-critical software, a call center company is trying to alleviate reservations about data protection and achieve the best possible acceptance among internal users. It therefore gives high priority to the quality characteristic of data minimization. At the same time, the company's business operations must be maintained at all times, even in the event of data loss. Accordingly, the quality characteristic of recoverability is also prioritized. Therefore, backups containing sensitive personal data of call center agents are indispensable for its fulfillment. This illustrates at least one conflict of objectives—data minimization vs. recoverability—which may be supplemented by interactions with other quality characteristics such as transparency or intervenability for the employees involved.

As a solution to this conflict of objectives, it was decided to develop a detailed backup and deletion concept for the various backup generations and to implement corresponding deletion routines at the technical level. To implement this procedure successfully, it is also necessary to plan and implement complementary activities in the company's organizational and social sphere. One starting point, for example, is to define appropriate operating instructions at the organizational level: What data are stored where and for how long in which backups? Are the backups encrypted? Who is allowed to access them? A criteria catalog or corresponding guiding questions can provide support here, such as "Is the number of backup systems required specified?," "In the case of additional redundant backup systems, have the redundancy mechanisms been specified?," and "Has the way in which the backups are created been determined?."

Corresponding measures should always be taken with the involvement of the works council or employee representatives, who should ideally be involved in resolving the conflict of objectives from the very beginning. Here, it is important to comply with the existing law, which stipulates a duty of co-determination as part of the introduction of technology as soon as there is a risk that employers could control the performance and behavior of their employees. Furthermore, it should be checked whether additional works council agreements are needed in which employees agree to the temporary storage of their sensitive personal data. In addition, all affected employees should be made aware of the measures (e.g., the backup and archiving systems from the given example) at an early stage and trained in their operation.

Thus, recognizing the issue allows a company to find a balance of interests between all those involved already during the development phase and to maintain it more easily during the implementation and operation phases.

Involving the works council or employee representatives is also a first step toward addressing the social level in the company. However, this alone is not sufficient to achieve high acceptance of the new technical solution. The first step is to raise employees' awareness of the need for the new technical solution and to make clear the importance of their contribution to data protection and security through the successful introduction of this new technology. Internal information events to which management invites the employees are first step of doing this. This highlights the relevance of the development project for the company and the role of managers as good role models who support the project. In addition, employees should be regularly informed about the progress and kept up to date. To this end, the appropriate communication channels and formats must be selected, which may vary depending on the company. Another step is to participate employees in the early phases of technology introduction, for example, in requirements analysis (see also the chapter "Achieving Usable Security and Privacy Through Human-Centered Design"). Furthermore, it is essential that employees receive training on the use of the new digital solution well in advance of the go live. Well-structured training should show both the general scope of functions and their limitations as well as the specific procedure in practical use cases. This will ensure that employees are not initially overwhelmed by the use of the new technology and the resulting changes in workflows.

Further general measures for maintaining a high level of data protection are the establishment of organizational regulations (e.g., locking one's screen when leaving the desk) and raising awareness for behavioral manipulation similar to social engineering attacks. Regarding social engineering, there are special kinds of training that expose employees to a trap, such as a pretend phishing email. Trapped employees are then informed about countermeasures. The German Federal Office for Information Security (BSI) provides current examples of phishing attacks and informs about countermeasures [12].

6 Summary

The introduction of new technologies or processes in a company is often subject to reservations and conflicts. In the case of data protection, this is particularly challenging due to the criticality, sensitivity, and legal requirements in this area. In this chapter, we therefore first looked at the challenges that must be considered when introducing corporate data protection measures. In particular, a lack of attention to the interactions between the technical, organizational, and social spheres of a company can lead to unintended interactions, up to and including rejection of the new technology and harmful behavior of employees. We presented possible solutions as to how a holistic approach considering all three spheres can contribute to successful technology introduction.

Acknowledgments This work is funded by the German Federal Ministry of Education and Research (grant numbers 16KIS1506K, 16KIS1507, and 16KIS1509).

References

1. Alipour, J.-V., Falck, O., & Schüller, S. (2020). Homeoffice während der Pandemie und die Implikationen für eine Zeit nach der Krise. ifo Institut – Leibniz-Institut für Wirtschaftsforschung an der Universität München (Vol. 73, pp. 30–36).
2. Anderson, R. (2008). *Security engineering: A guide to building dependable distributed systems*. Wiley
3. Barki, H., & Hartwick, J. (1994). Measuring user participation, user involvement, and user attitude. *MIS Quarterly, 18*(1), 59–82.
4. Barnes, S. (2006). A privacy paradox: Social networking in the United States. *First Monday, 11*(9).
5. Beal, G., & Rogers, E. (1960). The adoption of two farm practices in a central Iowa community. Special report 26, Iowa Agricultural and Home Economics Experiment Station Publications, Iowa.
6. Blumberg, V., & Kauffeld, S. (2020). Anwendungsszenarien und Technologiebewertung von digitalen Werkerassistenzsystemen in der Produktion. *GIO Gruppe. Interaktion. Organisation. Zeitschrift für Angewandte Organisationspsychologie, 51*(1), 5–24.
7. Bosse, C. K., Dietrich, A., Kelbert, P., Küchler, H., Schmitt, H., Tolsdorf, J., & Weßner, A. (2020). Beschäftigtendatenschutz: Rechtliche Anforderungen und Technische Lösungskonzepte. In W. Kummer, F. Schweighofer, E. Hötzendorf (Eds.), *Conference volume of the 23rd Edition of the Conference International Legal Informatics Symposions IRIS*. Vachendorf.
8. Bosse, C. K., Dietrich, A., & Schmitt, H. (2021). IT-Rahmenwerk für den eschäftigtendatenschutz. Technologieeinführung aus rechtlicher und arbeitswissenschaftlicher Perspektive. *Informatik, 2020*, 815–828.
9. Bosse, C. K., Dietrich, A., & Weßner, A. (2021). Selbstbewertungsinstrument für den betrieblichen Datenschutz. Unterstützung für die Umsetzung des Beschäftigtendatenschutzes in KMU. *Datenschutz und Datensicherheit-DuD, 45*(1), 23–27.
10. Bosse, C. K., Hellge, V., Schröder, D., & Dupont, S. (2019). Digitalisierung im Mittelstand erfolgreich gestalten. In C. Bosse & K. Zink (Eds.), *Arbeit 4.0 im Mittelstand. Chancen und Herausforderungen des digitalen Wandels für KMU* (pp. 13–34). Springer-Gabler.
11. Brignull, H. (2011). Dark patterns: Deception vs honesty in UI design. https://alistapart.com/article/dark-patterns-deception-vs.-honesty-in-ui-design/
12. Bundesamt für Sicherheit in der Informationstechnik (BSI). (2011). Aktuelle Beispiele für Phishing-Angriffe. https://www.bsi.bund.de/DE/Themen/Verbraucherinnen-und-Verbraucher/Cyber-Sicherheitslage/Methoden-der-Cyber-Kriminalitaet/Spam-Phishing-Co/Passwortdiebstahl-durch-Phishing/Aktuelle-Beispiele-fuer-Phishing/aktuelle-beispiele-fuer-phishing_node.html
13. Bruhn, P. (2020). IT-Sicherheit und Datenschutz. In *Homeoffice und mobiles Arbeiten im Team effektiv umsetzen. Essentials*. Springer Vieweg.
14. Bösch, C., Erb, B., Kargl, F., Kopp, H., & Pfattheicher, S. (2016). Tales from the dark side: Privacy dark strategies and privacy dark patterns. In *Proceedings on Privacy Enhancing Technologies* (pp. 237–254).

15. Caraban, A., Karapanos, E., Gonçalves, D., & Campos, P. (2019). 23 ways to nudge: A review of technology-mediated nudging in human-computer interaction. In *Proceedings of the 2019 CHI Conference on Human Factors in Computing Systems*, CHI '19 (pp. 1–15). Association for Computing Machinery.
16. Choong, Y.-Y., & Theofanos, M. (2015). What 4,500+ people can tell you—employees' attitudes toward organizational password policy do matter. In T. Tryfonas & I. Askoxylakis (Eds.), *International Conference on Human Aspects of Information Security, Privacy, and Trust* (pp. 299–310). Springer.
17. Cialdini, R. B. (2009). *Influence: The psychology of persuasion, revised edition*. Harper Business.
18. CPRA. (2019). The California Privacy Rights Act of 2020. https://oag.ca.gov/system/files/initiatives/pdfs/19-0021A1. Consumer Privacy—Version 3.
19. Davis, F. (1985). *A technology acceptance model for empirically testing new end-user information systems—theory and results*. PhD thesis, Massachusetts Inst. of Technology.
20. Davis, F., Bagozzi, R., & Warshaw, P. (1989). User acceptance of computer technology: A comparison of two theoretical models. *Management Science, 35*(8), 982–1003.
21. Dietrich, A., Bosse, C. K., & Schmitt, H. (2021). Kontrolle und Überwachung von Beschäftigten. *Datenschutz und Datensicherheit-DuD, 45*(1), 5–10.
22. Feth, D., & Schmitt, H. (2020). Requirement and quality models for privacy dashboards. In *2020 IEEE 7th International Workshop on Evolving Security & Privacy Requirements Engineering (ESPRE)* (pp. 1–6). IEEE.
23. Garfinkel, S. (2005). *Design principles and patterns for computer systems that are simultaneously secure and usable*. PhD thesis, Massachusetts Institute of Technology.
24. GDPR. (2016). Regulation (EU) 2016/679 of the European Parliament and of the Council of 27 April 2016 on the protection of natural persons with regard to the processing of personal data and on the free movement of such data, and repealing Directive 95/46/EC (General Data Protection Regulation. https://eur-lex.europa.eu/eli/reg/2016/679/oj
25. Gerber, P., Volkamer, M., & Gerber, N. (2017). Das Privacy-Paradoxon–Ein Erklärungsversuch und Handlungsempfehlungen. In *Dialogmarketing Perspektiven 2016/2017* (pp. 139–167). Springer Gabler.
26. Good, N. S., & Krekelberg, A. (2003). Usability and privacy: A study of Kazaa P2P file-sharing. In *Proceedings of the SIGCHI Conference on Human Factors in Computing Systems*, CHI '03 (pp. 137–144). Association for Computing Machinery.
27. Hasenbein, M. (2020). *Der Mensch im Fokus der digitalen Arbeitswelt*. Springer.
28. Inglesant, P. G., & Sasse, M. A. (2010). The true cost of unusable password policies: Password use in the wild. In *Proceedings of the SIGCHI Conference on Human Factors in Computing Systems*, CHI '10 (pp. 383–392). Association for Computing Machinery.
29. ISO 9241-11. (2018). Ergonomics of human-system interaction—part 11: Usability: Definitions and concepts (ISO 9241-11:2018).
30. Jasperson, J., Carter, P. E., & Zmud, R. W. (2005). A comprehensive conceptualization of post-adoptive behaviors associated with information technology enabled work systems. *MIS Quarterly, 29*(3), 525–557.
31. Kanzenbach, K. (2020). Rechtliche Grundlagen zum Homeoffice und der Telearbeit. In *DGUV forum* (Vol. 8, pp. 18–24).
32. Kirrane, S., Fernández, J. D., Dullaert, W., Milosevic, U., Polleres, A., Bonatti, P. A., Wenning, R., Drozd, O., & Raschke, P. (2018). A scalable consent, transparency and compliance architecture. In *European semantic web conference* (pp. 131–136). Springer.
33. Lo Iacono, L., Smith, M., von Zezschwitz, E., Gorski, P. L., & Nehren, P. (2018). Consolidating principles and patterns for human-centred usable security research and development. In *European workshop on usable security*.
34. Matzutt, R., Müllmann, D., Zeissig, E.-M., Horst, C., Kasugai, K., Lidynia, S., Wieninger, S., Ziegeldorf, J. H., Gudergan, G., Spiecker gen. Döhmann, I., Wehrle, K., & Ziefle, M. (2017). myneData: Towards a trusted and user-controlled ecosystem for sharing personal data. In *INFORMATIK 2017*.

35. Moorstedt, M. (2020). Tracking von Mitarbeitern. In der Krise boomt auch die Überwachung durch den Chef. https://www.sueddeutsche.de/digital/home-office-ueberwachung-tracking-chef-zoom-1.4868739
36. KES Online. (2022). Jeder zweite Angestellte umgeht Security-Lösungen. https://www.kes.info/archiv/schlaglichter/schlaglicht/?tx_ttnews%5Byear%5D=2022&tx_ttnews%5Bmonth%5D=06&tx_ttnews%5Bday%5D=10&tx_ttnews%5Btt_news%5D=228&cHash=9153da146aff24a9c080c4347cdb1fc8
37. Osranek, R., & Staat, P. (2020). *Moderne Führung als Ausdruck neuer Werte* (2nd ed.). Ayway media GmbH.
38. Poikola, A., Kuikkaniemi, K., & Honko, H. (2015). MyData. A Nordic model for human-centered personal data management and processing. Finnish Ministry of Transport and Communications.
39. Polst, S., & Feth, D. (2020). Privacy ad absurdum-how workplace privacy dashboards compromise privacy. In *Mensch und Computer 2020-Workshopband.*
40. Polst, S., Kelbert, P., & Feth, D. (2019). Company privacy dashboards: Employee needs and requirements. In *International Conference on Human-Computer Interaction* (pp. 429–440). Springer.
41. Rogers, E. (2003). *Diffusion of innovations* (5th ed.). Free Press.
42. Rudolph, M., Feth, D., & Polst, S. (2018). Why users ignore privacy policies—a survey and intention model for explaining user privacy behavior. In *International Conference on Human Aspects of Information Security, Privacy, and Trust.*
43. Schaper, N. (2019). Arbeitsgestaltung in Produktion und Verwaltung. In G. Schaper, N. Nerdinger, & F.W. Blickle (Eds.), *Arbeits-und Organisationspsychologie* (pp. 411–434). Springer.
44. Schemm, M. (2020). Bußgeld wegen Datenschutzverstößen bei H&M. https://datenschutz-hamburg.de/pressemitteilungen/2020/10/2020-10-01-h-m-verfahren
45. Schepers, J., & Wetzels, M. (2007). A meta-analysis of the technology acceptance model: Investigating subjective norm and moderation effects. *Information & Management, 44*(1), 90–103.
46. Schmitt, H., Bosse, C. K., Dietrich, A., & Polst, S. (2021). Wie ich an deine Daten kam oder Dark Patterns und Phishing im Beschäftigtenkontext. In *Jusletter IT 27.*
47. Schmitt, H., & Groen, E. C. (2021). Qualitätsmodell zur Förderung des Beschäftigtendatenschutzes. *Datenschutz und Datensicherheit-DuD, 45*(1), 28–32.
48. Sharma, R., & Yetton, P. (2007). The contingent effects of training, technical complexity, and task interdependence on successful information systems implementation. *MIS Quarterly, 31*(2), 219–238.
49. Thaler, R., & Sunstein, C. (2008). *Nudge: Improvising decisions about health, wealth, and happiness.* Yale University Press.
50. Tolsdorf, J., Bosse, C. K., Dietrich, A., Feth, D., & Schmitt, H. (2020). Privatheit am Arbeitsplatz. Transparenz und Selbstbestimmung bei Arbeit 4.0. *Datenschutz und Datensicherheit-DuD, 44*(3), 176–181.
51. TrUSD Project Consortium. (2022). TrUSD—Transparente und selbstbestimmte Ausgestaltung der Datennutzung im Unternehmen. https://www.trusd-projekt.de/
52. Venkatesh, V., & Bala, H. (2008). Technology acceptance model 3 and a research agenda on interventions. *Decision Sciences, 39*(2), 273–315.
53. Venkatesh, V., & Davis, F. D. (2000). A theoretical extension of the technology acceptance model: Four longitudinal field studies. *Management Science, 46*(2), 186–204.
54. Visser, L., Voigt, P., & Vraetz, M. (2021). *Das Recht auf Homeoffice in der Pandemie* (1st ed.). Baden-Baden.
55. Weinmann, M., Schneider, C., & Brocke, J. v. (2016). Digital nudging. *Business & Information Systems Engineering, 58*, 433–436.
56. Whitten, A., & Tygar, J. D. (1999). Why Johnny can't encrypt: A usability evaluation of PGP 5.0. In *8th USENIX Security Symposium (USENIX Security 99)*, Washington, D.C. USENIX Association.

57. Wieseke, J., Kraus, F., & Rajab, T. (2010). Ein interdisziplinärer Ansatz zur Überwindung von Technologieadaptionsbarrieren. *Zeitschrift für betriebswirtschaftliche Forschung, 62*(7), 822–859.
58. Zink, K., Schröder, D., Hellge, V., & Bosse, C. (2019). Zukunft der Arbeit = Arbeit 4.0? In K.J. Zink (Ed.), *Arbeit und Organisation im digitalen Wandel* (pp. 53–93). Baden-Baden.

Data Cart: A Privacy Pattern for Personal Data Management in Organizations

Jan Tolsdorf and Luigi Lo Iacono

1 Introduction

The entry into force of the General Data Protection Regulation (GDPR) [29] in the European Union (EU) in 2016 has had a lingering impact worldwide on how individuals' personal data are processed. Essentially, entities that determine the purpose and/or process personal data are held more accountable than before for protecting the privacy of individuals. For instance, these entities are obligated to implement individuals' rights to transparency and intervention (Art. 12–21 GDPR), as well as to take measures for upholding the GDPR's principles for the privacy preserving and secure processing of personal data (Art. 5 GDPR). To reduce the risks of privacy violations and data breaches, the GDPR obligates these entities to implement Technical and Organizational Measures (TOMs) *"to ensure and to be able to demonstrate that [personal data] processing is performed in accordance with"* the GDPR (Art. 24 GDPR). For example, TOMs can include, but are not limited to, organizational measures such as risk assessments, implementation of a privacy policy, and awareness training for employees, as well as technical measures such as encryption and pseudonymization or tools to enforce the data protection policy. Among other things, this has caused organizations to (1) reorganize their business processes, (2) implement data protection management, (3) redesign their privacy policies, and (4) train their employees involved in personal data processing [76]. Failure to comply with the GDPR, such as not implementing the rights of individuals or insufficient protection of personal data, has already resulted in heavy fines for organizations [67]. Similar to the GDPR, other data protection laws around the world now also impose sanctions for these types of breaches, including

J. Tolsdorf (✉) · L. Lo Iacono
Hochschule Bonn-Rhein-Sieg, Sankt Augustin, Germany
e-mail: jan.tolsdorf@h-brs.de; luigi.lo_iacono@h-brs.de

N. Gerber et al. (eds.), *Human Factors in Privacy Research*,
https://doi.org/10.1007/978-3-031-28643-8_18

the CPPA in California [75] and the APPI in Japan [57]. This development has also influenced academic discourse in the disciplines of Computer Science, Information Systems, and Human–Computer Interaction (HCI) for quite some time. In this context, related work on human factors in privacy has focused almost exclusively on the needs of individuals whose personal data are being processed, i.e., on the needs of data subjects. Among other things, these works include (1) examining the effectiveness and behavioral impact of transparency enhancing tools with respect to legal requirements [43, 52, 69, 81], (2) studying tools that provide data subjects with the ability to intervene and consent as required by law [27, 48, 80], (3) examining the compliance of transparency and intervention mechanisms with the GDPR's demand to provide information on personal data processing to data subjects *"in a concise, transparent, intelligible and easily accessible form, using clear and plain language"* (Art. 12) [42, 54, 79], (4) studying individuals' perceptions of their (new) rights introduced by the GDPR [3, 61], and (5) designing (new) transparency and intervention tools that comply with both legal and individuals' privacy requirements [9, 27, 53, 68, 81].

However, the current focus of research on human factors under contemporary data protection laws neglects the fact that privacy protection remains highly dependent on the privacy-compliant processing of personal data within organizations through the "correct" application and use of TOMs by employees responsible for personal data processing [12]. Following the notion of Human-Centered Design (HCD), the design of an organizations' internal TOMs must therefore account for the needs and capabilities of data processing employees. TOMs that are simply implemented without considering these factors are likely to be ineffective and even harmful to the organization. For example, previous work has found that data processing employees are not fully familiar with the essential terminology, concepts, and basic rules of the GDPR, which increases the risk of non-compliance [78]. In addition, TOMs may impose a burden on established business routines and increase the workload of data processing employees [36]. In this regard, industry reports indicate that up to 90% of all data breaches are caused by some form of human error [37]. Particular problems are both the accidental processing of data without permission and the forwarding of data to the wrong recipients. For example, this is reportedly true for 39% of incidents in the USA in 2019 [60] and for two-thirds of incidents in the Netherlands in 2020 [62]. Reasons include negligence of employees [66], high stress levels at work, and overladen communication channels (e.g., email) [11]. Half of the incidents resulted in disciplinary or other professional consequences for the employees [30]. The GDPR in particular has therefore increased the pressure on organizations and their data processing employees to comply with the regulation's strict rules.

The obvious solution is to provide data processing employees with TOMs that fulfill usability [28] criteria when it comes to the privacy compliant handling of personal data. However, stakeholders involved in the design and development of TOMs, e.g., employers and IT engineers, often face the challenge of translating complex legal, technical, and human requirements into concrete design and architectural decisions. In particular, the development from scratch and going through a complete HCD process can be extremely resource intensive [49]. To speed up the development process of TOMs and keep it cost-efficient, it may be advisable to use privacy design strategies and privacy patterns. These represent existing and proven concepts for the implementation of TOMs. In this chapter, we introduce a privacy pattern for the implementation of TOMs for data processing employees.

The remainder of this chapter is structured as follows: Sect. 2 provides some overall background information relevant for the implementation of TOMs using privacy patterns. Section 3 then provides a brief outline of the HCD development process of our own privacy pattern, including the requirements elicited. Next, our privacy pattern is presented in Sect. 4, followed by insights gained in our evaluation in Sect. 5. We then conclude this chapter in Sect. 6 by summarizing our approach.

2 Background

This section provides background information on the implementation of TOMs using privacy patterns under the GDPR. Section 2.1 provides a brief overview of the key principles set out in the GDPR that must be adhered to when processing personal data and that TOMs should help comply with. Section 2.2 outlines the principles of the design philosophy Privacy by Design (PbD) to be considered when implementing TOMs. At last, Sect. 2.3 describes how privacy patterns can be leveraged to implement TOMs that comply with these principles.

2.1 GDPR Principles

Generally speaking, the implementation of TOMs is supposed to help entities who process personal data to comply with the GDPR's foundational principles put forward in Art. 5 of the regulation. In the following, we provide an overview of the different principles and briefly explain their implications for the development of TOMs aimed at assisting data processing employees in the privacy-compliant handling of personal data.

- *Lawfulness, fairness, and transparency* denote (1) that personal data processing must be based on a valid legal basis prior to processing, (2) that personal data are not processed in a manner that is unjustifiably harmful, unlawfully discriminatory, unexpected, or deceptive to data subjects, and (3) that personal data processing is transparent, open, and clear to data subjects. The design of TOMs should generally help ensure that the processing of personal data by data processing employees complies with these principles. For example, depending on the situation, TOMs should help data processing employees understand whether the processing of personal data is based on an organization's legitimate interests or must be based on the data subject's consent. TOMs should also help inform data subjects about the nature and scope of the processing.

- *Purpose limitation* denotes that personal data may only be obtained for specific, explicit, and legitimate purposes. The data must not be processed in a way that is incompatible with the purposes for which they were obtained. TOMs should therefore ensure that data processing employees process personal data only for specified purposes to perform a specific job task.
- *Data minimization* refers to only processing personal data that are adequate, relevant, and limited to what is necessary for a given purpose. Thus, TOMs should facilitate limiting data collection to personal data that are necessary for a purpose associated with the job tasks of data processing employees.
- *Accuracy* indicates that the personal data processed are accurate and up to date and that reasonable efforts are made to erase or rectify inaccurate data in relation to a specific purpose. TOMs should therefore help data processing employees ensure that the personal data they process meet these characteristics.
- *Storage limitation* denotes that the processing of personal data does not allow identifying data subjects for longer than is required for the original purpose or to comply with legal obligations. TOMs should therefore delete personal data or make personal data inaccessible to data processing employees after a job task has been completed, and no legal regulations prescribe longer storage.
- *Integrity and confidentiality* require the implementation of appropriate technical and organizational safeguards to ensure personal data security, including safeguards against unauthorized or unlawful processing, accidental loss, destruction, or damage. Accordingly, TOMs should only grant access to personal data if data processing employees are authorized and the job task requires the personal data processing. TOMs should further support data processing employees in storing and processing personal data in a suitably protected manner.
- *Accountability* means that controllers, i.e., entities who define the purposes for personal data processing, ensure and are able to demonstrate compliance with the aforementioned principles. This generally requires controllers to ensure and be able to demonstrate that their data processing employees' actions comply with these principles. This may include providing privacy policies based on an inventory of processing records, documenting and tracking processing activities, and creating data protection awareness among data processing employees.

2.2 Privacy and Data Protection by Design

The GDPR requires that the implementation of TOMs takes into account the principles of *data protection by design and by default* (Art. 25 GDPR). These principles build upon the design philosophy of Privacy by Design (PbD) [16]. PbD advocates that *"privacy must be incorporated into networked data systems and technologies, by default. Privacy must become integral to organizational priorities, project objectives, design processes, and planning operations"* [16]. PbD provides seven principles on how to integrate privacy [16]:

1. *Proactive not reactive; preventative not remedial*—all privacy policies and mechanisms must be in place prior to processing so that privacy issues can be resolved before they become real problems.
2. *Privacy as the default*—the default case guarantees integrity of privacy and provides fair processing of personal data. This includes, but is not limited to, purpose limitation, data minimization, transparency, and intervention capabilities.
3. *Privacy embedded into design*—privacy protection should not be considered an "add-on" but an integral part of information systems and business practices. It requires considering the broader context and all stakeholder views for finding the best solution.
4. *Full functionality*—PbD means promoting privacy as a complement, not a trade-off, and provides for innovative and creative solutions, which take into account all legitimate interests.
5. *End-to-end security*—privacy requires consideration of the entire processing chain, from collection to destruction of personal data ("cradle to grave").
6. *Visibility and transparency*—controllers should meet their accountability obligations by demonstrating compliance and providing truthful information about the processing.
7. *Respect for user privacy*—data protection should reflect the interests and needs of data subjects and requires user-oriented approaches in the design of tools, information systems, and business processes.

In 2010, the International Conference of Data Protection and Privacy Commissioners recognized PbD *"as an essential component of fundamental privacy protection"* and promoted its widespread adoption in legislation [63]. However, the translation of its principles into specific guidelines for action is a major practical problem [6, 24, 39, 73, 74]. PbD is frequently linked to Privacy Enhancing Technologies (PETs, see also the chapter "Acceptance Factors of Privacy-Enhancing Technologies on the Basis of Tor and JonDonym") because their development usually implicitly takes into account some PbD principles, in particular, *privacy by default* and *end-to-end security* [24]. However, PbD has always taken a holistic view and must be seen as a kind of lesson from the past, showing that implementing privacy by means of technology is only part of the answer toward more privacy, but not the answer itself [50]. That is, PETs should be understood as an integral part of PbD but must be accompanied by complementary measures that respect the privacy implications at the design stage of the technology.

Moreover, implementing PbD using a purely legally oriented process promotes the manifestation of one-size-fits-all solutions, which are detrimental to effective privacy protection because they disregard the nature of privacy, which is individualistic, contextual, diverse, and multifaceted [44, 51]. That said, PbD itself already takes this issue very much into account, promoting the principle of *respect for user privacy—keep it user-centric.* It essentially requires human factors of privacy to be incorporated in every IT system and business process [16, 17]. In particular, it emphasizes on the need for privacy controls to be *"human-centered, user-centric, and user-friendly so that informed privacy decisions may be reliably exercised"* [16]. As such, there are increasing efforts to reinforce this principle in TOM development [32] and to expand the implementation of PbD to a human-centric process which accounts for this need [7, 31, 51, 71].

2.3 Privacy (Design) Patterns

Privacy patterns are design patterns used to translate the abstract principles of PbD and *data protection by design and by default* into practical advice for developing privacy-friendly systems and processes. In the following, we first briefly introduce the idea behind design patterns in general and then provide an overview of the use of privacy patterns in system design, business process design, and in HCI.

Design Patterns

Design patterns are proven solutions to known and recurring problems in a specific domain that are systematically recorded and documented [35]. The pattern approach was first developed and introduced in the field of urban and building architecture to document proven architectural designs in a standardized structure [2]. Later, the concept of design patterns became particularly popular in software engineering [35] and was eventually adapted to related fields, such as human–computer interaction [22] and cybersecurity [83]. Since the design of complex systems usually involves a wide range of recurring problems, engineers also usually need to draw on different design patterns to implement system requirements. To facilitate access to various design patterns, they are commonly organized in *pattern catalogs*. A pattern catalog represents a collection of design patterns that systematically classifies design patterns into different categories [14]. Its underlying systematization can be informal or based on formal pattern taxonomies. Pure pattern catalogs often consider patterns in isolation and ignore the fact that design patterns are frequently interdependent with other design patterns. For example, a design pattern may represent, among other things, an aggregation or specialization of other design patterns. Therefore, if a pattern catalog contains a sufficiently large number of design patterns, it may be useful to convert it into a *pattern system* capable of describing these dependencies [14]. Pattern systems, also known as *pattern languages*, describe dependencies between individual design patterns based on a predefined set of relationship types, as well as guidelines and rules for their implementation [15].

Privacy Pattern Collections

The concept of design patterns from software development was later extended to security [83] and privacy [65, 70]. Continuous efforts by the research community have resulted in a comprehensive collection of privacy patterns being available today, covering a multitude of topics including but not limited to anonymity [70] and pseudonymity [34], the development and application of privacy-enhancing technologies [40], as well as issues targeting human-computer interaction [25, 33, 38] with an emphasis on transparency [72]. The privacy patterns mainly support designers and developers in identifying privacy requirements for their system or process, provide suggestions for a suitable system architecture, or provide concrete design and implementation guidelines [47]. To this end, the pattern descriptions are often accompanied by conceptual representations, UML diagrams, sequence diagrams, and screenshots. Many of the privacy patterns available have further been documented in a repository that is maintained by a collaboration of international researchers.[1] The patterns have also been organized into catalogs targeting specific domains, such as the online context [4, 65] and the Internet of Things [55]. In addition, some catalogs categorized patterns according to the principles of the privacy framework in ISO/IEC 29100 with the aim of further simplifying the application of privacy patterns to comply with international standards and privacy laws [4, 26]. Meanwhile, there are first proposals for privacy pattern systems [19, 20, 40], as well as proposals for a suitable modeling language to concisely describe dependencies between privacy patterns [15].

Privacy Design Strategies and Tactics

Privacy design strategies allow a mapping between legal requirements and system requirements and are suitable for specifying clear objectives related to PbD in order to achieve a certain level of privacy protection [21]. For better distinction and labeling, privacy patterns are often classified according to eight privacy design strategies [41]: (1) *Minimize* the amount of personal data that are processed (2) *Hide* personal data and their interrelationships from plain view (3) *Separate* the processing of personal data into compartments (4) *Aggregate* personal data to the highest level and with the least possible detail (5) *Inform* data subjects about personal data processing (6) *Control* over personal data processing by data subjects (7) *Enforce* privacy policies compatible with legal requirements (8) *Demonstrate* compliance with privacy policies and legal requirements

A recent literature survey revealed that about half of the privacy patterns published in peer-reviewed articles focus on the strategies *hide* and *separate*, which are usually strongly characterized by the use of TOMs [47]. In addition, various tactics are available for implementing each data protection strategy. A tactic

[1] https://privacypatterns.org/

represents a homogeneous set of privacy patterns and summarizes their underlying main concept [21]. Tactics provide a useful intermediate level of abstraction for modeling systems and processes because they are more fine-grained than privacy strategies, but more abstract than privacy patterns.

Patterns for Business Processes and Workflows

Akin to design patterns for system design and architecture, there also exist patterns for modeling business processes to include obligations imposed by privacy laws [1, 5, 8, 13, 18, 64]. Patterns in this category support organizations in modeling their high-level architecture and business processes while incorporating PbD. Some approaches employ enterprise architecture model description languages to make the interdependence of systems and the associated data flows transparent and understandable [18]. This also allows determining which components must be added or implemented in order to comply with privacy principles or regulatory requirements [10]. Other approaches employ description languages for business process models to incorporate privacy principles and regulatory-mandated organizational measures into business processes by default [1, 5, 8, 13, 64].

However, the scope covered by the approaches varies; some works focus on patterns covering the standard cases of data protection law, in particular those of the GDPR. Cases covered include controllers' obligations and data subjects' rights [1, 18, 64]. They may be used as templates by organizations and architects to avoid having to model standard processes themselves. Second, there are methodologies available for modeling legal requirements and creating patterns using standard modeling languages [8, 13, 64]. They support organizations and architects in documenting their own patterns and processes in a comprehensible and consistent manner. Third, some works present more specific patterns for business processes in certain contexts that help to reduce the level of abstraction of the former two approaches [5].

Usable Privacy and Interaction Patterns

Privacy patterns focus not only on technical and architectural aspects but also on usability aspects, i.e., designing privacy protection in a human-centered manner to make it efficient, effective, and satisfying for the respective user group. To this end, numerous so-called *HCI (privacy) patterns* have been proposed to provide usable interfaces for PETs [33, 38]. In particular, several patterns have been proposed under the design strategy *inform*, which are commonly referred to as privacy transparency patterns and are suitable for implementing data subjects' information rights [33, 72].

Independent of the topic of privacy, design patterns that define problems and solutions targeting the perceived interaction behavior are generally referred to as *interaction design patterns* [22]. The term emerged in the HCI community to clearly distinguish design patterns with a focus on interaction behavior from design patterns for the realization of interfaces in software engineering. Interaction design patterns are usually the result of a HCD process in which the pattern was developed and evaluated together with the affected stakeholders [33, 56].

3 Privacy Pattern Development

In this section, we outline the development process of our privacy pattern *Data Cart*. Generally speaking, stakeholders involved in the design and development process of tools that adhere to PbD need a deep understanding of (1) the situation and context *in which* the tools will be used, as well as (2) the personal data processing activities *for which* the tools will be used [58]. To incorporate these aspects early in the design process for a privacy pattern for data processing employees that supports them in the data protection compliant handling of personal data, we applied a User-Centered Design (UCD) approach (see the chapter "Achieving Usable Security and Privacy Through Human-Centered Design") with data processing employees from two public institutions in Germany. In the following, we first provide a brief overview of the UCD study in Sect. 3.1. Then, in Sect. 3.2, we outline the main requirements identified for tools to support our stakeholders in managing personal data in a privacy-compliant manner. The detailed study procedure, elicited requirements, and development process are available elsewhere [77].

3.1 User-Centered Design Study

A total of 19 data processing employees participated in our UCD study. A summary of their demographics is available in Table 1. Overall, our sample was highly educated, as all participants held a university degree. At the time of participation, they had been in their job and with the organization for between 1 and 19 years (median = 3 years and mean = 5.4 years). In most cases, our participants held multiple job roles, including research officer, third-party funding officer, legal officer, team assistant, network manager, and innovation manager. Their tasks included consulting and coaching activities, guiding and supporting grant applications or patent approvals, and monitoring ongoing projects or start-ups. In these activities, they primarily processed personal data of employees of the organization. The data typically included personnel data, contact data, and demographic data. In addition, our participants often processed classified information (e.g., patents). Other tasks include public relations and marketing as well as networking, which includes the regular planning and hosting of events. These activities require extensive processing of private and professional contact data, as well as image recordings. In all of these activities, participants regularly cooperated and communicated with their colleagues and other departments or with external organizations such as project sponsors and funding agencies. Particularly often, our participants were in contact with the HR

Table 1 Participant demographics

ID	Sex[S]	Age (years)	Education (highest)	Job description (primary)	Job tenure (years)
P01	f	35–44	PhD	Research Funding Officer	6–10
P02	f	35–44	PhD	Research Promotion Officer	1–5
P03	m	25–34	PhD	Research Officer	1–5
P04	f	45–55	Master's degree	Research Officer	1–5
P05	f	45–55	Master's degree	Research Officer	1–5
P06	f	45–55	Master's degree	Research Officer	6–10
P07	f	35–44	Master's degree	Research Officer	1–5
P08	f	35–44	Master's degree	Network Manager	16–20
P09	m	25–34	Master's degree	Innovation Manager	1–5
P10	f	55–65	State exam	Research Officer	16–20
P11	f	35–44	State exam	Legal Officer	1–5
P12	f	25–34	Master's degree	Third-Party Funding Coordinator	1–5
P13	f	35–44	Master's degree	Research Officer	6–10
P14	f	35–44	PhD	Research Officer	1–5
P15	f	35–44	State exam	Third-party Funding Coordinator	1–5
P16	f	45–55	Master's degree	Research Officer	1–5
P17	f	45–55	Master's degree	Research Officer	6–10
P18	f	35–44	PhD	Research Officer	6–10
P19	f	45–55	Bachelor's degree	Team Assistant	6–10

Note. [S]Options were diverse, female, male, prefer not to say

Department to request personal data instead of obtaining them directly from the data subjects. In most cases, their tasks require sharing (personal) data with others or using the data to generate statistics and reports. Thirteen participants self-reported processing personal data very frequently or regularly, while six participants reported processing such data occasionally.

The UCD study consisted of a series of eight workshops to investigate the data processing employees' needs for assistance in handling personal data and to evaluate potential solutions. An overview of the full development process is given in Fig. 1. In the first workshop, we adopted a concept of Polst et al. [59] in order to familiarize ourselves with the stakeholder group and their everyday work. In the subsequent workshops, we elicited common problems that our participants encountered when processing personal data. We explicitly addressed data protection concerns and asked as to how they envision a redesign of the processes to improve privacy. Based on the obtained feedback, we developed a concept and refined it in several sessions with UX designers and usable privacy and security experts. We then evaluated the concept using pen and paper mockups to conduct a pluralistic walkthrough [82] with our participants. From the results, we compiled a list of final requirements and drafted a prototype. The prototype was implemented as a web application. It included several scenarios of our participants' everyday work. We ran formative usability tests to evaluate the prototype's usability and privacy-enhancing properties.

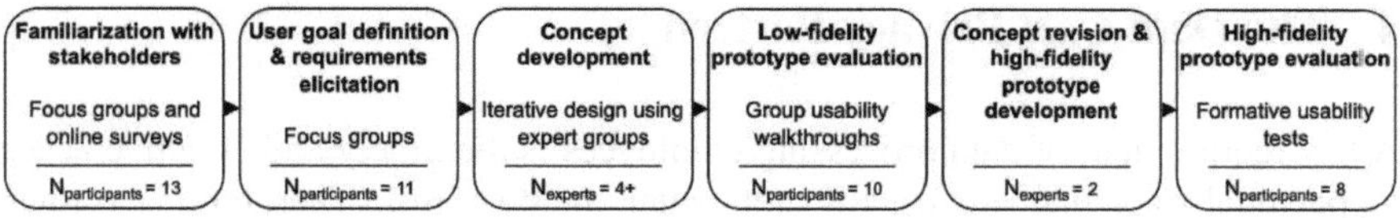

Fig. 1 Summary of the development process of *Data Cart*. Note that due to busy schedules of our participants and staff turnover in the departments, not all participants participated in all steps of the UCD study ($N_{participants\ total} = 19$)

We also discussed the extent to which the tool would change established work processes and the handling of personal data with our participants.

For the most part, we relied on focus groups because we expected our participants to enrich each other [45], but we also used interviews because both methods are well suited for both requirements elicitation and evaluation [46]. We either adapted existing workshop concepts to our needs or created our own study protocol in accordance with established guidelines. All study protocols were designed and reviewed by two subject-matter experts, as well as researchers from a larger research project team, and researchers with experience conducting user studies. Depending on the type of study, we piloted studies with members of our own institutions or other organizations.

3.2 Data Processing Employee Requirements

Participants identified numerous problems and opportunities to improve workflows and strengthen data protection within them. Major concerns were the inconsistent processes and decentralized infrastructure across different departments. This greatly affects the gathering of personal data and the handling of outdated data. Participants complained that much of their time was spent communicating with other departments, such as HR, or with the data subjects themselves when they needed data. Employees rated clearly identifiable responsibilities and fast, as well as complete, responses to their inquiries as essential factors for their job tasks, as they are often subject to tight deadlines.

Moreover, participants were well aware of their responsibility in dealing with personal and classified data. They assured that they strived to act to the best of their knowledge but expressed their uncertainties in practice. In particular, they felt insecure due to a lack of knowledge about data protection rules that apply to certain situations and data. They desired tools to keep them from committing unlawful actions and demanded clear instructions without any room for interpretation. Besides, participants showed concern about the lack of transparency of their processing activities to data subjects and were also unaware whether and how data subjects would have consented. Consequently, they asked to make the extent of processing and data flow transparent to data subjects.

4 The *Data Cart* Privacy Pattern

A key requirement of data processing employees is the timely and effective access to personal data that are usually not under their control. Therefore, a primary task is to assemble a set of different and varying personal data and data subjects from external sources that are needed for a particular business process. This may require initiating multiple data queries, keeping track of them, and processing the responses. Similar complexity in the compilation and tracking of different items and attributes is a well-known problem in online shopping. This is why we adapted the shopping cart[2] interaction pattern to our context and created the metaphor of a "data cart." The metaphor builds on data processing employees' existing knowledge of interaction concepts for complex processes where one first defines an output based on metadata, considers different statuses (e.g., availability), and only gains access after completing different tasks (e.g., payment, delivery). For this purpose, the steps necessary to model the processing of personal data in administrative tasks have been roughly mapped to an online shopping cart. The data cart metaphor serves two purposes. First, we used the metaphor in the context of our internal design and development cycle, as well as in internal communication within the research team. This allowed for a common understanding of the interaction concept among all researchers. Second, we also used the metaphor to break down the complexity of data protection for our participants and integrate privacy requirements into meaningful workflows that align with their needs.

Based on the data cart metaphor and taking into account legal concerns and stakeholder requirements, we developed an employee-centric solution that provides sufficient flexibility to meet various use cases of our stakeholders related to the processing of personal data. The solution basically provides for synchronizing the recurring tasks of retrieving and managing personal data with privacy obligations. The result is a harmonized combination of process flow and interaction concept, which we have documented as a privacy pattern. In the remainder of this section, we provide a basic description of the pattern following established templates [19].

Name Data Cart

Summary A single point of access for data processing employees to obtain and manage personal data in a data protection compliant manner.

Context This pattern applies to data processing employees working in organizations that elicit personal data as a part of an overarching business process and must share the personal data with other entities or departments as part of this business process. Elicitation is usually done in structured surveys through forms or by requesting the personal data from other departments within the organization. The pattern has been evaluated with data processing employees from public institutions who mainly process employee personal data for purposes such as academic services,

[2] https://web.archive.org/web/20211124013206/http://welie.com/patterns/showPattern.php?patternID=shopping-cart

consulting, and patent registration and exploitation. The personal data processed generally included information on contact, education, finances, professional activity, as well as pictures and personal identifiers.

Problem Data processing employees are frequently required to process personal data for (time-critical) job tasks, which necessitates extensive communication with an organization's employees, departments, and partners. In an organization, particularly heterogeneous business processes prevent effective data inquiries, either because the data received are incomplete and incorrect or because the correct contact person in other departments is unknown. In many of these cases, data processing employees perceive data protection as a burden because they are uncertain whether they act in compliance with data protection, or whether certain measures are necessary, and how they should put them into practice. In practice, data processing employees thus act with uncertainty and make efforts to protect themselves from misconduct that they do not know are necessary or even correct. As a result, employers, as data controllers thus liable for the actions of their employees, may subsequently fail to comply with their accountability obligations.

Solution Provide a privacy enhancing personal data management interface to personal data that (1) streamlines data collection processes in organizations and aligns them with data protection requirements, (2) standardizes access to personal data for data processing employees, (3) simplifies access to privacy policies for data processing employees, and (4) supports both controllers and data processing employees in demonstrating transparency and compliance by documenting processing activities.

GDPR Principles *Lawfulness*, *fairness*, and *purpose limitation* are addressed by reducing human error due to ignorance, since information about the legal basis and purpose become an integral part of any request for personal data; *data minimization* and *accuracy* are achieved through (1) centrally controlled access to personal data, (2) provision of meta-information about data, and (3) triggering of updates, and *storage limitation* and *integrity and confidentiality* are supported by incorporating *privacy by default* (e.g., encryption of exports) and data handling information. *Fairness*, *transparency*, and *accountability* are supported by the implicit documenting of requests. *Accountability* is further addressed by making data processing employees aware of personal data processing obligations through clear and uniform privacy notices.

Privacy Design Strategies [41]
Primary:

- *Enforce* privacy policies compatible with legal requirements
- *Demonstrate* compliance with privacy policies and legal requirements

Supports:

- *Minimize* the amount of personal data that are processed
- *Inform* data subjects about personal data processing
- *Control* over personal data processing by data subjects

4.1 Process Flow Model

In this section, we describe the process flow associated with the *Data Cart* solution outlined above. It shall serve IT architects, developers, and process managers as a means to understand and integrate the *Data Cart* pattern into their own systems and processes, respectively. The process flow divides into tasks to define a personal data processing activity, process personal data, and demonstrate compliance. The basic flow is outlined in Fig. 2 and divides into eight tasks. A detailed process flow diagram is shown in Fig. 3. The process flow starts by assuming that a data processing employee has a demand to process personal data and opens the *Data Cart* interface. The process flow is as follows:

1. The first process step requires data processing employees to model a data processing activity to be performed. For this purpose, they must choose a processing activity from the organization's records of processing activities for which they are authorized. According to Art. 30 GDPR, this directory must be maintained by all data controllers with regular processing activities and contains a list of all legitimate personal data processing activities. Each entry comprises a purpose, categories of personal data, categories of data subjects, categories of recipients, legal basis, and, if applicable, further information on technical and organizational measures. Upon selection, employees are provided with a summary of the processing record. This requires employees to become aware of the legal basis before processing begins. In the event that the personal data have already been collected via form, this can also be imported instead. In such a case, the appropriate processing record entry can be selected automatically.
2. In the second process step, data processing employees define tuples of required categories of personal data and data subjects. They may also add additional details, such as specific recipients, the version of personal data they require, or a personal message to the data source (e.g., data subject, department). Once

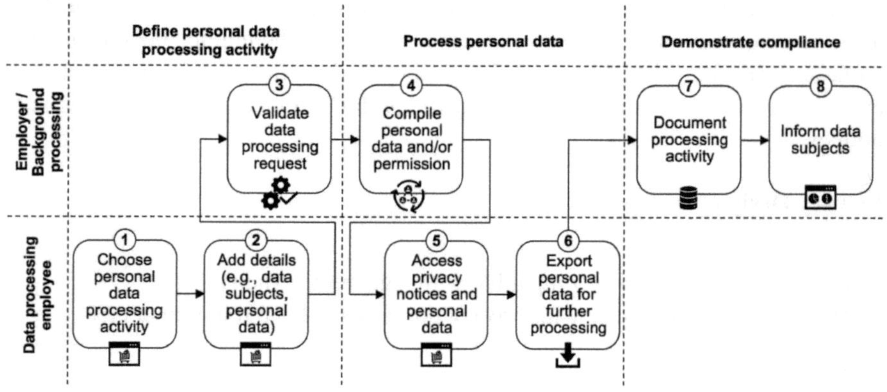

Fig. 2 Flow of the concept developed using the metaphor of a data cart

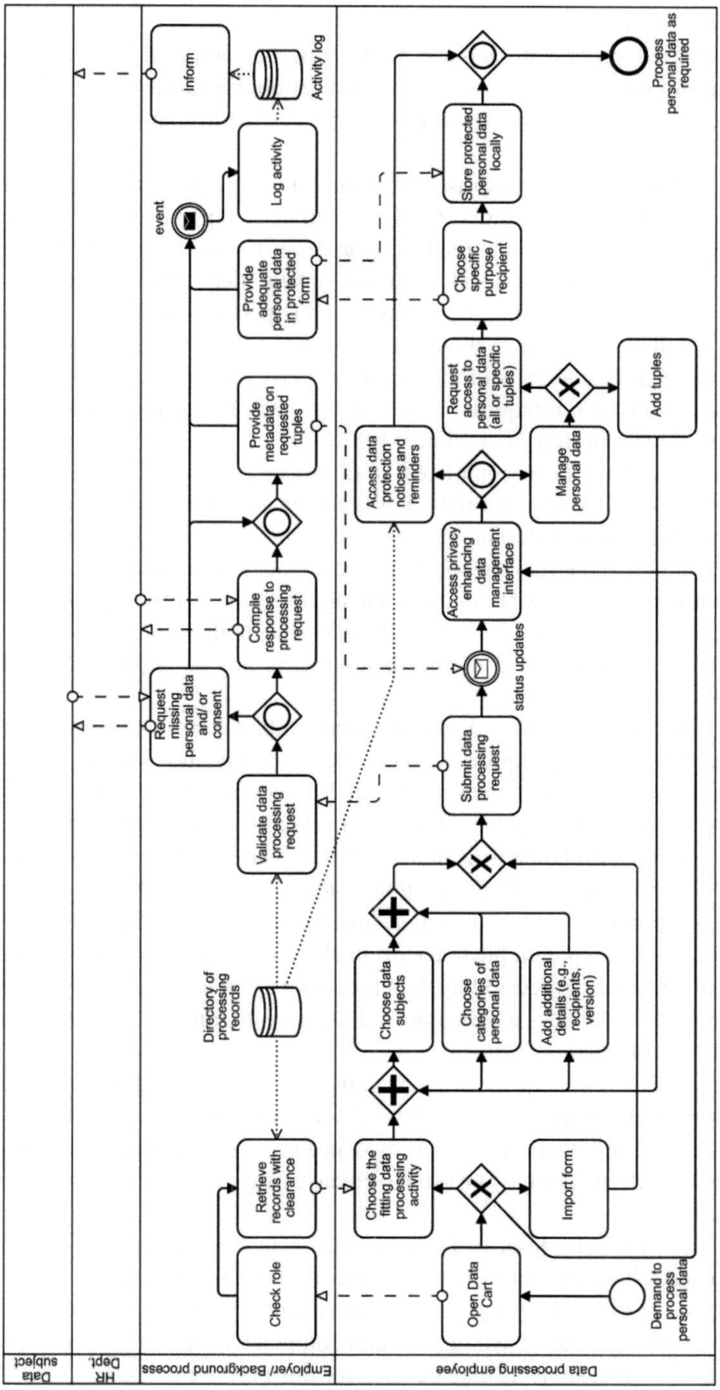

Fig. 3 Process flow diagram of the concept developed using the metaphor of a data cart

finished, the modeled processing activity is to be submitted as a new data processing request.
3. The submitted processing request must then be validated by verifying for lawfulness of processing against the processing policies extracted from the record of processing activities and by checking the availability and timeliness of the personal data requested.
4. The next process step comprises obtaining missing personal data and permissions. Depending on the processing activity, this may require initiating requests to the respective data subjects or departments to provide the missing data and approvals. It is critical from our stakeholders' point of view that the request be structured and that input validation is performed. Requests must also include detailed information about the requester and their legitimacy, as well as procedural and legal aspects of the underlying processing. Our own pattern does not specify how such a request should be designed, but privacy patterns similar to *informed consent* may be used here [33].
5. After all tasks have been completed, data processing employees get access to a privacy enhancing personal data management interface. It provides access to metadata of the data processing request, including status information and details about the tuples requested. In addition, it provides access to contextual privacy policies and reminders extracted from the organizations' directory of processing records. Furthermore, the interface provides the ability to request additional combinations of personal data and data subjects and to request access to the personal data (e.g., exports).
6. To access raw personal data, data processing employees must choose a specific purpose for which they require the data. Based on this, they should be provided with an export of the personal data, which contains only the data authorized for the purpose and recipients. The export should be adequately protected by default, as our stakeholders do not have the necessary knowledge to do this themselves. All exports should further contain a copy of the data protection information provided in the data management interface, as well as an ID to ensure traceability of the exported file to the original request. The exported personal data then shall be further processed by data processing employees as required. Based on stakeholder feedback, we recommend using common data exchange formats (e.g., MS Excel).
7. All actions, including requests for data and data exports, are logged to document all personal data processing activities. After completing a processing activity, requests can be archived and serve as evidence for later audits and traceability. In addition, the activity log may be used to create a usage history for data processing employees.
8. Furthermore, the here described concept advocates transparency and conceptually provides that data subjects are informed about the processing carried out on the basis of the activity log. This is not covered by our own pattern. Instead, depending on the needs, existing tools and components optimized for employees in their role as data subjects may be used for this purpose [23, 68].

4.2 Interaction Concept

Based on the process flow outlined above, we developed a corresponding user interaction concept that reflects our stakeholders' point of view. The interaction concept including a mapping to the requirements elicited is shown in Fig. 4. The interaction concept is divided into five parts.

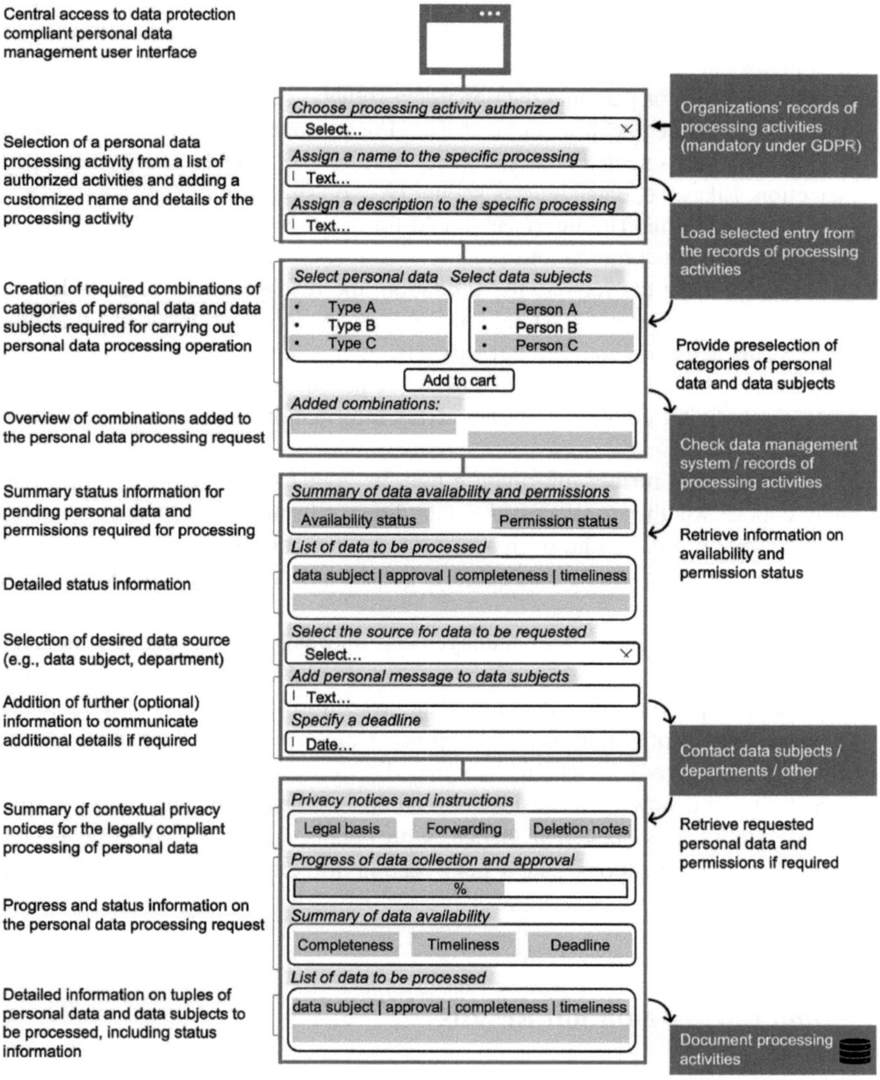

Fig. 4 Basic interaction concept designed following the data cart metaphor

1. First, data processing employees should be offered a personal data management tool that provides for centralized access to personal data and enforces consistency of the full data management process.
2. To model a data processing request, data processing employees should be provided with a preloaded list of processing activities for which they are authorized. Upon selection, employees should be provided with a summary of the processing record. In addition, the planned processing must be given a name and a description. These steps require employees to become aware of the legal basis before processing begins. At the same time, the interaction concept provides for contextual support, such as providing templates and contextual information. Templates may be based on previous requests, too.
3. To define tuples of personal data and data subjects, data processing employees should be provided with predefined lists. For personal data, these lists may be derived from the selected processing record entry and should be offered as a pre-selection. Likewise, data subjects should be accessible from a list of employees in the organization. The interface should further support the iterative adding of multiple different combinations.
4. When submitting the request for validation, the results should be provided for review in an overview. It should include status information on whether the processing activity can start immediately after submission of the processing request or whether additional actions are required, such as collecting personal data or obtaining consent. Detailed status information should be accessible as needed.

   At this point, further information may be added to the request. Employees may choose whether to request the data directly from the data subjects, via an administrative department, or in a customized manner. They may also compose individual messages to the data subjects and set a deadline for responding to the request.
5. The privacy enhancing data management interface should provide detailed information on the status of pending requests. In addition, it provides frequently needed or important information on data protection tailored to its users' needs. This includes information on allowed processing operations, whether processing has been approved, to whom data may be disclosed, deletion periods, data sensitivity, and how data must be safeguarded. In general, the interface aims to provide such notices at a glance, with details accessible when necessary. Additional visualizations and a help section for questions accompany detail views.

5 *Data Cart* Evaluation Results

In this section, we report on results obtained in the UCD study by evaluating our *Data Cart* mock-ups. We report on our participants' perceptions and understanding of the "data cart" metaphor in Sect. 5.1. We then present our participants' feedback on *Data Cart*'s properties for data protection in Sect. 5.2, followed by limitations and open issues in Sect. 5.3.

5.1 Metaphor and Concept Understanding

Overall, we found that the data cart metaphor was helpful in outlining the basic assumptions and processes of the *Data Cart* concept to data processing employees. In particular, we found that the data cart metaphor supports data processing employees in understanding that a personal data processing operation always requires the definition of a tuple consisting of one or multiple purposes, data subjects, and data categories, but without the need to understand the details of the GDPR. In this context, the data cart metaphor was useful to explain the basics of a directory of processing activities, since we found that data processing employees in our UCD study were generally unfamiliar with this concept and its meaning. Only one participant indicated that they knew their organization maintained such a directory.

5.2 Data Protection Properties of Data Cart

In total, we identified five themes on data protection in our participants' feedback on *Data Cart*. The themes are summarized in Table 2.

Desire for Systematic Data Protection by Design In general, the *Data Cart* concept encouraged our participants to discuss their need for systematic data protection that integrates with work processes, rather than always being added as an additional expense and interfering with work. Participants pointed out that the correct handling of personal data *"is too often overlooked in everyday life, and the use of a such a tool would, on the one hand, simplify this and, on the other hand, somehow make you aware of the relevance of data protection and data"* (P06). Furthermore, our participants praised the PbD approach taken by *Data Cart*, because *"[personal data] would be handled in a more sensitive way without making it [(data protection)] too much of an issue"* (P04). In addition, the approach to systematic data protection in the form of *Data Cart "creates legal certainty and can somehow take away uncertainty"* (P05) when dealing with personal data: *"Well, basically, because everything is already predefined [. . .] I think you feel a bit safer, because you can make fewer mistakes yourself, because it is automated or because hints are given"* (P03) and *"because I don't have to worry at all about whether the person consents or not, because it is all there"* (P03).

Central Source of Information for Data Protection Our participants positioned *Data Cart* as a central information platform for data protection topics, which *"compiles the information quite well, so you don't have to go through the hassle of*

Table 2 Summary of participant feedback related to data protection properties of *Data Cart*

Theme	Description
Desire for systematic data protection	▷ Establishing data protection by design ▷ Enabling efficient, effortless, and secure handling of personal data
Central source of information for data protection	▷ Eliminating non-uniform handling of data protection rules by providing clear and understandable instructions on data protection ▷ Keeping data privacy information available and allowing quick access to "important" information
Raising awareness of data protection	▷ Sensitizing data processing employees for data protection ▷ Allowing sensitization of data subjects ▷ Correcting and aligning interindividual understanding of "sensitive data"
Integration limits as a barrier for data protection	▷ Transitions between processes and systems are critical for data protection compliance ▷ Processing of data remains unaffected without adaptation of processes
Consequences of systematic data protection as an obstacle to work	▷ Conflicting with established work practices and procedures

finding out how to proceed with it [(personal data)]" (P03). Particularly important was quick access to important information, i.e., that one can *"immediately see which data I'm allowed to pass on externally or internally, I think that's pretty good"* (P05) *"because you're simply dealing with sensitive data, and you don't always know whether you're allowed to [process data] or not"* (P01).

Raising Awareness of Data Protection *Data Cart* is seen as a driver of awareness for both data processing employees and data subjects. Our participants particularly welcomed the sensitization for legally compliant data processing: *"Otherwise, you are just less aware of it, so I think it makes you more aware that these are all very important data and that they must also be specially protected"* (P04). Here, too, PbD played a role: *"Because otherwise it's like this in the everyday handling of data: I don't even think about what people have approved, what they haven't approved"* (P08), but *"just by having this tool at your disposal, you're more likely to even think about 'do I need to pay attention to anything right now?'."* At the same time, documentation and communication through *Data Cart* allows data processing employees to fulfill their desire to inform data subjects: *"I find this tool quite good for that. That I can then write to [those] whose data I process [. . .] and make them aware that their data are being processed and whether they agree to it at all"* (P06).

Integration Limits as a Barrier for Data Protection Our participants noted that tools like *Data Cart* cannot solve all privacy issues. Especially if tools are introduced as a supplement to existing processes or current workflows, *"because* *then the data are accessible again: I have to archive them for later auditing [...] and then, of course, these sensitive data are stored there. That's a place where everyone has access"* (P04). Further problems arise from the lack of digitalization, since requests for project proposals are often made via traditional means of communication not under control of *Data Cart*, yet they may already contain critical data: *"But I wonder what happens when you simply receive data. So just in everyday work, one simply gets some kind of data by email"* (P15).

Consequences of Systematic Data Protection as an Obstacle to Work It becomes clear that the handling of personal data enforced by *Data Cart* creates new obstacles: *"Because if we use this here, we make the request, it gets approved, so the data have to be checked first [...] At that moment, we can't continue at that point. And that delays some workflows"* (P01). In particular, lack of or denial of approval is perceived as the biggest obstacle: *"If someone's data are not approved, then I can't continue processing. Of course, we don't have this situation now because no one knows that the data are being used"* (P06).

5.3 Limitations and Open Issues

Based on our analysis of *Data Cart*, we identified several further topics and issues related to TOMs like *Data Cart* from the perspective of data processing employees [77]. For example, there are possible integration barriers, especially if tools are introduced as a supplement to existing processes or current workflows. Further problems may arise from a lack of digitalization in organizations, which could cause a significant overhead in both the integration and operation. Further issues may result from the consequences of systematic data protection, as it enforces a specific way of working that might need additional change management efforts. These potential issues should definitely be considered when implementing tools based on *Data Cart* and will require further investigation in the future.

6 Conclusion

Data processing employees have always played an important role in putting privacy goals into practice. To assist them in the privacy-compliant handling of personal data, TOMs must be designed to align with their needs and capabilities. To this end, this chapter introduced and presented the privacy pattern *Data Cart*, consisting of a process flow model and interaction concept. *Data Cart* offers a practical solution to stakeholders involved in privacy research or engineering for the human-centered design of TOMs under the GDPR. It (1) streamlines data management processes and brings them in line with data protection requirements, (2) standardizes access to personal data, (3) facilitates employee access to privacy policies, and (4) enables documentation of personal data processing. In general, we found that *Data Cart* addresses data processing employees' desire for systematic data protection, i.e., data protection that integrates with work processes, rather than always being added as an additional expense and interfering with work. In this context, a PbD approach seems to be valued for implicitly enforcing data protection in the handling of personal data by designing the entire process from the perspective of data processing employees. By mapping the organization's requirements directly into the process and interface design, data processing employees benefit by focusing more on the essential process and being less exposed to uncertainty when processing personal data. In our UCD study, data processing employees perceived *Data Cart* as a relief because it reduces the manual compliance effort on their end. *Data Cart* may be adapted in the future to meet participants' demands for more comprehensive solutions and become an integral part of standard software or its own class of standard software for privacy management used in organizations.

Acknowledgments This chapter is derived in part from an article published in Behaviour & Technology 2022 ©Taylor & Francis, available online: https://www.tandfonline.com/10.1080/0144929X.2022.2069596. This research was supported by the German Federal Ministry of Education and Research (BMBF) under the contract numbers 16KIS0899 and 16KIS1508.

References

1. Agostinelli, S., Maggi, F. M., Marrella, A., & Sapio, F. (2019). Achieving GDPR compliance of BPMN process models. In *Proceedings of the CAiSE Forum as part of the 31st International Conference on Advanced Information Systems Engineering (CAiSE Forum)* (pp. 10–22).
2. Alexander, C., Ishikawa, S., Silverstein, M., Jacobson, M., Fiksdahl-King, I., & Shlomo, A. (1977). *A pattern language: Towns, buildings, construction*. OUP.
3. Alizadeh, F., Jakobi, T., Boden, A., Stevens, G., & Boldt, J. (2020). GDPR reality check—claiming and investigating personally identifiable data from companies. In *Proceedings of the IEEE European Symposium on Security and Privacy Workshops (EuroSPW)* (pp. 120–129).
4. Aljohani, M., Blustein, J., & Hawkey, K. (2018). Toward applying online privacy patterns based on the design problem: A systematic review. In *Proceedings of the 7th International Conference on Design, User Experience, and Usability (DUXU)* (pp. 608–627).
5. Aljohani, M., Hawkey, K., & Blustein, J. (2016). Proposed privacy patterns for privacy preserving healthcare systems in Accord with Nova Scotia's personal health information act. In *Proceedings of the 4th International Conference on Human Aspects of Information Security, Privacy and Trust (HAS)* (pp. 91–102).
6. Alshammari, M., & Simpson, A. (2017). Towards a principled approach for engineering privacy by design. In *Proceedings of the 5th Annual Privacy Forum (APF)* (pp. 161–177).
7. Ayalon, O., & Toch, E. (2021). User-centered privacy-by-design: Evaluating the appropriateness of design prototypes. *International Journal of Human-Computer Studies, 154*, 102641.
8. Barati, M., & Rana, O. (2021). Design and verification of privacy patterns for business process models. In S. Patnaik, T.-S. Wang, T. Shen, & S. K. Panigrahi (Eds.), *Blockchain technology and innovations in business processes* (pp. 125–139). Springer.
9. Bier, C., Kühne, K., & Beyerer, J. (2016). PrivacyInsight: The next generation privacy dashboard. In *Proceedings of the 4th Annual Privacy Forum (APF)* (pp. 135–152).
10. Blanco-Lainé, G., Sottet, J.-S., & Dupuy-Chessa, S. (2019). Using an enterprise architecture model for GDPR compliance principles. In *Proceedings of the 12th IFIP Working Conference on the Practice of Enterprise Modeling (PoEM)* (pp. 199–214).

11. Brackenbury, J., & Bailey, R. (2020). 2020 Outbound Email Security Report | Egress. https://www.egress.com/newsroom/2020-outbound-email-security-report
12. Brodie, C., Karat, C.-M., Karat, J., & Feng, J. (2005). Usable security and privacy: A case study of developing privacy management tools. In *Proceedings of the 1st Symposium on Usable Privacy and Security (SOUPS)* (pp. 35–43).
13. Buchmann, E., & Anke, J. (2017). Privacy patterns in business processes. In *Proceedings of the 47th Jahrestagung der Gesellschaft für Informatik (INFORMATIK)* (pp. 793–798).
14. Buschmann, F., Meunier, R., Rohnert, H., Sommerlad, P., & Stal, M. (1996). *Pattern-oriented software architecture—a system of patterns* (Vol. 1). Wiley.
15. Caiza, J. C., Martín, Y.-S., Del Alamo, J. M., & Guamán, D. S. (2017). Organizing design patterns for privacy: A taxonomy of types of relationships. In *Proceedings of the 22nd European Conference on Pattern Languages of Programs (EuroPLoP)* (pp. 1–11).
16. Cavoukian, A. (2011). *Privacy by design the 7 foundational principles implementation and mapping of fair information practices*. Brochure, Information and Privacy Commissioner of Ontario Canada.
17. Cavoukian, A., Shapiro, S., & Cronk, R. J. (2014). Privacy engineering: Proactively embedding privacy by design. White paper, Information and Privacy Commissioner of Ontario Canada.
18. Coelho, M. D., Vasconcelos, A., & Sousa, P. (2021). Privacy by design enterprise architecture patterns. In *Proceedings of the 23rd International Conference on Enterprise Information Systems (ICEIS)* (pp. 743–750).
19. Colesky, M., & Caiza, J. C. (2018). A system of privacy patterns for informing users: Creating a pattern system. In *Proceedings of the 23rd European Conference on Pattern Languages of Programs (EuroPLoP)* (pp. 1–11).
20. Colesky, M., Caiza, J. C., Del Álamo, J. M., Hoepman, J.-H., & Martín, Y.-S. (2018). A system of privacy patterns for user control. In *Proceedings of the 33rd Annual ACM Symposium on Applied Computing (SAC)* (pp. 1150–1156).
21. Colesky, M., Hoepman, J.-H., & Hillen, C. (2016). A critical analysis of privacy design strategies. In *Proceedings of the IEEE Security and Privacy Workshops (SPW)* (pp. 33–40).
22. Dearden, A., & Finlay, J. (2006). Pattern languages in HCI: A critical review. *Human–Computer Interaction, 21*(1), 49–102.
23. Dehling, F., Feth, D., Polst, S., Steffes, B., & Tolsdorf, J. (2021). Components and architecture for the implementation of technology-driven employee data protection. In *Proceedings of the 18th International Conference on Trust, Privacy and Security in Digital Business (TrustBus)* (Vol. 12927, pp. 99–111).
24. Domingo-Ferrer, J., Hansen, M., Hoepman, J.-H., Le Métayer, D., Tirtea, R., Schiffner, S., Danezis, G., European Union, & European Network and Information Security Agency. (2014). Privacy and Data Protection by Design - from Policy to Engineering. Report, European Union Agency for Cybersecurity (ENISA).
25. Doty, N., & Gupta, M. (2013). privacy design patterns and anti-patterns—patterns misapplied and unintended consequences. In *Proceedings of the 1st Trustbusters for User Interfaces Workshop* (pp. 1–5).
26. Drozd, O. (2016). Privacy pattern catalogue: A tool for integrating privacy principles of ISO/IEC 29100 into the software development process. In *Proceedings of the 10th IFIP International Summer School on Privacy and Identity Management* (pp. 129–140).
27. Drozd, O., & Kirrane, S. (2020). Privacy CURE: Consent comprehension made easy. In *Proceedings of the 35th IFIP International Conference on ICT Systems Security and Privacy Protection (IFIP SEC)*.
28. EN ISO 9241-11:2018. Ergonomics of Human-System Interaction Part 11: Usability: Definitions and Concepts. International Standards. International Organization for Standardization.
29. European Union. (2016). General Data Protection Regulation. Regulation (EU) 2016/679.
30. Evdokimov, A., Reva, A., & Maris, K. (2020). Taking care of corporate security and employee privacy. Survey, AO Kaspersky Lab.
31. Feth, D., Maier, A., & Polst, S. (2017). A user-centered model for usable security and privacy. In *Proceedings of the 5th International Conference on Human Aspects of Information Security, Privacy and Trust (HAS)* (pp. 74–89).

32. Fischer-Hübner, S., & Berthold, S. (2017). Privacy-enhancing technologies. In *Computer and Information Security Handbook* (pp. 759–778). Elsevier.
33. Fischer-Hübner, S., Köffel, C., Pettersson, J. S., Wolkerstorfer, P., Graf, C., Holtz, L. E., König, U., Hedbom, H., & Kellermann, B. (2010). HCI pattern collection—version 2. Deliverable D4.1.3, PrimeLife.
34. Gabel, A., & Schiering, I. (2019). Privacy patterns for pseudonymity. In *Proceedings of the 13th IFIP International Summer School on Privacy and Identity Management* (pp. 155–172).
35. Gamma, E., Helm, R., Johnson, R., & Vlissides, J. (1995). *Design patterns: Elements of reusable object-oriented software*. Addison-Wesley.
36. Gan, M. F., Chua, H. N., & Wong, S. F. (2019). Privacy enhancing technologies implementation: An Investigation of its impact on work processes and employee perception. *Telematics and Informatics, 38*, 13–29.
37. Goodman, S. (2020). Human Error to Blame for 9 in 10 UK Cyber Data Breaches in 2019. https://www.cybsafe.com/press-releases/human-error-to-blame-for-9-in-10-uk-cyber-data-breaches-in-2019/
38. Graf, C., Wolkerstorfer, P., Geven, A., & Tscheligi, M. (2010). A pattern collection for privacy enhancing technology. In *Proceedings of the 2nd International Conferences on Pervasive Patterns and Applications (PATTERNS)* (pp. 21–16).
39. Gürses, S., Troncoso, C., & Diaz, C. (2011). Engineering privacy by design. In *Proceedings of the 4th Conference on Computers, Privacy & Data Protection (CPDP)* (pp. 1–25).
40. Hafiz, M. (2013). A pattern language for developing privacy enhancing technologies. *Software: Practice and Experience, 43*(7), 769–787.
41. Hoepman, J.-H. (2014). Privacy design strategies. In *Proceedings of the 29th IFIP International Conference on ICT Systems Security and Privacy Protection (IFIP SEC)* (pp. 446–459).
42. Johansen, J., & Fischer-Hübner, S. (2020). Making GDPR usable: A model to support usability evaluations of privacy. In *Proceedings of the 14th IFIP International Summer School on Privacy and Identity Management* (pp. 275–291).
43. Karegar, F., Pulls, T., & Fischer-Hübner, S. (2016). Visualizing exports of personal data by exercising the right of data portability in the data track—are people ready for this? In *Proceedings of the 11th IFIP International Summer School on Privacy and Identity Management* (pp. 164–181).
44. Knijnenburg, B. P., Page, X., Wisniewski, P., Lipford, H. R., Proferes, N., & Romano, J. (2022). Introduction and overview. In B. P. Knijnenburg, X. Page, P. Wisniewski, H. R. Lipford, N. Proferes, & J. Romano (Eds.), *Modern Socio-Technical Perspectives on Privacy* (pp. 1–11). Springer.
45. Krueger, R. A., & Casey, M. A. (2015). *Focus groups: A practical guide for applied research* (5th ed.). Sage.
46. Lazar, J., Feng, J. H., & Hochheiser, H. (2017). *Research methods in human computer interaction* (2nd ed.). Elsevier.
47. Lenhard, J., Fritsch, L., & Herold, S. (2017). A literature study on privacy patterns research. In *Proceedings of the 43rd Euromicro Conference on Software Engineering and Advanced Applications (SEAA)* (pp. 194–201).
48. Machuletz, D., & Böhme, R. (2020). Multiple purposes, multiple problems: A user study of consent dialogs after GDPR. *Proceedings on Privacy Enhancing Technologies, 2020*(2), 481–498.
49. Mathis, F., Vaniea, K., & Khamis, M. (2021). Prototyping usable privacy and security systems: Insights from experts. *International Journal of Human–Computer Interaction. 38*(5), 468–490.
50. Morton, A., & Sasse, M. A. (2012). Privacy is a process, not a PET: A theory for effective privacy practice. In *Proceedings of the Workshop on New Security Paradigms (NSPW)* (pp. 87–104).
51. Mulligan, D. K., & King, J. (2012). Bridging the gap between privacy and design. *University of Pennsylvania Journal of Constitutional Law, 14*(4), 1–46.
52. Murmann, P., & Fischer-Hübner, S. (2017). Tools for achieving usable ex post transparency: A survey. *IEEE Access, 5*, 22965–22991.

53. Murmann, P., Reinhardt, D., & Fischer-Hübner, S. (2019). To be, or not to be notified: Eliciting privacy notification preferences for online mhealth services. In *Proceedings of the 34th IFIP International Conference on ICT Systems Security and Privacy Protection (IFIP SEC)* (pp. 99–114).
54. Nouwens, M., Liccardi, I., Veale, M., Karger, D., & Kagal, L. (2020). Dark patterns after the GDPR: Scraping consent pop-ups and demonstrating their influence. In *Proceedings of the CHI Conference on Human Factors in Computing Systems* (pp. 1–13).
55. Papoutsakis, M., Fysarakis, K., Spanoudakis, G., Ioannidis, S., & Koloutsou, K. (2021). Towards a collection of security and privacy patterns. *Applied Sciences, 11*(4), 1396.
56. Pauwels, S. L., Hübscher, C., Bargas-Avila, J. A., & Opwis, K. (2010). Building an interaction design pattern language: A case study. *Computers in Human Behavior, 26*(3), 452–463.
57. Personal Information Protection Commission Japan. (2020). Amended act on the protection of personal information.
58. Piras, L., Al-Obeidallah, M. G., Pavlidis, M., Mouratidis, H., Tsohou, A., Magkos, E., Praitano, A., Iodice, A., & Crespo, B. G.-N. (2020). DEFeND DSM: A data scope management service for model-based privacy by design GDPR compliance. In *Proceedings of the 17th International Conference on Trust, Privacy and Security in Digital Business (TrustBus)* (pp. 186–201).
59. Polst, S., Kelbert, P., & Feth, D. (2019). Company privacy dashboards: Employee needs and requirements. In *Proceedings of the 1st International Conference on Human-Computer Interaction for Cybersecurity, Privacy and Trust (HCI-CPT)* (pp. 429–440).
60. Privacy Rights Clearinghouse (PRC). (2020). PRC Data Breach Chronology. Database 1.13.20, Privacy Rights Clearinghouse.
61. Presthus, W., & Sørum, H. (2019). Consumer perspectives on information privacy following the implementation of the GDPR. *International Journal of Information Systems and Project Management, 7*(3), 19–34.
62. Rapportage Datalekken 2020. (2020). Technical report, Autoriteit Persoonsgegevens.
63. Resolution on Privacy by Design. (2010). Technical report, *32nd International Conference of Data Protection and Privacy Commissioners*.
64. Robak, M., & Buchmann, E. (2020). How to extract workflow privacy patterns from legal documents. In E. Ziemba (Ed.), *Information Technology for Management: Current Research and Future Directions* (pp. 214–234). Springer.
65. Romanosky, S., Acquisti, A., Hong, J., Cranor, L. F., & Friedman, B. (2006). Privacy patterns for online interactions. In *Proceedings of the 13th Conference on Pattern Languages of Programs (PLoP)* (pp. 1–9).
66. Rosen, E. (2015). Human error biggest cause of data breach: Survey. *Bloomberg Law*.
67. Runte, C., & Kamps, M. (2021). *GDPR enforcement tracker report: Executive summary* (2nd ed.). CMS Law-Now.
68. Sahqani, W., & Turchet, L. (2021). Co-designing employees' data privacy: A technology consultancy company use case. In *Proceedings of the 28th Conference of Open Innovations Association (FRUCT)* (pp. 398–406).
69. Schufrin, M., Reynolds, S. L., Kuijper, A., & Kohlhammer, J. (2021). A visualization interface to improve the transparency of collected personal data on the internet. *IEEE Transactions on Visualization and Computer Graphics, 27*(2), 1840–1849.
70. Schumacher, M. (2003). Patterns and security standards—with selected security patterns for anonymity and privacy. In *Proceedings of the 8th European Conference on Pattern Languages of Programms (EuroPLoP)* (pp. 1–11).
71. Senarath, A., Arachchilage, N. A. G., & Slay, J. (2017). Designing privacy for you: A practical approach for user-centric privacy. In *Proceedings of the 5th International Conference on Human Aspects of Information Security, Privacy and Trust (HAS)* (pp. 739–752).
72. Siljee, J. (2015). Privacy transparency patterns. In *Proceedings of the 20th European Conference on Pattern Languages of Programs (EuroPLoP)* (pp. 1–11).
73. Spiekermann, S., & Cranor, L. F. (2009). Engineering privacy. *IEEE Transactions on Software Engineering, 35*(1), 67–82.

74. Stark, L., King, J., Page, X., Lampinen, A., Vitak, J., Wisniewski, P., Whalen, T., & Good, N. (2016). Bridging the gap between privacy by design and privacy in practice. In *Extended Abstracts of the CHI Conference on Human Factors in Computing Systems (CHI EA)* (pp. 3415–3422).
75. State of California. (2018). California Consumer Privacy Act. Assembly Bill No. 375.
76. Tikkinen-Piri, C., Rohunen, A., & Markkula, J. (2018). EU general data protection regulation: Changes and implications for personal data collecting companies. *Computer Law & Security Review, 34*(1), 134–153.
77. Tolsdorf, J., Dehling, F., & Lo Iacono, L. (2022). Data cart—designing a tool for the GDPR-compliant handling of personal data by employees. *Behaviour & Information Technology, 41*(10), 2070–2105.
78. Tolsdorf, J., Dehling, F., Reinhardt, D., & Lo Iacono, L. (2021). Exploring mental models of the right to informational self-determination of office workers in Germany. *Proceedings on Privacy Enhancing Technologies, 2021*(3), 5–27.
79. Tolsdorf, J., Fischer, M., & Lo Iacono, L. (2021). A case study on the implementation of the right of access in privacy dashboards. In *Proceedings of the 9th Annual Privacy Forum (APF)* (pp. 23–46).
80. Utz, C., Degeling, M., Fahl, S., Schaub, F., & Holz, T. (2019). (Un)informed consent: Studying GDPR consent notices in the field. In *Proceedings of the ACM SIGSAC Conference on Computer and Communications Security (CCS)* (pp. 973–990).
81. Veys, S., Serrano, D., Stamos, M., Herman, M., Reitinger, N., Mazurek, M. L., & Ur, B. (2021). Pursuing usable and useful data downloads under GDPR/CCPA access rights via co-design. In *Proceedings of the 17th Symposium on Usable Privacy and Security (SOUPS)* (pp. 217–242).
82. Wilson, C. (2014). Pluralistic usability walkthrough. In *User interface inspection methods* (pp. 81–97). Elsevier.
83. Yoder, J., & Barcalow, J. (1997). Architectural patterns for enabling application security. In *Proceedings of the 4th Conference on Patterns Language of Programming (PLoP)* (pp. 1–31).

Index

N. Gerber et al. (eds.), *Human Factors in Privacy Research*,
https://doi.org/10.1007/978-3-031-28643-8

MIX
Papier aus verantwortungsvollen Quellen
Paper from responsible sources
FSC® C105338

If you have any concerns about our products,
you can contact us on
ProductSafety@springernature.com

In case Publisher is established outside the EU,
the EU authorized representative is:
Springer Nature Customer Service Center GmbH
Europaplatz 3, 69115 Heidelberg, Germany

Printed by Libri Plureos GmbH
in Hamburg, Germany